Methods in Cell Biology

VOLUME 56
Video Microscopy

Series Editors

Leslie Wilson
Department of Biological Sciences
University of California, Santa Barbara
Santa Barbara, California

Paul Matsudaira
Whitehead Institute for Biomedical Research and
Department of Biology
Massachusetts Institute of Technology
Cambridge, Massachusetts

Methods in Cell Biology

Prepared under the Auspices of the American Society for Cell Biology

VOLUME 56
Video Microscopy

Edited by

Greenfield Sluder and David E. Wolf

Department of Cell Biology
Worcestor Foundation for Experimental Biology
Shrewsbury, Massachusetts

ACADEMIC PRESS

San Diego   London   Boston   New York   Sydney   Tokyo   Toronto

Front cover: A digital image of a PtK1 epithelial cell arrested in mitosis with the microtubule depolymerizing drug nocodazole. A differential interference contrast image (green/blue) was overlaid with an immunofluorescence image of the spindle assembly checkpoint component Mad2. Localization of Mad2 to kinetochores that are not properly attached to spindle microtubules (red/orange) is thought to generate a signal that inhibits enaphase onset. Photograph courtesy of Jennifer C. Waters and E. D. Salmon.

This book is printed on acid-free paper.

Academic Press
a division of Harcourt Brace & Company
525 B Street, Suite 1900, San Diego, California 92101-4495, USA
http://www.apnet.com

Academic Press Limited
24-28 Oval Road, London NW1 7DX, UK
http://www.hbuk.co.uk/ap/

International Standard Book Number: 0-12-564158-3

Printed and bound by CPI Group (UK) Ltd, Croydon, CR0 4YY

Transferred to Digital Print 2012

CONTENTS

CONTRIBUTORS

Numbers in parentheses indicate the pages on which the authors' contributions begin.

Eric J. Alm (91), Department of Biology, The University of California, Riverside, California 92521

Keith Berland (19), Lineberger Comprehensive Cancer Center, University of North Carolina at Chapel Hill, Chapel Hill, North Carolina 27599

Keith M. Berland (277), IBM, Yorktown Heights, New York 10598

Kerry Bloom (185), Department of Biology, University of North Carolina, Chapel Hill, North Carolina 27599

Richard A. Cardullo (91), Department of Biology, The University of California, Riverside, California 92521

Richard W. Cole (253), Division of Molecular Medicine, Wadsworth Center for Laboratories and Research, New York State Department of Health, Albany, New York 12201-0509

Chen Y. Dong (277), Laboratory for Fluorescence Dynamics, Department of Physics, University of Illinois at Urbana-Champaign, Urbana, Illinois 61801

Kenneth Dunn (217), Department of Medicine, Indiana University Medical Center, Indianapolis, Indiana 46202-5116

Todd French (19), Sandia National Laboratories, Nanostructure and Semiconductor Physics Department, Albuquerque, New Mexico 87185-1415

Todd French (277), LJL Bio-Systems, Sunnyvale, California 94089

Neal Gliksman (63), Universal Imaging Corporation, West Chester, Pennsylvania 19380

Enrico Gratton (277), Laboratory for Fluorescence Dynamics, Department of Physics, University of Illinois at Urbana-Champaign, Urbana, Illinois 61801

Edward H. Hinchcliffe (1), Department of Cell Biology, University of Massachusetts Medical Center, Worcester, Massachusetts 01655

Jan Hinsch (147), Leica, Inc., Allendale, New Jersey 07401

Ted Inoué (63), Universal Imaging Corporation, West Chester, Pennsylvania 19380

Ken Jacobson (19), Lineberger Comprehensive Cancer Center, Department of Cell Biology and Anatomy, University of North Carolina at Chapel Hill, Chapel Hill, North Carolina 27599

H. Ernst Keller (135), Carl Zeiss, Inc., Thornwood, New York 10594

Paul S. Maddox (185), Department of Biology, University of North Carolina, Chapel Hill, North Carolina 27599

Frederick R. Maxfield (217), Department of Biochemistry, Cornell University Medical College, New York, New York 10021

Masafumi Oshiro (45), Hamamatsu Photonic Systems, Division of Hamamatsu Corporation Bridgewater, New Jersey 08807

Conly L. Rieder (253), Division of Molecular Medicine, Wadsworth Center for Laboratories and Research, New York State Department of Health, Albany, New York 12201-0509; Department of Biomedical Sciences, State University of New York, Albany, New York 12222; Marine Biology Laboratory, Woods Hole, Massachusetts 02543

E. D. Salmon (153,185), Department of Biology, University of North Carolina, Chapel Hill, North Carolina 27599

Sidney L. Shaw (185), Department of Biology, University of North Carolina, Chapel Hill, North Carolina 27599

Randi B. Silver (237), Department of Physiology and Biophysics, Cornell University Medical College, New York, New York 10021

Greenfield Sluder (1), Cell Biology Group, Worcester Foundation for Biomedical Research, Shrewsbury, Massachusetts 01545

Peter T. C. So (277), Department of Mechanical Engineering, Massachusetts Institute of Technology, Cambridge, Massachusetts 02139

Phong Tran (153), Department of Biology, University of North Carolina, Chapel Hill, North Carolina 27599

Yu-li Wang (305), Cell Biology Group, Worcester Foundation for Biomedical Research, Shrewsbury, Massachusetts 01545

Clare M. Waterman-Storer (185), Department of Biology, University of North Carolina, Chapel Hill, North Carolina 27599

Jennifer Waters (185), Department of Biology, University of North Carolina, Chapel Hill, North Carolina 27599

David E. Wolf (117), Department of Physiology, Biomedical Imaging Group, University of Massachusetts Medical School, Worcester, Massachusetts 01655

Elaine Yeh (185), Department of Biology, University of North Carolina, Chapel Hill, North Carolina 27599

PREFACE

On July 4, 1997, as we were putting the finishing touches on this book, we, along with the rest of the world, watched in amazement as the Pathfinder spacecraft sent back its first pictures of the surface of Mars. This opening of a previously inaccessible world to our view is what we have come to expect from video imaging. In the past three decades video imaging has taken us to the outer planets, to the edge of the universe, to the bottom of the ocean, and into the depths of the human body. Video microscopy has also grown in scope and importance in the past 20 years. Three technological revolutions have literally "rocketed" video microscopy to its current level: the development of high-speed microcomputers, the development by the defense industry of low-light-level video cameras, and the development of image processing algorithms and hardware. Where are these developments taking us? Like the Pathfinder, they are taking us to new worlds. Video imaging can take us not only to other planets but also to the interior of living cells. It can reveal not only the forces that created and altered other planets but also the forces of life on earth.

Video technology in microscopy has evolved rapidly from a laboratory curiosity that was a poor substitute for conventional photomicrography into a central player in the remarkable renaissance of light microscopy as a premier tool for the study of cells. The combination of sophisticated image analysis software with refined instrument-grade cameras has transformed video microscopy into a powerful analytical tool that allows researchers to conduct quantitative biochemistry with individual living cells, not populations of dead cells. Importantly, video technology has provided researchers with a new dimension, that of time. We can now characterize changes in cellular physiology in "real time," the time frame of the cell.

In many ways this volume is a direct outgrowth of the Analytical and Quantitative Light Microscopy course (AQLM) that we give each spring at the Marine Biological Laboratory in Woods Hole, Massachusetts. Our experience in this course has led us to deliberately avoid the obvious temptation of publishing exhaustive technical treatments of only the newest cutting-edge technology and applications. Instead, we have aimed the content of this book at the sorts of students who come to our program, namely, established researchers who need to apply video microscopy to their work yet have little experience with the instrumentation. We have assumed that the readership of this volume will be largely composed of biologists, not physicists or engineers. Thus, we have asked the authors to present the basics of the technology, providing very practical information and in particular helping the reader avoid common mistakes that lead to artifacts. Our rationale has been that video microscopy requires an integrated system composed of several sophisticated instruments that all must be coordinately optimized. Failure to properly control any of the links inevitably leads to the loss of specimen detail.

As a consequence, we start with chapters on some of the basics, such as microscope alignment, mating cameras to microscopes, and elementary use of video cameras and monitors. We then provide chapters on the properties of various camera systems, how to use them in quantitative applications, and the basics of image processing. Finally, we present chapters covering some of the more sophisticated applications.

The preparation of a book of this sort cannot be done in isolation, and many people have made direct and indirect contributions to this volume. First, we thank our contributors for taking the time to write articles covering the basic and practical aspects of video microscopy. Although this may be less glamorous than describing the latest and greatest aspects of their work, we expect that their efforts will provide a valuable service to the microscopy community. We also thank the members of the commercial faculty of our AQLM course for their years of patient inspired teaching of both students and us. Through their interactions with students in various microscopy courses, these people, principals in the video and microscopy industries, have defined how the material covered in this book is best presented for use by researchers in biology. We are grateful that some of them have taken the time from their busy schedules to write chapters for this book. We especially thank Ms. Cathy Warren here at the Worcester Foundation for her tireless assistance in the preparation of the manuscripts for this volume and her invaluable help with the myriad administrative details. Also, we thank Ms. Jennifer Wrenn at Academic Press for her extraordinary patience and cheerful willingness to work with us.

Finally, we acknowledge three individuals, who are no longer with us, for their contributions both to the AQLM course and to the field of video microscopy: Fredric Fay of the Department of Physiology, UMASS Medical Center; Gerald Kleifgen of Dage MTI; and Philip Presley of Carl Zeiss. These people are greatly missed not only for their intellectual contributions, but also because they were wonderful people who helped define the interactive environment that is the field of video microscopy. We respectfully dedicate this volume to their memories.

Greenfield Sluder
David E. Wolf

CHAPTER 1

Video Basics: Use of Camera and Monitor Adjustments

Greenfield Sluder and Edward H. Hinchcliffe

Cell Biology Group, Worcester Foundation for Biomedical Research,
Shrewsbury, Massachusetts 01545; and Department of Cell Biology,
University of Massachusetts Medical Center, Worcester, Massachusetts 01655

I. Introduction

In this chapter we discuss the practical aspects of adjusting the video camera and the monitor to prevent the loss of specimen image gray level information, a subject that is often treated in a most cursory fashion in the owners manuals for cameras and monitors. The problem one faces in coordinating camera and monitor adjustments is that there are too many interacting variables and no standard against which to evaluate the performance of the system. One can vary the intensity of the optical image on the photosensitive surface of the camera, camera settings, and monitor settings. All three appear interactive and can be used individually to seemingly optimize the quality of the image to some extent. When done in a haphazard fashion, empirical "knob twiddling" opens the real possibility of loosing specimen information, typically in the brightest and/or darkest portions of the image. Also when recording the results of an experiment, one can find that the quality of the recorded image is not as good as the image one saw on the monitor. This "discovery" is often made after the experiment is over and it is too late to remedy the situation.

II. The Black-and-White Video Signal

Here we review, in a simplified fashion, a few aspects of the monochrome composite video signal, produced by tube-type cameras and conventional CCD cameras. For a complete description of the composite video signal and how it is displayed on the monitor the reader is referred to S. Inoue's 1986 book "Video Microscopy," which is currently available and the second edition of this book by S. Inoue and K. Spring, which will be published in late 1997.

The camera converts the optical image on its photosensitive surface, or faceplate, into a sequence of electrical signals which comprise the video signal. For cameras using the North American (NTSC) format the image at the camera faceplate is sampled from top to bottom in 525 horizontal scan lines every 1/30th of a second. Within each scan line, the faceplate light levels are converted into a voltage that is proportional, within limits, to light intensity. Shown in Fig. 1A is a schematic representation of a number of scan lines through the image of a "cell" that is darker than the background. Figure 1B shows the voltage output of the camera as a function of position along each of these representative scan lines. When displayed one above the other in sequence, as occurs in the monitor, the scans recreate the two dimensional intensity profile of the specimen

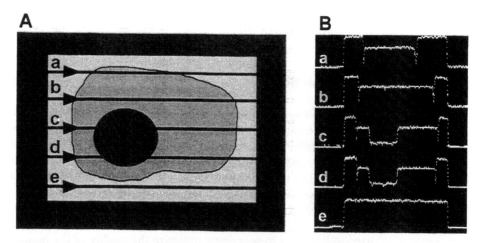

Fig. 1 Video signals corresponding to scanlines through the image of a "cell." (A) Schematic drawing of a cell with rectangular black border as imaged on the faceplate of the video camera. The horizontal black lines labeled a–e are representative scan lines through the image with the scans going from left to right and progressing from top to bottom. (B) Actual camera output for the scan lines showing camera output voltages as a function of position along the scan lines.

image. In reality each frame (one complete image) is not scanned sequentially in 525 lines. Rather, every other scan line (together forming a "field") is transmitted in 1/60th of a second and then the remaining scan lines are transmitted in the next 1/60th of a second. This is known as the "interlace" and serves to eliminate an annoying vertical sweeping of the image.

Figure 2 shows a portion of the signal coming from the camera; in this case two scan lines are represented of a specimen image that is brighter in the center of the image than at the periphery. Starting at the left is the H-sync pulse of negative voltage which serves to coordinate the monitor scan with that of the camera. The voltage then returns to 0 volts (part of the H-blanking pulse that blanks the signal while the electron beam in the camera returns from the end of one scan line to the start of the next scan). The voltage then rises to 0.05–0.075 volts, a value known as the "setup" or "pedestal," which is the picture black level. Thereafter the voltage rises and falls in relation to the image intensity along the scanline. At the end of the active picture portion of the scan the voltage returns to the pedestal value and then drops to 0 volts at the beginning of the

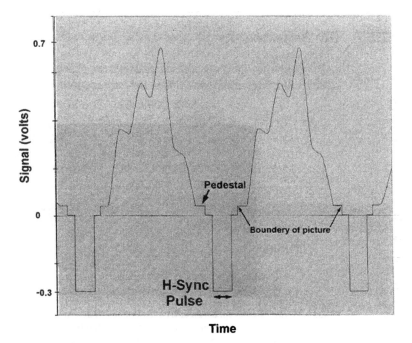

Fig. 2 The composite video signal for two scanlines of a specimen image that is brighter in the center of the image than at the periphery. The active picture portion of the signal is indicated for the second scan and the pedestal voltage is indicated at the end of the first scan. Between scans is the H-blanking pulse (0 volts) upon which is superimposed the H-sync pulse (negative voltage). The full range of signal amplitude for specimen image light intensities is encoded in slightly less than 0.7 volts.

H-blanking pulse, upon which is superimposed the negative H-sync pulse. The full range of signal amplitude for specimen image light intensities is encoded in slightly less than 0.7 volts.

Figure 3 (lower portion of the figure) shows the step gray scale test target or "specimen" that will be used for the rest of this chapter. In this and subsequent figures, the grayscale specimen is viewed by a camera and the resulting image on the monitor is shown in the lower portion of the figure. The upper part of the figure shows the camera output voltages for a scanline running across the grayscale as displayed by a Dage MTI, "RasterScope" device. The camera output voltage rises in a stepwise fashion from just above 0 volts for the black band to just under 0.7 volts for the white band with a proportionate increase in camera output voltage for each step in image intensity. We have not used quite the full 0.7-volt range of camera output in order to display the first and last steps. With the monitor properly adjusted all intensity values in the specimen are represented in the monitor image. The purpose in using this grayscale is that one can easily identify the loss of grayscale information when either the camera or monitor are improperly adjusted.

III. Adjusting the Camera and Video Monitor

To set up the video system one must do two things. First, the camera is adjusted so that all image intensity values are represented in its output signal using the

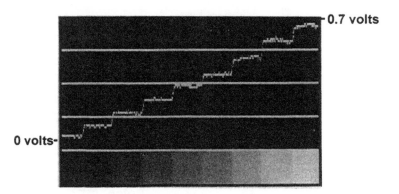

Fig. 3 Grayscale test specimen and camera output. The test specimen used throughout this chapter consists of a step grayscale running from black on the left to white on the right. In this and subsequent figures, the grayscale specimen is viewed by a camera and the resulting image on the monitor is shown in the lower portion of the figure. The upper part of the figure shows the camera output voltages for a scanline running across the grayscale as displayed by a Dage MTI RasterScope. Black is represented by a voltage just above 0 volts and white is represented by a voltage just under 0.7 volts. We have used slightly less than the full dynamic range of the camera in order to show the black and the white steps on the voltage profile. Here the camera response is linear; for each increase in specimen image intensity there is an equal increase in camera output voltage (gamma = 1).

full 0.7-volt camera output to include the highest and lowest image intensity values. Second, the monitor is adjusted to display all gray values present in the camera output. Proper adjustment of the camera is of paramount importance, because loss of information at the camera level cannot be later restored. Also, note that image storage devices record just what the camera provides, not the optimized image on the monitor screen.

Below we discuss what the camera/monitor controls mean, how one rationally uses them, and some of the consequences of their improper use.

A. Camera Controls

1. Gain

Camera gain is an amplifier setting that modifies the amplitude of the camera output voltage for any given intensity of light input. Sometimes this control is labeled as "contrast." Shown in Fig. 4B is a scanline through our grayscale specimen; the voltage profile shows that the darkest specimen detail (black) is just above 0 volts and the brightest specimen detail (white) is 0.7 volts. The image of the grayscale on a properly adjusted monitor is shown immediately below.

When the gain is set too high, as shown in Fig. 4A, the camera output for each step is relatively larger but the lighter grays in the specimen are driven to the full 0.7 volts and become saturated at white. Thus, the practical consequences of too high a gain setting are improved contrast between the darker gray steps and the complete loss of the specimen information contained in the lighter gray values. Obviously this lost information cannot be restored by monitor adjustment or image processing.

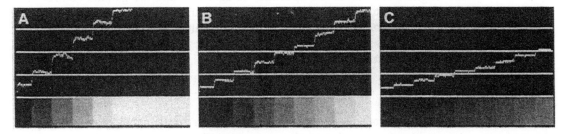

Fig. 4 Camera gain. (B) The camera and monitor are adjusted to capture and display respectively all grayscale values. (A) The results of raising the camera gain while leaving the monitor settings constant. For the camera output trace note that the black level remains at 0 volts, the height of each step is increased, and the lighter gray values in the specimen image are driven to 0.7 volts, resulting in an irretrievable loss of highlight information. Also note that the contrast between the darker gray values is increased. (C) The results of lowering the camera gain. Black remains at 0 volts, the step size between gray levels is reduced, and the white portion of the specimen is represented by less than 0.7 volts. Although the monitor representation of the specimen contains all grayscale information, the contrast between the gray steps is low.

When the gain is lowered below the proper setting (Fig. 4C), the output of the camera is reduced for each step of specimen image intensity. The black value remains essentially 0 volts and the white portion of the specimen is represented by a voltage less than the full 0.7 volts. When viewed on the monitor, the black band of the target remains black while the white band is represented as a gray. The practical consequence of too little gain is, therefore, a low-contrast image. Although this image in principle retains all the gray level information, small differences in specimen image intensity are lost in the noise and can be effectively lost.

2. Linearity of Camera Gain: Gamma

The gamma of a camera is the slope of the light transfer characteristics curve plotted as log signal output as a function of log illuminance of the faceplate. The gamma (κ), the slope of the curve, is the exponent of the relationship of faceplate illuminance to signal output: $I/I_D{}^\kappa = i/i_d$, where I is the illuminance, i the signal current, and I_D the illuminance that would give a signal current equal to the dark current i_D (Inoue, 1986).

The discussion thus far has been predicated on the assumption that the camera gain is the same for all specimen image intensities, or for when gamma is 1.0 (Fig. 5A). For cameras that allow control of gamma the linearity of the amplifier response can be adjusted to accentuate contrast in desired ranges of specimen image brightness. For example, setting gamma to <1 causes a greater amplification of the lower specimen intensities relative to the higher ones (Fig. 5B), resulting in the accentuation of contrast in the dimmer portions of the specimen. Conversely, setting gamma to >1 amplifies the brighter values more than the dimmer ones (Fig. 5C). The result is the selective enhancement of contrast in the brighter regions of the specimen image. In using the gamma control of a camera we must bear in mind, however, that one is effectively suppressing the specimen information contained in the gray values that are less amplified. For a more complete description of the use of the gamma function to selectively manipulate contrast, see Inoue (1986).

3. Automatic Gain

Often cameras have an automatic gain feature to accommodate varying specimens and specimen illumination intensities. For many cameras this feature can be turned on and off deliberately. For some cameras the automatic gain, when engaged, partially modulates the manual gain setting. In its simplest incarnation the automatic gain circuit raises or lowers the highest voltage in the image signal and sets it to 0.7 volts. In theory the very brightest portion of the specimen image would serve as the reference even if it corresponded to a minute bright spot in some portion of the image. In practice, however, the response of the automatic gain circuit is dampened to an extent that varies with camera model

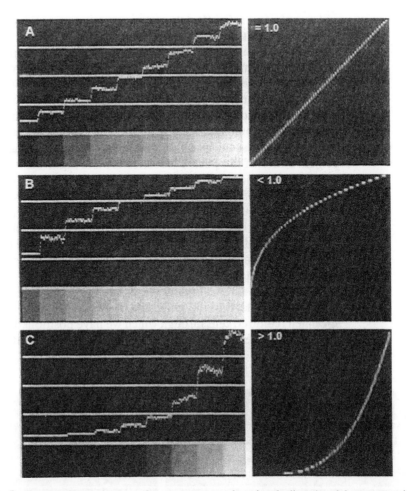

Fig. 5 Gamma. Shown here are the consequences of varying the linearity of the camera gain; that is, varying gamma. For each of the three panels, the left side shows the camera output and the monitor image of the specimen. The right side shows the camera output (ordinate) as a function of light input (abscissa). When the camera gamma is 1.0 (A), the camera gain is equal for all light intensities in the specimen image. When the camera gamma is set <1.0 (B), the gain is higher for the lower light intensities. All grayscale values are captured and displayed, but there is higher contrast between the darker gray levels than the lighter ones. (C) Camera gamma set higher than one; the gain is higher for the brighter specimen intensities. Although all grayscale values, in principle, are represented in the camera output, the contrast between the brighter specimen gray values is greater than that for the darker gray steps. Here the loss of the darker gray values in the monitor representation of the specimen is an artifact.

and manufacturer. In effect the automatic camera gain responds to a percentage of the image that has bright features.

As a consequence, caution should be exercised in using the automatic gain feature in certain applications such as fluorescence microscopy where a small percentage of the image area is composed of high-intensity values on a dark or black background. The large amount of dark area leads to a rise in the camera gain which can cause the brighter values in the light-emitting portion of the specimen to become saturated. In order to best differentiate between brightnesses within the fluorescent specimen one should manually adjust the camera gain. A demonstration of this phenomenon is shown in Fig. 6. Here we have put a narrow strip of the grayscale on a neutral gray background (Fig. 6A, lower portion) and adjusted to camera so that all specimen image intensity values are represented in the camera output (Fig. 6A, upper). When the neutral gray background is replaced with a black surround, the camera gain automatically increases (Fig. 6B, upper) thereby driving the lighter gray values to white (Fig. 6B, lower). Note that the black level remains unchanged and the contrast between the darker gray steps is enhanced.

Figure 7 shows how bringing the image of the microscope field stop into the image area can lead to the same sort of problem. Shown here is a low-contrast specimen (cultured cells through which we have positioned the cursor or horizontal sampling line) with the image of the microscope field stop lying outside of

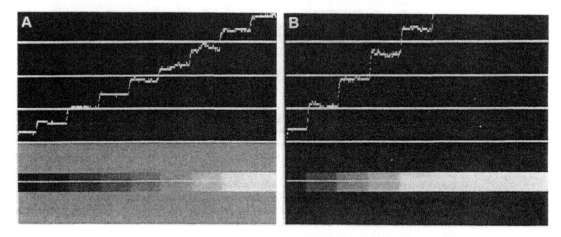

Fig. 6 Autogain and average image intensity. (A) The lower portion shows the monitor display of the specimen used; a strip of grayscale on a surround of neutral gray. The white horizontal line (cursor) through the grayscale shows the scanline for which the camera output voltages are shown immediately above. (B) With the camera in the autogain mode the gray surround is replaced with a black surround leaving camera and monitor adjustments constant. The large area of black in the specimen leads the autogain circuit to increase the camera gain causing a loss of lighter gray specimen detail, albeit with an increase in contrast between the darker gray values of the specimen.

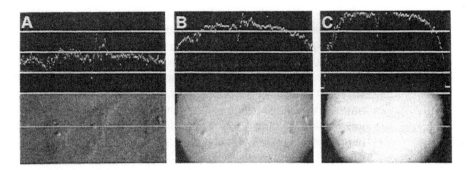

Fig. 7 Autogain and field stop position. (A) The lower portion shows a field of cells with the horizontal white cursor line. The camera output voltage profile along this line is shown in the upper portion of this figure; no attempt was made to utilize the full 0.7-volt camera output. (B and C) As the image of the microscope field stop is increasingly brought into the field seen by the camera, the autogain circuit increases the camera gain. As a consequence, lighter gray values are driven to white and specimen information is lost.

the camera's field of view (Fig. 7A, lower). When the field stop is closed down, the automatic increase in camera gain drives the lighter specimen grays toward saturation an information can consequently be lost (Figs. 7B and 7C).

The use of the automatic gain feature is also not indicated for quantitative applications with specimens whose intensity varies with time. For example, when measuring the bleaching of a fluorescent specimen, the automatic gain circuit will attempt to compensate for loss of specimen intensity leading to a progressively noisier but constant-intensity image.

4. Specimen Illumination Intensity

To optimize the amount of specimen gray level information included in the camera output attention must be paid to the intensity of the specimen image projected on the camera faceplate. When the camera is used in the automatic gain mode, this may not be critical in that the camera can deal with a range of light intensities. However, information can be lost if the average image intensity is close to the limits which the camera can accommodate. Clearly insufficient image intensity will, in the extreme, produce no signal from the camera compared to noise. At the other extreme, an image that is too bright will saturate the faceplate of the camera and again no image is transmitted from the camera. Typically one has enough control over the intensity of the specimen image to avoid such gross problems. An overly bright image is most easily controlled either by reducing the output of the microscope illuminator or by introducing neutral density filters into the illumination pathway. However, in some applications, such as polarization or high extinction differential interference contrast microscopy, one may be forced to work at the low end of intensities that the camera can use.

Figure 8 shows the loss of specimen gray level information when the intensity of the specimen image on the camera faceplate is raised or lowered. Although this demonstration was conducted with the camera in the fixed-gain mode, the phenomena shown here can occur when the specimen image intensity exceeds the range of values that the automatic gain control can accommodate. At proper specimen image intensity (Fig. 8B), all gray levels are represented in the camera output. At inappropriately high specimen image intensity (Fig. 8A) the lighter gray values become saturated and are thus displayed as white. There is some increase in contrast between the darker gray values. At inappropriately low specimen image intensity (Fig. 8C) the contrast between specimen gray values is reduced and the darker gray values go to black. The white portion of the specimen is displayed as gray.

5. Black Level

For each scan line of the video signal (Fig. 2) the black level is at a voltage that is slightly higher than the 0-volt blanking level. This 0.05- to 0.075-volt offset is called the "setup" or "pedestal." For cameras that have black level control the lower image intensity represented by zero camera output (i.e., black) can be manually controlled by varying this offset. Sometimes this camera control is labeled "brightness."

Figure 9A shows the grayscale specimen with the camera gain and black level set properly. Figures 9B and 9C show the results of two increased levels of the black level control with the gain fixed. As the black level is increased, more of the darker gray level steps are set to 0 volts. The whole output curve for the camera is shifted downward to lower voltages with the step size (gain) remaining the same. As a result, the lower gray values are lost and the white value is

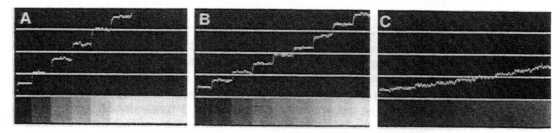

Fig. 8 Specimen image intensity and camera response with fixed gain. (B) Camera output voltage profile and monitor image of the grayscale specimen at adequate illumination with camera and monitor adjusted to capture and display all gray levels. (A) With camera and monitor settings held constant, specimen illumination is increased. The result is the saturation of the light gray values and an increase in the contrast between darker gray values. (C) A decrease in specimen illumination intensity results in smaller changes in camera output for each step of the grayscale. Thus, the image of the grayscale on the monitor is of low contrast, ranging from black to dark gray.

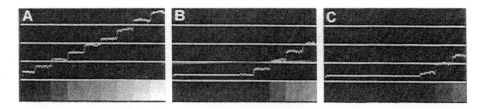

Fig. 9 Black level control. (A) The camera and monitor are adjusted to capture and display respectively all grayscale values. Leaving camera gain and the monitor settings constant, the black level control for the camera is increased in B and C. As this control is progressively increased, more of the darker gray values are set to black. Note that the contrast between remaining steps does not change and that the white specimen step is progressively brought to lower camera output voltages; hence the white portion of the specimen is displayed as gray on the monitor.

reduced to gray. In practice, this means that the indiscriminate use of the black level control can lead to the loss of specimen detail.

Some cameras have an automatic black level option that, in principle, sets the lowest signal voltage of the image to the setup or pedestal level (0.05–0.075 volts). In practice, however, autoblack circuits typically do not detect only the very lowest voltage in the field because insignificant objects such as small black granules in the specimen and imaged dirt particles would serve as the reference for the black level. As with the autogain function, the autoblack circuit is dampened so that it monitors the lowest voltages in a percentage of the field or responds to dark objects of some threshold size. The extent to which the autoblack circuit is dampened can vary with camera model and manufacturer. Also, the degree of averaging may not be the same for autogain and autoblack. For demanding applications it may be best to use the manual black level control, because small areas of dark gray specimen detail can be driven to black and the information lost when the camera is used in the autoblack mode.

Obviously one would not want to indiscriminately use the black level control for images that have true blacks, because one could lose the darker specimen detail. However, the great value of this control becomes apparent when working with low-contrast specimens that are represented (under conditions of adequate illumination) by only lighter gray values. By setting the black level we can make the darkest gray become black and then raise the gain to amplify the lightest gray to white. What started as a low-contrast image, represented by only higher camera output voltages, now is represented by the full 0.7-volt range of the camera output. In effect we have enhanced the contrast of the image retaining all relevant gray level information. This operation is illustrated in Figs. 10A–10C in which we have sought to increase the contrast between the midlevel gray steps 4–7 (of course mindful of the irretrievable loss of higher and lower specimen image intensities). Figure 10A shows the camera set to display all gray levels. For Fig. 10B we have increased the black level to drop gray level 4 to black and

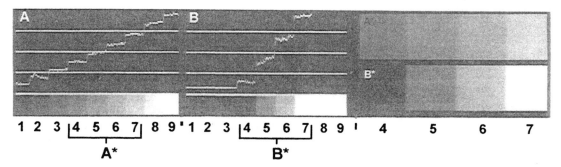

Fig. 10 Use of black level and gain controls to enhance contrast. (A) The camera output voltages and the monitor representation of the grayscale when the camera and monitor are set to display all gray levels (here numbered for reference). For B the black level control was increased to set gray level 4 to almost black and the gain was increased to set gray level 7 to almost white. The result is a marked increase in the contrast between gray levels 4–7. (C) A direct comparison between gray levels 4–7 taken from A and B. The use of black level and gain controls in this fashion is useful when one has a low-contrast specimen or when one wishes to selectively enhance the contrast for desired portions of the specimen image.

then raised the gain to display gray level 7 as white. The contrast enhancement for gray levels 4–7 is shown in Fig. 10C.

This strategy can be implemented deliberately with the use of manual black level and gain controls especially when one is working with a demanding specimen or is interested in optimizing the visualization of only a subset of the specimen detail. For routine work or for specimens that are of relatively uniform intensity across the field of the microscope, the use of autoblack in combination with autogain may perform this operation satisfactorily. However, note that when the camera is used with both autoblack and autogain, bringing the image of the microscope field stop into the field seen by the camera can lead to a marked loss of specimen contrast. The camera black level control will take the image of the field stop to be the darkest specimen value leaving the real specimen values at light grays. In addition, the presence of a large area of black may lead to an automatic increase in gain which can drive the light gray values to white.

B. Monitor Controls

Black-and-white monitors typically have "brightness" and "contrast" knobs to modify the displayed image characteristics. We emphasize that these controls modify just the monitor response to the signal from the camera and thus exert no control over the characteristics of the image being captured by a recording device. This means that monitor controls should be used to optimize the display only after the camera is properly adjusted.

1. Brightness Control

This varies the current of the electron beam that scans the phosphor surface at the face of the picture tube. At the minimal setting the screen may be black and show no image. At maximum setting portions of the image representing zero camera input appear gray and the brighter gray values may be represented with low contrast. Figure 11 shows the consequences of setting the monitor brightness control either a bit too low or a bit too high in relation to the proper setting. Figure 11B shows the grayscale with the monitor properly adjusted and the camera output trace showing that all specimen intensity values are represented in the camera signal. Figures 11A and 11C show the results of setting the brightness control too high and too low respectively without any change in camera output. When set too high, blacks becomes gray and the lighter gray values are driven to white. When set too low, the white value goes to gray and the dark gray values are shown as black. In both cases information that is present in the camera output is functionally lost in the image displayed on the monitor.

2. Contrast Control

This is a gain control that varies the change in electron beam current relative to changes in the input voltages from the camera. The higher the "contrast" setting the more the camera signal is amplified. Shown in Fig. 12B is the grayscale with the monitor properly adjusted and the camera output trace showing that all specimen intensity values are represented in the camera signal. Figures 12A and 12C show the results of setting the contrast control too high and too low respectively with no change in camera output. At too high monitor gain the blacks stay black and the light gray values are driven to white. With a real specimen on the microscope this sometimes leads to "blooming" of the highlights.

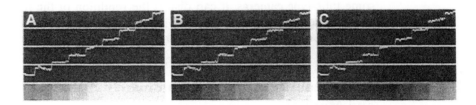

Fig. 11 Monitor settings: brightness. For this exercise the camera was properly adjusted to capture all gray levels in the specimen as is shown in the camera voltage output profiles in all panels. With the monitor properly adjusted (B) all gray levels are displayed on the face of the monitor. When the brightness control is set too high (A) the black step is displayed as a gray and the lighter gray values are saturated and thus white. When the brightness control is set too low (C) the darker gray values become black and the lighter gray values become darker. In both cases, the improper adjustment of monitor brightness leads to the functional loss of specimen information that is present in the camera signal.

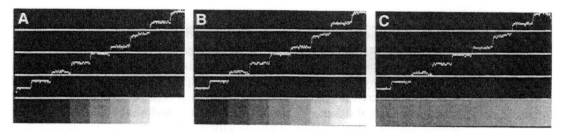

Fig. 12 Monitor settings: contrast. For this exercise the camera was properly adjusted to capture all gray levels in the specimen as is shown in the camera voltage output profiles in all panels. When contrast is properly set, all gray values are displayed, as shown in B. When monitor contrast is increased (A), the lighter gray values are driven to saturation and become white and displayed contrast between the remaining gray levels is increased. In the monitor display there is a loss of the brighter specimen information present in the camera signal (upper portion of this panel). When monitor contrast is lowered (C), the grayscale displayed is of low contrast without true black or true white.

Lower than proper monitor contrast settings yield an image of low contrast with the white value going to gray and at the extreme no image shown on the monitor screen.

3. Monitor Response and Contrast Enhancement

Since monitors typically have a gamma between 2.2 and 2.8, the brightness of the image on the monitor is not linearly proportional to the input voltage from the camera. The contrast between lighter gray values is greater than the contrast between the darker gray values because the lighter values are amplified to a greater extent. Although this high gamma compensates for the low gamma of some tube-type cameras, it can distort the rendition of the image produced by linear cameras such as CCDs. This can be controlled with an image processor that allows one to manipulate the gamma of the camera output signal (see Fig. 5).

IV. Practical Aspects of Coordinately Adjusting Camera and Monitor

When examining a specimen that has known gray level characteristics, such as the grayscale we have used as a test target, adjusting the camera and monitor is relatively straightforward because we know what specimen information needs to appear on the monitor image. However, when working with a real microscope image we obviously do not have the luxury of knowing what to look for. The difficulty is that the camera and monitor controls appear interactive and we have no good reference from which to work. How then does one properly adjust the camera and the monitor to ensure that we do not inadvertently lose specimen information?

Unfortunately, there is no simple answer to this question nor any foolproof protocol in the absence of a means to evaluate camera output independently of the monitor response. An added complication is that some but not all cameras have features that allow the operator to monitor their output to some extent. Thus, the particulars of coordinately adjusting camera and monitor controls will depend upon the specific instrumentation available to the operator. Nevertheless, the principles of coordinately adjusting the camera and monitor are discussed below.

As mentioned earlier, one must first ensure that the camera output contains the specimen gray levels that are relevant and adjust the signal so that these gray levels are represented over as much of the 0.7-volt camera output as possible. The only way to do this knowledgeably and effectively is to have the capability of measuring the camera signal output independently of the monitor response. Without such capability we have found that it is virtually impossible to precisely or repeatedly optimize the output of the camera. Given the cost of a scientific-grade video system and the importance of maximizing the amount of specimen information that can be captured, the relatively low cost of obtaining some means of monitoring camera output is a good investment.

Traditionally the way of monitoring camera output has been to use an oscilloscope to examine the waveform of the live composite video signal as illumination intensity, gain, and black level controls are manipulated. This works well if one has an appropriate oscilloscope that can be dedicated for use with the video equipment and one becomes adept at displaying the waveform for the portion of the specimen that is of interest. Given that many researchers do not have an easy familiarity with the use of oscilloscopes and that these instruments are bulky, there is a great temptation not to use them. As an alternative, some image processors offer the ability to set a cursor line through the image of a specimen and then quantify the pixel values along that line. By positioning the cursor through the brightest and darkest portions of the image one can determine the characteristics of the camera output. This works best if the processor allows one to use a live image rather than a static captured image; the live image enables one to immediately see the results of camera control manipulations. For the figures shown in this chapter we used a RasterScope made by Dage MTI. This easy-to-use instrument, hooked up between the camera and the monitor, displays the camera output profile for a positionable cursor line through the live image of the specimen. The graphic representation of the camera output is superimposed on the image of the specimen. Once the camera is properly adjusted, the image of the camera output can be turned off.

1. If one has the means to monitor camera output, the coordinate adjustment of camera and monitor is relatively easy. First, set the camera and monitor controls to their "home" or neutral settings. Adjust specimen illumination so

that a reasonable image appears on the monitor. With the camera in manual or automatic gain modes, reduce the camera black level control (sometimes called "brightness") until the darkest portion (of interest) of the specimen along the cursor line is set to just above 0 volts. Then the gain control is raised until the brightest portion (of interest) in the specimen is set at just 0.7 volts. If the specimen illumination is a bit too low, one will not be able to raise the highlights to the full 0.7 volts. If so, repeat this sequence with increased illumination. Once these adjustments are made one can be assured that all the specimen gray levels of interest are included in the output of the camera and that the contrast is maximized.

To adjust the monitor first set the monitor to underscan the video signal thereby providing a reference background outside the image field. Then set monitor brightness at maximum and back this control off until the portions of the field outside the scanned area are just black. The monitor contrast control is then adjusted so that the lightest values are white with no "blooming" or spreading of the whites into adjacent areas.

2. If one does not have the means to assess camera output independently of the monitor image, there is no way to reliably ensure that the camera is optimally adjusted. Nevertheless, the empirically derived protocol that seems to work best for us is described below. Although this sequence can be used with the camera set in manual or automatic gain, the latter setting may be helpful.

First set the camera and monitor controls to their "home" or neutral settings. Then adjust the monitor brightness control so that the areas outside the scanned field just become black. Next adjust the camera black level control until the darkest portion of interest in the specimen image appears black. Readjust the monitor brightness control, if necessary, to reestablish a true black for the monitor. Then adjust the camera gain control until the lightest portions of the specimen image are just under saturation using blooming of the whites as an indication of overdriving the camera. It is all right if the whites appear a light gray on the monitor at this point. Next adjust the monitor gain control until the lightest values are just white without any blooming into adjacent areas. If the image on the monitor seems to be of unusually low contrast and noisy when the camera gain is set to its highest value, try raising the illumination intensity for the specimen and repeating the set up procedure.

When we evaluated the results of these empirical camera adjustments with the RasterScope device we found that we sometimes set the black level slightly too low leading to a loss of the very darkest grays. Also, the camera gain was often not set high enough so that we were not using maximum possible camera output for the brightest features of the specimen image. As a consequence, we often set the monitor gain a bit too high. In addition, we were not readily able to tell when the specimen illumination was slightly too low; under such conditions we were not using quite the full output of the camera and compensating for this with monitor gain.

In conclusion, these exercises revealed that we were consistently able to obtain better images of the specimen on the monitor when we had a means to directly evaluate the camera output. The implications of these findings for the use of image recording devices should be evident.

Acknowledgments

We thank Paul Thomas and Dean Porterfield of the Dage MTI, Inc. for the kind loan of a RasterScope during the preparation of this chapter. We also thank David Wolf, Ken Spring, Paul Thomas, and Ted Salmon for critical readings of the manuscript.

References

Inoue, S. (1986). "Video Microscopy." Plenum Press, New York.
Inoue, S., and Spring, K. R. (1997). "Video Microscopy, The Fundamentals." Plenum Press, New York.

CHAPTER 2

Electronic Cameras for Low-Light Microscopy

Keith Berland and Ken Jacobson★

Lineberger Comprehensive Cancer Center
★ Department of Cell Biology and Anatomy
University of North Carolina at Chapel Hill
Chapel Hill, North Carolina 27599

Todd French

Sandia National Laboratories
Nanostructure and Semiconductor Physics Department
Albuquerque, New Mexico 87185-1415

METHODS IN CELL BIOLOGY, VOL. 56

═══ I. Introduction

A. Low-Light Imaging in Biology

The demand for low-light-level cameras in biology is mainly in fluorescence imaging. Fluorescence microscopy normally produces images which are much dimmer than images from other common methods (e.g., phase contrast, DIC, and bright field), and high performance cameras are often necessary to get reliable low-light detection. Modern electronic cameras have become very sensitive and easy to use, facilitating their use in a wide variety of biological applications, ranging from imaging rapid changes in ion concentration in live cells to the distribution of labeled molecules in fixed specimens (Shotton, 1995; Wang and Taylor, 1989; Arndt-Jovin *et al.,* 1985). Sensitive cameras allow observation of cells labeled with a variety of fluorescent probes at illumination levels which cause insignificant perturbation of cellular function over extended periods, providing a means to study real-time changes in structure, function, and concentration. Detection with low illumination levels can also help reduce photobleaching or "fading" of fluorescent probes to manageable levels. Moreover, low-light electronic cameras are replacing the traditional role of film cameras for microscopic imaging.

There are many different types of electronic cameras and a wide variety of requirements in biological imaging. Thus, it can be difficult to determine which cameras best fit a particular application. The aim of this chapter is to provide an introduction to electronic cameras, including a discussion of the parameters most important in evaluating their performance, and some of the key features of different camera formats. This chapter does not present a comprehensive list of all available devices, a difficult task given the rapid and continuous changes in camera technology. The best sources of up-to-date information are the camera manufacturers. Rather, it is hoped that this chapter will provide a basic understanding of how electronic cameras function and how these properties can be exploited to optimize image quality under low light conditions.

B. The Importance of Proper Optical Imaging

Broadly speaking, most fluorescence imaging can be characterized as "low light," and highly sensitive cameras are invaluable for recording quality images. However, one should not overlook the microscope itself, since the actual image intensity which must be measured by the camera depends critically on the microscope and imaging optics. The goal in light-limited imaging will be to optimize the collection efficiency of the microscope (i.e., to get as much light as possible from a luminescent specimen to the camera). Choosing appropriate optical components for each application such as objectives, filters, and dichroic mirrors is thus very important. This may require some extra work, but is easy to do and well worth the effort since poor optical design can result in significant light loss. Using a high-quality microscope together with the proper optical components should result in substantially higher light levels reaching the camera. More light translates into better images, even for the most sensitive cameras.

There are several important factors which determine the collection efficiency of a microscope. First, the fraction of fluorescence emission that can be collected by the objective lens is determined by its numerical aperture (NA). Higher NA lenses collect more light. A second factor is the throughput (transmission) of the objective, which varies for different lens designs and is also dependent on lens materials and coatings (Keller, 1995). Manufacturers can be consulted regarding which lenses have higher throughput. Filters and dichroic mirrors also reduce the collection efficiency, although they are necessary in order to attenuate the illumination relative to the fluorescence emission. These elements should be carefully chosen to optimize the signal relative to the background, based on the excitation and emission spectra of the particular fluorescent molecule to be imaged. Finally, microscopes often have multiple mirrors, lenses, and windows between the back aperture of the objective and the camera, resulting in light loss. Microscopes with multiple camera ports do not usually send the same amount of light to each port, and one should determine which receives the most light. When low-light imaging is a priority, it may prove fruitful to upgrade the microscope itself since many designs now have a direct optical path from the back of the objective to the camera which can improve throughput substantially.

Table I shows the total fluorescence detection efficiency for two different cameras and highly optimized optics. The total collection efficiency of about 10% represents an optimistic upper limit on the collection efficiency. Although it may be difficult to actually reach 10% with any given microscope or fluorescent probe molecule, it should be possible to get reasonably close, say, within a factor of 2 to 10. On the other hand, it is also not uncommon to collect much less light than the numbers shown in the table. For example, use of low-NA objectives or a camera port that receives, for example, 50% (or less) of the light will both reduce the efficiency. Other potential sources of light loss include: the use of too many or inappropriate filters; use of DIC microscopes without the slider or analyzer removed; dirty optics, perhaps covered with dust or immersion oil; and optical misalignment. Without proper attention to optical imaging, the total collection efficiency could easily be much less than 1%, or even less than 0.1%, throwing away much needed light!

C. Detection of Low Light

The total detection efficiency can be used to estimate the camera signal amplitude for various applications, beginning with a single fluorescent molecule. If we estimate that with moderate illumination a single probe will emit about 10^5 photons/sec, the microscope and two cameras from Table I would detect 10,000 and 1000 photons, respectively, for a 1-sec exposure (neglecting photobleaching)

Table I
Factors Which Determine the Efficiency of a
Fluorescence Microscope

Microscope elements	Optimized collection efficiency
Objective lens: Numerical aperture[a]	.3
Objective lens: Transmission	.9
Dichroic mirror	.8
Microscope throughput[b]	.8
Filters[c]	.6
Total collection efficiency	~10%
Total detection efficiency (Camera = 50% quantum efficiency)	5%
Total detection efficiency (Camera = 10% quantum efficiency)	1%

Note. The total collection efficiency is given by the product of the different factors shown. Some of the values shown will be difficult to achieve, as described in text. The total detection efficiency represents the percentage of emitted photons detected by the camera.

[a] Represents the fraction of emitted light that is collected by the objective.

[b] The light transfer efficiency from the objective to the camera, excluding other listed elements.

[c] Filter transmission can be higher if excitation and emission wavelengths are well separated.

(Tsien and Waggoner, 1995; So *et al.*, 1995).[1] On the other hand, a shorter exposure, say, one video frame (1/30 second exposure), would detect 330 and 33 photons, respectively. A very important point one should notice from these numbers is that it is possible to measure reasonably large numbers of photons, especially with a high quantum efficiency camera, even from a single fluorophore provided it does not photobleach too rapidly (Funatsu *et al.*, 1995; Schmidt *et al.*, 1995). The microscope magnification and camera pixel size will determine how many pixels the image from a single fluorophore will be spread across, as discussed below. In any case, the technology is available to reliably measure fluorescence even from only a small number of probe molecules, particularly when the optical collection efficiency is also optimized. The remainder of this chapter is devoted to the discussion of electronic camera technology.

[1] This photoemission rate can vary greatly depending on the photophysics of the specific probe and the illumination level, but 10^5 is a reasonable estimate for this illustration. In fact, this number can realistically, in some cases, be 10 to 1000 times larger or as low as desired if illumination is reduced. Photobleaching will normally limit the total number of photons emitted, since fluorescent molecules often will photobleach after 10^3–10^5 excitations. Thus, the rates of 10^5 or higher may not be sustainable, depending on the exposure time. If long or multiple exposures are needed, one would often use lower-intensity illumination to limit photobleaching. One might also use lower illumination to limit perturbation of live cell preparations.

II. Parameters Characterizing Imaging Devices

The fundamental purpose of an electronic camera is to generate an electrical signal which encodes the optical image projected on its surface. When the light intensity is reasonably high, there are a number of cameras which perform very well. As the image intensity declines, it becomes increasingly more difficult to obtain high-quality images. Applications with the lowest light levels generally require some compromises between spatial resolution, temporal resolution, and signal-to-noise levels. However, as cameras become more sophisticated the required compromises are less severe. Ideally, a camera would record the arrival of every photon with high spatial and temporal precision, without introducing any extraneous signal, or noise. Of course, ideal detectors are hard to come by, and it is thus important to understand how real detectors perform. In the following section we discuss the key parameters which determine a camera's characteristic response. Most modern low-light cameras use solid state, charge coupled device (CCD) chips as the light-sensing detector (Hiraoka et al., 1987; Aikens et al., 1989). Therefore, the following sections will concentrate on CCD camera performance, although the basic ideas should be valid for any camera type. There are many excellent references available for readers interested in more details about vacuum tube-based cameras as well as solid state cameras and microscopy in general (Inoué, 1986; Bright and Taylor, 1986; Tsay et al., 1990).

A. Sensitivity and Quantum Efficiency

Camera sensitivity refers to the response of the camera to a given light level. A convenient measure of this quantity is the quantum efficiency (QE), which refers to the fraction of photons incident on the detector surface that generate a response (photoelectrons). Different CCD materials and designs can have vastly different quantum efficiencies, from a few percentages to above 80% in some devices. The higher quantum efficiency cameras are generally more sensitive, although a camera's minimum noise level must also be considered to determine how "sensitive" it is. Camera sensitivity is thus sometimes expressed in terms of the minimum amount of light which can be reliably measured, often called the noise equivalent floor. In this case, a lower sensitivity rating means the camera can detect dimmer light (i.e., is more sensitive). To be completely clear, it is best to specify the signal-to-noise ratio (see below) which can be achieved for a given light level. The use of different definitions in reporting the sensitivity can lead to confusion, and the manufacturer should be consulted when data sheets are unclear.

1. Spectral Response

The quantum efficiency of an image detector is wavelength dependent. This property is called the spectral response and is due to the nonuniform response of the device materials to different colors of light. Some typical curves of quantum efficiency versus wavelength are shown in Fig. 1 for three different CCDs. Most CCDs have their peak response in the visible wavelengths and are normally more sensitive to red light than blue. Special coatings are sometimes added to the detector surface to alter the spectral response, particularly for work in the ultraviolet range. It is important to verify that a camera has reasonably high sensitivity to the wavelengths of light that are to be imaged in a particular experiment.

2. Units

Camera sensitivity is also often reported in units of radiant sensitivity, which is equivalent to the quantum efficiency, but expressed in units of amps per watt rather than electrons per photon. One can easily switch between quantum efficiency and radiant sensitivity using:

$$ QE = 1240 \, \frac{R_S}{\lambda} , $$

where R_S is the radiant sensitivity in amps per watt and λ is the wavelength in nanometers. Both quantum efficiency and radiant sensitivity are radiometric units.

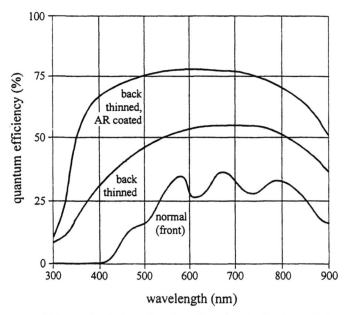

Fig. 1 Quantum efficiency of a CCD as a function of wavelength, showing typical curves for three devices: normal (front face), back-thinned, and back-thinned with anti-reflective coating.

When the sensitivity is reported as the smallest resolvable signal, as mentioned above, it is usually expressed in photometric units such as lux, lumen, or candela. Quoting sensitivity in lux is ambiguous unless the light spectrum is specified. Photometric units are adjusted for the response of the average human eye to different colors of light and thus differ from radiometric units such as energy or number of photons. The integrated product of the radiant quantity and the efficiency of the eye yields the corresponding photometric (or sometimes called luminous) quantity. The luminous efficiency function of the eye is called the CIE $V(\lambda)$ curve (CIE publication, 1990). Note that one lumen is equivalent to 1/683 watts (~1.5 milliwatts) of monochromatic light whose frequency is 540×10^{12} Hz (a wavelength of about 555 nm in air). Figure 2 shows the conversion between radiometric and photometric units as a function of wavelength for monochromatic light.

3. Fill Factor (CCD Format)

The sensitivity of a CCD detector is also affected by its geometry. Not all the light reaching the CCD falls on active pixel areas, as some of the surface may be masked (covered) as part of the device design for transferring stored photo-electrons. The surface electrodes also obscure some incoming light. Interline transfer CCDs, for example, have inactive space between each row of pixels for charge transfer, resulting in reduced collection efficiency. The fill factor refers to the fraction of the detector that is active, determined mainly by the surface area of the pixels. For interline transfer CCDs, the fill factor can be as low as 25%. This loss is sometimes recovered by positioning microlenses over each pixel to collect more light, increasing the effective active pixel area and thus improving the fill factor. Other CCD formats, such as frame transfer CCDs, do not have masked areas between pixels and have higher fill factors, often quoted as 100% although some light may still be obscured by electrodes and surface electronics.

B. Camera Noise and the Signal-to-Noise Ratio

Electronic imaging is complicated by the presence of noise, which degrades image quality. If the response of a camera to a given light level is below the noise level, the image will not be resolved. The most important measure of a camera's response is therefore not the sensitivity itself, but rather the signal-to-noise ratio which can be achieved for a given light level reaching the detector. The signal-to-noise ratio (S/N) is defined as the signal level generated by the camera's response to light, divided by the average noise level. The higher the signal-to-noise ratio, the better the image quality. The signal-to-noise ratio is often expressed in alternate units, either decibels (dB) or bits. To convert the signal-to-noise ratio to decibels, the following formula is used:

$$(S/N)_{dB} = 20 \log (S/N).$$

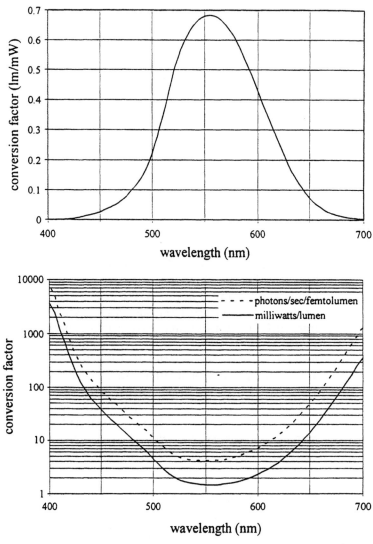

Fig. 2 Factors for converting between radiant flux (milliwatts or photons/second) and luminous flux (lumens) for monochromatic light. The two panels are shown for reader convenience, although they contain the same information.

The equivalent number of bits can be found by calculating the base 2 logarithm of the signal-to-noise ratio or dividing $(S/N)_{dB}$ by 6, since 1 bit is equivalent to 6 dB.

Noise plays an important role in determining the performance of imaging devices, particularly in low-light-level imaging where the signal is small by definition (see Section III, D about amplifying the signal). To achieve a good signal-to-noise ratio therefore requires a camera that has very low noise. Knowledge of noise sources and how their effects can be minimized is thus valuable in understanding how various cameras are designed and in determining which camera types are most appropriate for a given application.

1. Readout Noise

Intimately coupled to the design of image detectors is the method they use to transfer the recorded image from the detector to a useable electronic format. This process, called readout, will introduce noise to the signal. The noise level is independent of the amount of light reaching the camera, and high readout noise will reduce the signal-to-noise ratio, particularly for low-light images. The readout noise level is highly dependent on the readout rate, increasing approximately as the square root of the readout rate. Thus, cameras with lower readout rates will have better signal-to-noise ratios than cameras with higher readout rates, other things being equal. The cost of a lower readout rate is reduced frame rate. For high-end video CCDs, read noise can be as low as 50 electrons/per pixel, increasing to several hundred electrons per pixel for lower-end devices. The readout noise level is determined by the electronics of each device, but some typical numbers for high-end electronics are 5 electrons per pixel at 50-kHz, 20 electrons at 1-MHz, and 50 electrons at 20-MHz readout rates.

2. Shot Noise

Using an electronic camera to detect an image is essentially an experiment in counting the number of photoelectrons stored in the pixel wells. The electron counting is a random process and will follow Poisson statistics. The uncertainty in the measured result, called shot noise or photon noise, will be the familiar "square root of N," where N, in this case, is the number of electrons collected in each pixel. Shot noise is intrinsic to the photon statistics of the image and cannot be reduced. As the light level increases, the shot noise increases, eventually exceeding the level of other noise sources (this is a good thing). When shot noise is the dominant noise source for an image, the signal-to-noise ratio grows as the square root of the number of photons measured at any pixel. This condition is the best obtainable; that is, the signal-to-noise ratio cannot exceed the shot noise limit. It will vary between applications whether the camera noise sources are most important or are insignificant when compared with shot noise.

3. Background and Noise

There are a number of sources of background signal in electronic imaging. Thermally generated electrons get collected in the same pixels as the photoelectrons. These electrons, called dark charge, are generated even when the CCD is not illuminated.[1] The associated noise, called dark noise, will grow as the square root of the dark charge, as does shot noise. Other factors, such as Rayleigh and Raman scattered illumination or stray light leaking through to the camera, can also contribute to the background signal, again contributing to the noise as the square root of the number of electrons generated. The background level ultimately lowers the dynamic range (discussed below) of the camera and raises the noise. With image processing, the average background level can be subtracted from an image, although this subtraction will not reduce the noise. While it is valuable to understand the various noise sources, there is no need in actual experiments to distinguish between shot noise and various other background noise sources. Rather, the total noise (excluding readout noise) can be calculated directly as the square root of the total number of electrons collected in any pixel.

Generally, for video rate applications, the dark noise is not as important as readout and shot noise. On the other hand, for longer image acquisition times, dark noise can be significant at room temperature. By cooling the camera with thermoelectric coolers or liquid nitrogen, dark charge, and thus dark noise, can be drastically reduced. High-quality cooled scientific cameras can achieve extremely low dark charge levels of less than a single electron per pixel per hour. This is in contrast to the dark charge of a few hundred electrons per pixel per second for standard, uncooled cameras.

Cooling is not the only strategy used to reduce dark charge. Some camera chips have multipinned phasing (MPP), which helps reduce collection of surface electrons on the CCD chip during image integration. MPP devices can reduce dark charge several hundred times. The major drawback of MPP is that it usually results in a reduction of pixel well capacity and thus limits the dynamic range.

4. Example Calculation

Assume a cooled slow-scan CCD camera has quantum efficiency of 50%, read noise of $N_r = 10$ electrons per pixel, and is cooled such that much less than one thermal electron is generated per pixel during a 1-sec exposure. What will be the signal-to-noise ratio when the number of photons per pixel per second reaching the camera is 100, 1000, and 10,000? First, the number of electrons generated is given by the number of photons multiplied by the quantum efficiency; or 50, 500, and 5000. The shot noise, N_s, is then 7, 22, and 71, respectively. The total noise is given by $N = \sqrt{N_r^2 + N_s^2 + N_d^2}$, where the dark noise, N_d, is negligible for this example. Total noise is thus 12, 24, and 71 electrons, with corresponding signal-to-noise ratio of 4, 19, and 70, respectively, or 12, 26, and 37 dB.

How would the signal-to-noise ratio be changed using a 1-sec or 1-min exposure if the same camera is illuminated by 1000 photons per pixel per second and also has a dark charge of 50 electrons per pixel per second? The total number of

---

[1] Camera data sheets often specify a dark current, in amps per square centimeter, rather than dark charge, in electrons per second per pixel. The conversion between the two is determined by the size of each pixel and the charge of the electron, which is 1.6×10^{-19} Coulombs.

photons reaching the detector would be 1000 (or 60,000 for the long exposure) generating 500 (or 30,000) electrons. The total dark charge for the two integration times is 50 (or 3000) electrons. The sum of dark noise and shot noise would then be 23 (or 182). Adding in read noise results in noise of 25 (or 182) electrons. The signal-to-noise ratio is then 20 (26 dB) or 165 (44 dB). Note that the shot noise limit for these two cases is a signal-to-noise ratio of 22 (27 dB) for the 1-sec exposure and 173 (45 dB) for the 1-min exposure. The signal-to-noise ratio can of course be calculated in one step rather than the two shown, but this method nicely illustrates how read noise becomes less significant for long exposures. Figure 3 further illustrates this point with plots of the signal-to-noise ratio as a function of signal intensity for three different values of the read noise.

C. Spatial Resolution

Electronic cameras form images by sampling the optical image in discrete pixels, as shown in Fig. 4. The spatial resolution of a camera system is determined by the pixel spacing and is independent of the optical resolution of the microscope, although limited by it (Castleman, 1993). In order to resolve image features of a given spatial frequency, the camera must sample the image with at least twice that frequency. This requirement can be roughly understood as simply stating that image details will be resolved only when the pixel spacing is smaller than the scale over which image features change. As an extreme but intuitive example, one can imagine a "camera" with only a single large pixel, for example, a photomultiplier tube, which can measure light intensity, but has no spatial

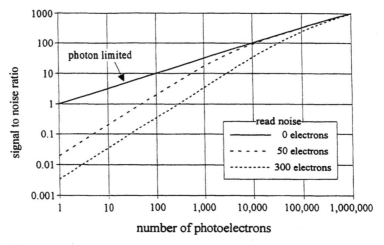

Fig. 3 The signal-to-noise ratio as a function of image intensity, assuming negligible dark noise. The three curves correspond to three different values for the read noise.

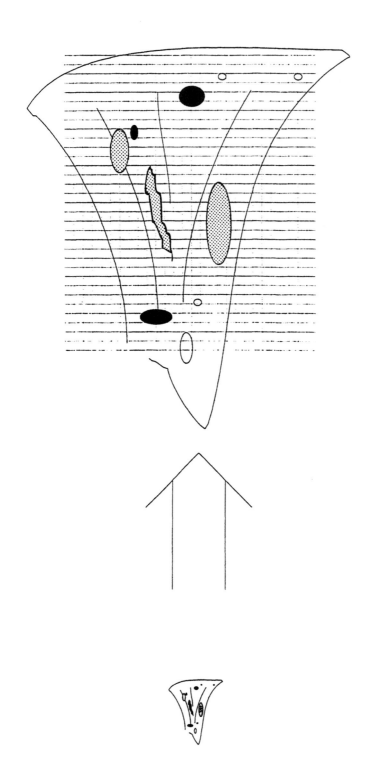

Specimen

Microscope magnification

Magnified image projected onto the camera

Fig. 4 Spatial resolution in electronic cameras is determined by the pixel size as described in the text. The magnified image is projected onto the camera, where it is sampled in discrete pixels. Image features which change over distance scales smaller than two pixels are not resolved in the electronic image. A real camera would have many more pixels than this cartoon.

resolution. By dividing the detector into pixels, the image will begin to be resolved and as the number of pixels is increased, higher levels of detail will become apparent, up to the diffraction-limited resolution of the microscope optics. CCDs generally have a pixel spacing from about 2–40 μm. The smallest resolvable details in the magnified image on the detector face plate are therefore about 4–80 μm in size, due to the sampling requirements stated above.

Most image detectors are physically small devices, typically 1-inch square for image tubes and even less for many CCD chips. Choosing how much an image is expanded or compressed in projecting the image on the device will involve a tradeoff between the resolution and the field of view. Image magnification also involves a tradeoff between spatial resolution and image brightness—the more an image is magnified, the more the light from each fluorophore is spread out, reducing the signal in any given pixel and thus the signal-to-noise ratio (Shotton, 1993).

1. Example

The size of a microscope image is governed by the object size and the microscope magnification, the smallest size being limited by the point spread function of the microscope (diffraction limited resolution). The spatial resolution of an imaging system can easily be calculated if the camera pixel size and magnification are known. Consider a CCD camera with 10×10-μm square pixels on a microscope with a 63X objective (NA = 1.4) and no further magnification. For each dimension of the camera chip, the 10-μm pixel at the camera corresponds to 159 nm (10 μm ÷ 63) in the specimen plane. This camera/lens combination would thus correctly sample image details down to 318 nm (2×159). This would not, however, fully exploit all the available image detail. The Rayleigh criteria for the optical resolution limit is given by 0.61λ/NA, which specifies the minimum separation at which two point objects can be resolved (Castleman, 1993). Assuming we have 500 nm light, this limit is 218 nm for a 1.4-NA objective. To correctly recover all the available image detail would require sampling at 109 nm (218 ÷ 2) pixel spacings. This could in fact be achieved with additional magnification of approximately 1.5×, which would sample the image every 106 nm. This extra magnification can be added to the microscope or sometimes to the camera itself.

D. Linearity and Uniformity

Generally, cameras are designed to be linear, meaning the detector has a linear relation between input signal intensity and output signal. This feature is essential for quantitative imaging and must be verified for each camera. A CCD is a linear detector, although linearity is lost if the detector is overexposed. In this case, photoelectrons from adjacent pixels overflow into neighboring wells causing blooming. Antiblooming circuitry and the improved design of modern CCDs reduces this problem. Also, the camera response across the detector surface is not always uniform and leads to image distortion or shading. The camera response is not the only source of this problem, since the illumination itself is often not uniform over the entire field of view. Shading (flat field) correction can be added to some cameras to electronically correct their response or the image may be corrected at a later time using image processing provided that the necessary correction is known from imaging a uniform sample.

Sometimes, video cameras are intentionally coupled to an output stage that is not linear. Two common sources of nonlinearity built into a video CCD camera are automatic brightness control and gamma correction. Automatic brightness control automatically changes the output preamplifier's gain when the number of stored charges in a pixel becomes too large. This change in gain is not normally under the camera operator's control and is therefore difficult to compensate for in quantitative imaging. The best solution is to turn off automatic brightness control. Many cameras (especially video CCD cameras) also have gamma-compensated output. Gamma correction is useful to properly transfer an image from the detector to a video monitor but is generally harmful to quantitative analysis. Monitors have a nonlinear response to the voltage applied to them. This nonlinearity is closely modeled by a power function:

$$\text{monitor\_brightness} = \text{monitor\_voltage}^{\gamma},$$

where the exponent gamma (γ) is about 2.5. In order to present a linear response, the signal applied to a monitor must be corrected with a inverse gamma exponent:

$$\text{monitor\_voltage} = \text{linear\_signal}^{(1/\gamma)}.$$

Cameras which perform gamma correction convert the linear CCD signal into a gamma-corrected one (using the second equation). This means a digitizer will quantize the data at nonlinear intensity levels. The advantage is that the first digitization controls the precision in the signal, and there is no further loss of precision when using a digital to analog converter for the video signal sent to the monitor. The disadvantage is that to make quantitative comparisons in the linear space of the original CCD signal, some of the original signal precision may be lost.[2]

E. Time Response

Cameras do not always respond immediately to a change in input intensity, introducing lag, and a decrease in temporal resolution. Lag can be introduced in several ways, including the time response of the phosphor screen in image intensifiers and the incomplete readout of all accumulated charge in any one frame of tube cameras. Lag is not significant in CCD cameras. If fast time resolution is of interest, it is very important to make sure the camera lag does not

---

[2] For more information about gamma and digital images see http://www.inforamp.net/~poynton/Poynton-color.html.

affect the measurement. For example, ratio imaging will likely contain artifacts if performed using a camera with lag, such as a silicon intensified target (SIT) camera.

F. Dynamic Range

There are a finite number of intensity levels, or gray levels, which can be detected by a given device, which can limit the detection of fine details. The range of levels is called the dynamic range. High dynamic range is important whenever there is a wide range of signal levels which must be recorded for a given imaging application. Dynamic range may be specified both within one frame (intrascene) and between frames (interscene). The dynamic range is determined by two factors: the overall operating range, or pixel well depth in the case of CCDs, and the average noise level, and is calculated by dividing the maximum signal level (for each pixel) by the minimum average noise level (excluding shot noise).

For example, a CCD with pixel capacity of 200,000 electrons and an average noise floor of 50 electrons has a dynamic range of 4000:I. Normally, like the signal-to-noise ratio, the dynamic range is expressed in either decibels or the equivalent number of bits. For this example, 4000 gray levels is equivalent to 72 dB or 12 bits.

Another way to think about dynamic range is that it specifies the smallest difference in signal level that can be resolved when the camera is operating at full range; that is, what is the smallest difference in image brightness that can be observed on top of a bright background. While this second definition is technically not quite correct, it is useful in as much as it gives some intuitive feeling for what the dynamic range is. The signal-to-noise ratio and dynamic range should not be confused. While the dynamic range describes the camera's ability to respond over a range of signal levels, the signal to noise is dependent on the conditions for each image acquired, comparing actual signal levels and camera noise. The signal-to-noise ratio is always less than the dynamic range.

The dynamic range of a camera can be limited by the choice of digitizer. For most video-grade CCDs this is not a problem, since they often have a dynamic range of only 40 dB or less, which is less than a dynamic range of an 8-bit digitizer (256:1). On the other hand, scientific-grade CCDs may have a dynamic range of greater than 100 dB (16 bits or 66,000:1). In this case, a digitizer with equivalent precision is required or digitization of the camera signal will reduce the dynamic range.

G. Gain

Gain refers to signal amplification by the camera to create an electrical signal which can be accurately measured. While sensitivity describes the input response of the device, gain determines the total output response. Gain stages can only increase the noise, but some gain is normally necessary to transit a signal which can be reliably read by the digitizing electronics. If the digitization is performed in close proximity to the detector, the gain requirements are less. Not all gain is provided by an electronic amplifier. Sometimes it is an inherent part of the photodetector, as in SIT cameras and image intensifiers, although these devices still have electronic gain as well.

When gain can be changed between frames, the camera can accommodate an interscene dynamic range that may be several orders of magnitude larger than the intrascene dynamic range. However, the response characteristics of a camera are normally not precisely known for different gain settings. For example, altering the gain may affect the slope of the camera's linear response or may even result in a nonlinear response. These differences must be measured and calibrated in order to use the detector with multiple gain settings. Without precise knowledge of the response of the detector at different gain settings it is impossible to make quantitative measurements that can be compared. Therefore, when the gain response is not well known, it is important that all the images of a time sequence, for example, use the same gain setting for proper quantitative analysis. Unfortunately, some images may have points that exceed the normal intrascene dynamic range. However, without changing the gain there is nothing which can be done, and the detector gain should be set to the best compromise. One exception is the gain of the output preamplifier of CCD cameras, which does not normally effect the response of the detector. However the effect of the system gain *must* be measured to verify linearity.

H. Frame Rate and Speed

Different applications require image acquisition at varying speeds or frame rates. The frame rate is determined by the pixel read rate and the number of pixels in the image. For a fixed number of pixels, the main drawback to higher frame rates is increased noise due to higher readout rates (faster readout also has less light per pixel per frame). Digitization is normally the rate-limiting step in image readout. For a certain level of noise and digitization depth (number of bits), the maximum digitization rate is consistent across all manufacturers. The frame rate is then determined by the number of pixels. Most video cameras have a fixed frame rate of 30 frames/sec. Scientific cameras and other specialized systems can offer variable frame rates. If a camera contains many pixels (one million or more), the frame rate will in almost all cases be lower than video rate, particularly for low-light applications. This is because the high read noise associated with fast readout would overwhelm the small signal levels inherent to low-light imaging, resulting in poor signal-to-noise ratios.

I. Resolution, Sensitivity, and Imaging Rate

A properly operated electronic camera is essentially an array of independent photodetectors. When the signal-to-noise ratio is reasonably high for most pixels, the result will be high-quality images. Various camera design choices employ different strategies to maximize signal-to-noise ratios, usually involving some compromise between spatial resolution, frame rate, and the signal-to-noise ratio. For example, when the signal per pixel is small compared with readout noise, one can simply acquire an image for a long time, provided the camera has low dark noise, in order to build up the required signal level. This describes the basic design of slow-scan CCD detectors, which increase the signal-to-noise ratio at the expense of time resolution. Another strategy is to use a camera with fewer (larger) pixels, effectively increasing the amount of signal per pixel. This will also increase the signal-to-noise ratio at the expense of spatial resolution. Conversely, for a given light level reaching the camera, increasing spatial resolution (more smaller pixels) will lead to a decrease in the signal-to-noise ratio as will increasing the frame rate. The required resolution and the image characteristics (intensity and spatial and temporal frequency) will determine which strategy and suitable camera design to employ for a given application.

Sometimes, when spatial resolution can be sacrificed, a technique called binning can be used to optimize the signal-to-noise ratio or the frame rate for a given application. Binning is an electronic technique, available only on certain camera models, that combines the stored charges of adjacent pixels on the CCD chip. Binning is effectively equivalent to using a camera with a smaller number of larger pixels than the actual camera in use—there is essentially no time loss associated with binning since combining electrons from multiple pixels can be completed much faster than pixel readout. There are several reasons one might choose to use binning. One is that there will be fewer pixels to digitize per frame, which will allow for higher frame rates. Another reason is to increase the signal level by increasing the active area of each effective pixel. This is especially useful for low-light situations where the number of photoelectrons collected by each single pixel is low compared to the readout noise. Getting more light to each pixel can also be accomplished using lower optical magnification, but electronic binning, when it is available, is certainly more convenient than changing optical hardware. The obvious drawback in either case is reduced spatial resolution. Another drawback of binning is that the full well capacity cannot always be used without saturation, since binning transfers all of the charge from multiple wells into one well before readout. This should not be a problem in most low-light imaging applications since the pixels are not full, and binning is used to increase the number of electrons to measurable levels.

III. Specific Imaging Detectors and Features

There are currently many options available in choosing electronic cameras, and devices with improved performance characteristics continue to evolve. More cameras are being manufactured with application-specific designs, resulting in exciting possibilities for biological imaging. In this section, we present some of the basic camera types which are commonly available and highlight the major features. For up-to-date information on the latest product specifications one should consult camera manufacturers. In choosing a camera for special applications, it is wise to evaluate several models, on a trial basis, in order to compare their performance and image quality using an investigator's actual samples.

A. Video CCD Cameras

Monochrome video CCDs are the most commonly available and least expensive CCD-based cameras. These are not considered low-ligh cameras, since they do not have particularly high quantum efficiency (10–20%). The typical minimum sensitivity of low-end devices is 0.5 lux (noise equivalent) which is barely low enough for microscopy. High-end video CCD cameras can have a sensitivity as low as 0.01 lux which can be used for fluorescence imaging of brightly labeled specimens. Readout noise is the main noise source since dark noise is usually minimal for video rate acquisition. The high readout noise and relatively low pixel capacity (36,000–100,000 electrons) limit the dynamic range. In the highest quality video CCDs, the dynamic range can be as high as 8 bits, although noise often limits the true dynamic range of many video CCDs to around 6 bits. Typical spatial resolution is approximately 700×500 pixels (interlaced) on a $\frac{1}{4}$- to $\frac{2}{3}$-inch chip. Video CCDs have a fixed frame rate of 30 Hz. These basic cameras have little flexibility, but are easy to use and compatible with standard video equipment.

Some video CCD cameras have been modified to enhance performance and increase flexibility. Circuits have been added which allow variable "on-chip" integration time, from a short 100 microseconds (300 times less than video rate exposures) up to several hundred seconds. Long integration times are very useful for low-light applications, increasing the total amount of light reaching the detector before it is read. On-chip integration has a major advantage over frame-averaging standard video camera output, since on-chip integration incurs read noise only once, rather than each time the image is read for averaging. Variable frame rate video CCDs incorporate cooling to reduce dark charge, which is important for the longer exposure times. These modified video CCDs are much more powerful and flexible than standard video cameras and provide an economical option for low to moderate light level imaging.

B. Slow-Scan CCD Cameras

Slow-scan, or scientific CCDs, are designed specifically for low-light applications, with the ideal attributes of both very high sensitivity and low noise. Slow-scan CCDs are currently the best available low-light cameras, provided fast time resolution is not required. These cameras are substantially more flexible than video cameras and this flexibility can be exploited to achieve superior image quality for many applications. They come in many different formats, and the particular CCD chip will determine the camera performance. The quantum efficiency of scientific CCDs ranges from 20% to above 80%, significantly better than video devices. The highest quantum efficiency is found in back-illuminated CCDs, which collect the image from the back surface of the chip rather than the front in order to reduce light losses incurred in passing through the electrodes and protective layers on the front surface of the chip. Applying antireflective coatings can also help achieve maximum quantum efficiency. Back-thinned CCDs are more expensive than other scientific CCDs, and the application will determine whether this added sensitivity is required and worth the investment.

In addition to high sensitivity, scientific CCD cameras also have very high spatial resolution, which is achieved by using CCDs with more pixels. Arrays may contain as many as 5000 × 5000 pixels. Some devices are built such that they can be placed side by side to create even larger pixel arrays. The growth in the number of pixels results in longer read times and thus slower frame rates. This should not pose a major problem unless fast time resolution is required, in which case the highest spatial resolution may be too much to hope for at present, at least for low-light applications. If only a small number of images are required, frame transfer CCDs can be used to acquire several images rapidly, with the images stored on inactive areas of the CCD and read out later at slow readout rates. This approach works because charge transfer to the inactive regions of the CCD chip can be completed much faster than the actual readout. As mentioned above, binning can also be employed to increase the frame rate.

Slow-scan cameras have very low background and, hence, low noise, allowing for long image acquisition times. The readout and thermal noise are minimized through slow readout rates and cooling or multipinned phasing. Readout noise as low as 5 electrons per pixel is common, and dark charge can be reduced to just a few electrons per pixel per hour. This results in the potential for very high signal to noise ratios in high light conditions or with long integration times for low light levels. Basically, using slow-scan CCDs, low-light images can be integrated (on-chip integration) for whatever time is necessary to achieve a reasonable signal-to-noise ratio, provided the sample is stable. The low noise also makes it possible to exploit the full dynamic range of each pixel. The electron well capacity of scientific CCDs is generally at least several hundred thousand, with some devices holding over one million electrons per pixel. The dynamic range of scientific cameras is 10–16 bits (1000:1–66,000:1), much greater than the 8-bit (maximum) video CCD.

C. Fast-Scan CCDs

When experiments require greater time resolution than slow-scan CCDs provide, fast-scan CCDs are another option. These are integrating cameras, like the slow-scan CCD, but have faster readout in order to achieve higher frame rates. Unfortunately, high readout rates result in increased noise, and the specimen must be fairly bright to acquire good images with these cameras. Fast-scan cameras generally sacrifice noise, dynamic range, or resolution for the higher frame rates. The advantage of the fast-scan camera over an intensified video camera (see below) is greater resolution, higher dynamic range, and ease of use (Ramm, 1994). These features can both be important if signal levels are highly variable. The fast scan camera also features variable integration times, providing flexibility for different applications. The high readout noise limits the dynamic range to 8–12 bits compared to the 10- to 16-bit resolution of comparable slow-scan cameras.

D. Intensified Cameras

In some imaging applications, the amount of light reaching the detector may be too low to acquire high-quality images in the requisite amount of time. Image intensifiers are designed to overcome this limitation by amplifying the amount of light reaching the camera. Figure 5 presents a schematic of an intensified CCD camera setup. In the nonintensified camera, the poor image quality is due to the low signal-to-noise ratio; that is, the low light levels do not result in a signal level sufficiently above the noise. Image intensifiers can increase the signal level substantially, without a corresponding increase in dark noise or readout noise, resulting in better signal-to-noise ratios. Image intensifiers do not reduce the shot noise inherent in the image and actually add some noise of their own to the system. If there were no dark noise or readout noise, the image intensifier would be of no help. The increase in signal-to-noise ratio, without the need for long integration times as with slow-scan cameras, makes intensified cameras very useful for low-light applications where fast time resolution is required, such as measuring the kinetics of rapid calcium changes.

A second major application of image intensifiers is to achieve very high time resolution through gating. Intensifiers can normally be turned on and off very

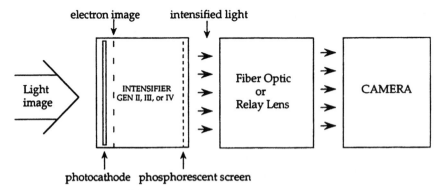

Fig. 5 Schematic of an intensified CCD camera. Light reaching the photocathode generates an electron image which generates an intensified light image upon striking the phosphorescent screen. The image on the screen is then projected to the camera.

rapidly, for some devices in less than 1 nsec. Thus, light can be collected from very short time windows to record transient events. Another application of high frequency gating is fluorescence lifetime imaging (Lakowicz and Berndt, 1991; Marriott *et al.,* 1991).

1. Intensifier Formats

There are several different types of image intensifiers. Gen I intensifiers employ vacuum tubes similar to standard tube-type cameras. Electrons generated at the photocathode (each by a single photon) are accelerated onto a phosphorescent screen. The electrons striking the screen produce multiple photons in proportion to the electron energy, thus amplifying the signal. The image produced on the phosphor screen is then viewed by a suitable camera. Typical gain for a Gen I intensifier may be as high as several hundred times. The quantum efficiency of the cathode is generally about 10%.

Gen II, Gen III, and Gen IV intensifiers employ a microchannel plate, which is a collection of small glass tubes, with metal surface coatings on the inside walls. Electrons collide with the surfaces as they are accelerated through the tubes, resulting in a cascade of electrons reaching the phosphorescent screen. The gain can be very high, up to a million times. The main differences between Gen II, Gen III, and Gen IV intensifiers are the efficiency of the cathode material and the method of construction. These intensifiers can have respectable quantum efficiencies, reaching 30–50% in high-quality devices and can sometimes have fairly low noise as well.

While intensifiers can greatly increase sensitivity, there are some drawbacks. First, their spatial resolution may be limited, depending on the size and spacing of the glass capillaries and the phosphor material. Although in some cases the intensifier may have comparable resolution to the camera itself, it is probably best not to use an intensifier if very high spatial resolution is required. The second drawback is that intensifiers often exhibit some nonuniformity and therefore require flat field correction. Intensifiers also often produce images with a lot of "snow," due to thermally generate background noise. Finally, the phosphor screen may introduce lag, due to long-lived phosphorescence. Proper choice of phosphor material can reduce lag at the cost of image intensity. If very fast time resolution is important in a particular application, the intensifier and phosphor material must be chosen appropriately.

Image intensifiers must be coupled to the camera in some manner. The two main methods are lens coupling and fiber optic coupling. The principle difference is in the coupling efficiency. Lens coupling is easy, but the transfer efficiency from the phosphor screen to the camera is typically less than 10%. Fiber coupling, on the other hand, results in 50–60% transfer efficiency. Lens coupling of image intensifiers is more convenient if the intensifier and camera need to be separated; for example, if either component fails or to use the camera alone without an intensifier.

2. Cameras for Image Intensifiers

All the camera types described above can be coupled to image intensifiers, although it is rare to couple them to slow-scan scientific cameras. Intensified video CCDs are very popular and reasonably affordable low-light cameras. The main drawbacks are high noise and low dynamic range (less than 8 bits). The low dynamic range may cause difficulty in trying to find the proper gain settings or may not be sufficient to resolve the details of images with wide variations in signal level. Fixed frame rate and lack of flexibility are limitations just as with the nonintensified video CCDs. An expensive but attractive option is to couple an image intensifier to a fast-scan scientific CCD. The major advantage of this pairing is the much larger dynamic range of 8–12 bits, which is useful when using high-end intensifiers. This configuration is most commonly seen in high-speed (faster than video rate) applications.

E. Color Video Cameras

Color cameras, designed primarily for the commercial and surveillance markets, are mass manufactured (found in the family camcorder) and often optimized to reproduce the response of the human eye to different colors of light. They are less sensitive than their monochrome counterparts because they must split the incoming light into its different colors, and the extra optics cause some light loss. Normally the light is sent through multiple beamsplitters; red, green and blue filters; or prisms to separate the color components. Each component is then imaged on a different region of a CCD or sometimes on three different CCDs. An alternate less-attractive strategy is to image the three colors sequentially using a filter wheel and then combine the three images to form a single color image.

Although color video cameras are also not sensitive enough for ultralow-light-level microscopy, they can sometimes be used with good results for conventional immunofluorescence and FISH. The best models have a minimum sensitivity reaching below 0.1 lux. Although most color cameras have fixed video format (30 Hz frame rate), some offer flexible exposure times, similar to the modified video cameras discussed above. These cameras are cooled to accommodate longer exposure times and may therefore prove quite useful in low- to moderate-light applications. The noise and dynamic range are comparable to monochrome video, as color video cameras use the same CCDs. The resolution is therefore also comparable (up to about 700 × 500 pixels), although a few higher-resolution models are available. Color cameras are easy to use and normally couple directly to monitors and video recording equipment.[4] Some also offer digital output which can be directly sent to a computer. The color separation schemes used in color cameras are not always optimal for fluorescence microscopy with various fluorescent probes. If multiple indicators must be imaged simultaneously, color cameras may be convenient, but for quantitative fluorescence imaging, it is often better to use monochrome cameras with customized filters.

---

[4] Standard video recording equipment has much lower resolution and dynamic range than the camera itself, except for professional video equipment.

1. Digital Still Cameras

A new alternative for capturing still images is the digital still camera. These cameras are almost always color cameras and are intended to replace traditional photographic cameras. They incorporate a CCD camera into a convenient camera body, often the same as film-based SLR cameras. Still cameras are easy to operate, commonly have high resolution (1000×1000 pixels), reasonable dynamic range (12 bits is typical), and may collect up to several images per minute. The high-sensitivity color versions may just be acceptable for fluorescence microscopy. However, these cameras can be very expensive, costing as much as intensified CCDs or low-end scientific CCDs.

F. Silicon Intensified Target Cameras

Silicon intensified target (SIT) cameras are built around an image tube rather than a solid-state detector. Photons striking a photocathode create photoelectrons, which are accelerated onto a silicon target. The accelerated electrons generate many electron-hole pairs in the target, resulting in the signal amplification of the SIT camera. Charges stored in the target are read out with a scanning electron beam, as in other TV tube cameras.

The SIT camera remains popular in video microscopy due to its reasonably high sensitivity. If low-light video rate imaging is required, the SIT camera can be a reasonable choice, although intensified CCDs should also be considered. Typical resolution is roughly comparable with video CCDs. These cameras have significant lag caused by incomplete charge readout of each frame, which effectively results in some degree of on-chip integration or image averaging between frames. This results in visually pleasing images, although it can also be a drawback if high time resolution is needed, such as in ratiometric calcium imaging, where lag produces significant artifacts. Although it has limited flexibility, the SIT is a very convenient camera, particularly for qualitative video-rate low-light imaging. When the SIT camera is coupled to a Gen I intensifier, it is called an ISIT camera. This camera has very high gain and is suitable for low-light level video, but as with the SIT camera, has little flexibility and significant lag.

G. CMOS Imagers

The promise of CMOS[5] imaging devices is a single integrated imaging sensor with on-chip processing and control (i.e., no external electronics). These devices are similar to CCDs but are far less common. The first CMOS imagers were charge injection devices (CID). The CID has an array of pixels, each independently addressed, and differs from CCDs in that electrons in each pixel are not removed by the readout process. The ability to randomly address pixels could be useful for imaging a small area of interest at higher rates. In addition, pixels of the CID can have different image integration times. This allows CID detectors the possibility of obtaining the maximum signal to noise for each pixel, even though

---

[5] CMOS stands for "complementary metal-oxide semiconductor."

there may be large differences in light intensity for various pixels. For example, pixels with low light levels can be averaged on chip for longer, while reading out bright pixels before overflow leads to blooming. CIDs have promising characteristics, although they often have high readout and dark noise. The high noise makes them difficult to use for low-light-level imaging.

Another CMOS imager is the active pixel device, developed primarily to overcome the high readout noise of CIDs. An active pixel sensor is a CID with an amplifier at each pixel. These active pixel sensors bring the readout noise to levels comparable to high quality CCDs (about 10 electrons/pixel). However, an active pixel sensor still has a large dark current ($\sim$5000 electrons/pixel/second), making them difficult to use for microscopy except for high-speed imaging. These devices have low power consumption and could be cheap to produce. In many respects (dynamic range, resolution, and addressability) CMOS imagers are comparable to or better than CCDs. However, the current sensors are not well suited to low-light-level imaging.

IV. Image Acquisition

Electronic cameras do not all have the same output format, and evaluating the image acquisition system is an important part of choosing a camera. The basic objective is to get an image into digital format and read it into a computer. For many cameras, such as slow-scan CCDs, cameras with variable integration modes, and cameras with other unique designs (multiple outputs, very high speeds), the manufacturer incorporates a dedicated digitizer as part of the camera system. This is the most desirable performance choice, and it should normally be easy to get the digital image into a computer for whatever processing is necessary. Video cameras, on the other hand, usually have an output which can be directly sent to standard equipment such as VCRs. When it is necessary to digitize the signal and import it into a computer for further processing, a frame grabber can be used. Frame grabbers can often be customized to capture analog signals from many different cameras (almost always fast-scan cameras).

Once in the computer, digital images need to be stored. Several custom formats are used by various image acquisition programs. For images with more than an 8-bit dynamic range, the most suitable nonproprietary format is portable network graphics (PNG)[6] This format also offers the advantages of image compression and metadata storage (useful for recording experimental details such as resolution, filters, labeling concentration, and so on).

There are numerous ways of processing an image once it is stored in digital format. Some basic capabilities can prove useful, such as flat field and background subtraction. Flat field correction is normally required even for cameras with uniform response because the illumination may not be uniform. Sometimes, additional processing is required just to visualize the data. For example, high-bit depth images do not typically display directly on monitors. They must be

---

[6] See the Portable Network Graphics homepage http://www.wco.com/~png/.

rescaled for the normal 8-bit (grayscale) display. Note that this is true even for a 24-bit display because there are 8 bits devoted to each of the red, green, and blue channels thus leaving only 256 possible gray levels displayable. Finally, the most useful data is not always the initial intensity images but rather a combination of images as in ratio imaging or fluorescence lifetime imaging. While basic image processing and storage is usually straightforward, it can become quite complicated if interfacing electronics and camera components are not correctly matched. It is therefore sensible to discuss output formats and image handling with manufacturers in order to avoid headaches after the camera is delivered.

V. Conclusions

The requirements for electronic camera performance are as varied as the many diverse applications of fluorescence microscopy. The information presented in this chapter will hopefully aid the reader in understanding how to appropriately apply the available technology to suit their imaging requirements. A good place to start when evaluating cameras for a particular biological application is to address the following questions:

1. What time resolution and frame rate are required?
2. Is high spatial resolution a priority?
3. How bright is a typical sample?

The answers to these questions can narrow down the appropriate choices considerably. For example, if time resolution and frame rate are of no concern, slow-scan CCDs certainly offer the best available performance both in terms of the signal-to-noise ratio and their spatial resolution. Slow-scan cameras are thus the first choice for experiments using fixed specimens, such as measurements using immunofluorescence and fluorescence *in situ* hybridization. On the other hand, if video-rate imaging is required, one need not evaluate slow-scan CCD cameras. A very basic video CCD may suffice if samples are heavily labeled or are not perturbed by high-intensity illumination. When video-rate imaging is required for very dim specimens an intensified CCD camera is probably most appropriate. A SIT camera may also be considered, provided its significant lag can be tolerated. The variable integration time video cameras are a very attractive option if one needs to acquire images at video rate acquisition as well as with longer integration times for less-bright samples. This flexibility can facilitate many diverse applications, with highly varied light levels. Whichever camera type is chosen, it is always best to evaluate the camera performance in the lab before making a purchase.

Finally, we reiterate that obtaining quality fluorescence images requires not only a good camera, but also optimal imaging. By maximizing the microscope light collection efficiency, one can substantially increase the amount of light reaching the camera. This is an important consideration, which can greatly benefit any application using fluorescence microscopy.

Acknowledgments

We gratefully acknowledge Ken Spring for helpful comments.

K.B. and K.J. were partially supported by NIH Grants GM41402 and GM35325. K.B. was also supported by NIH Grant NS10351.

T.F. was supported by the US Department of Energy under contract DE-AC04-94AL85000. Sandia National Laboratories is a multiprogram laboratory operated by Sandia Corporation, a Lockheed Martin Company, for the US Department of Energy.

References

Aikens, R. S., Agard, D. A., and Sedat, J. W. (1989). Solid-state imagers for microscopy. *In* "Fluorescence Microscopy of Living Cells in Culture. Part A: Fluorescent Analogs, Labeling Cells, and Basic Microscopy" (Y. L. Wang and D. L. Taylor, eds.), pp. 291–313. Academic Press, New York.

Arndt-Jovin, D. J., Robert-Nicoud, M., Kaufman, S. J., and Jovin, T. M. (1985). Fluorescence digital imaging microscopy in cell biology. *Science* **230**, 247–256.

Bright, G. R., and Taylor, D. L. (1986). Imaging at low light level in fluorescence microscopy. *In* "Applications of Fluorescence in the Biomedical Sciences" (D. L. Taylor, A. S. Waggoner, F. Lanni, R. F. Murphy, and R. R. Birge, eds.), pp. 257–288. Alan R. Liss, New York.

Castleman, K. R. (1993). Resolution and sampling requirements for digital image processing, analysis, and display. *In* "Electronic Light Microscopy" (D. Shotton, ed.), p. 71–94. Wiley-Liss, New York.

CIE publication 86 (1990). "Spectral luminous efficiency function for photopic vision."

Funatsu, T., Harada, Y., Tokunaga, M., Saito, K., and Yanagida, T. (1995). Imaging of single fluorescent molecules and individual ATP turnovers by single myosin molecules in aqueous solution. *Nature* **374**, 555–559.

Hiraoka, Y., Sedat, J. W., and Agard, D. A. (1987). The use of a charge-coupled device for quantitative optical microscopy of biological structures. *Science* **238**, 36–41.

Inoué, S. (1986). "Video Microscopy." Plenum, New York.

Keller, E. H. (1995). Objective lenses for confocal microscopy. *In* "Handbook of Biological Confocal Microscopy" (J. B. Pawley, ed.), 2nd edition, pp. 111–126. Plenum, New York.

Lakowicz, J. R., and Berndt, K. W. (1991). Lifetime-selective fluorescence imaging using an RF phase-sensitive camera. *Rev. Sci. Instrum.* **62**, 1727–1734.

Marriott, G., Clegg R. M., Arndt-Jovin, D. J., and Jovin, T. M. (1991). Time resolved imaging microscopy. *Biophys. J.* **60**, 1374–1387.

Ramm, P. (1994). Advanced image analysis systems in cell, molecular, and neurobiology applications. *J. Neurosci. Methods* **54**, 131–149.

Schmidt, T., Schutz, G. J., Baumgartner, W., Gruber, H. J., and Schindler, H. (1995). Characterization of photophysics and mobility of single molecules in a fluid lipid membrane. *J. Phys. Chem.* **99**, 17662–17668.

Shotton, D. (1993). An introduction to the electronic acquisition of light microscope images. *In* "Electronic Light Microscopy" (D. Shotton, ed.), pp. 1–38. Wiley-Liss, New York.

Shotton, D. M. (1995). Electronic light microscopy: Present capabilities and future prospects. *Histochem. Cell Biol.* **104**, 97–137.

So, P. T. C., French, T., Yu, W. M., Berland, K. M., Dong, C. Y., and Gratton, E. (1995). Time-resolved fluorescence microscopy using two-photon excitation. *Bioimaging* **3**, 49–63.

Tsay, T. T., Inman, R., Wray, B., Herman, B., and Jacobson, K. (1990). Characterization of low-light-level cameras for digitized video microscopy. *J. Microscopy* **160**(2), 141–159.

Tsien, R. Y., and Waggoner, A. (1995). Fluorophores for confocal microscopy. *In* "Handbook of Biological Confocal Microscopy" (J. B. Pawley, ed.), 2nd edition, 267–279. Plenum, New York.

Wang, Y. L., and Taylor, D. L. (eds.) (1989). "Fluorescence Microscopy of Living Cells in Culture. Part A: Fluorescent Analogs, Labeling Cells, and Basic Microscopy." Academic Press, New York.

CHAPTER 3

Cooled CCD Versus Intensified Cameras for Low-Light Video—Applications and Relative Advantages

Masafumi Oshiro

Hamamatsu Photonic Systems
Division of Hamamatsu Corporation
Bridgewater, New Jersey 08807

METHODS IN CELL BIOLOGY, VOL. 56

I. Overview

Imaging is becoming an important tool in the area of biological science and the advantages generate great benefits:

- The higher-sensitivity imaging detectors enable us to see very-low-light objects which otherwise are too dark to see.
- The spectral sensitivity of the human eye is limited to the range of 400 to 700 nm. The spectrum sensitivity range of imaging detectors is broader; from the range of x-ray to infrared.
- The dynamic range of the human eye is very poor. Special CCD detectors can achieve a 12- to 16-bit dynamic range.
- The human eye cannot follow an object if it moves too fast or too slow. Imaging detectors can capture images in the nanosecond range for high-speed objects. Time-lapse recording enables us to see slow-moving objects.
- Images can be analyzed using image processors to characterize the objects.
- Images can be stored into VCR, OMDR, and computers for future reference and further analysis.
- Images can be modified using image processors and these processed images generate new information which cannot be obtained with the human eye.

To respond to the requirements of biological science researchers, a variety of imaging detectors are available. A CCD (charge coupled device) camera is the most popular imaging detector for high-light applications. For low-light applications there are two types of CCD cameras available. One is the intensified CCD (ICCD) camera which uses an image intensifier (II) and a CCD camera (Fig. 1). The II intensifies low-light images and the intensified image is projected onto a CCD camera through relay optics such as a relay lens or fiber plate. This enables us to see a low-light image which cannot be seen by a CCD alone. The other is a cooled CCD (CCCD) camera which uses a similar CCD chip for high-light imaging. The CCCD reduces camera noise by cooling and slowly reading out the signal (Fig. 2). The reduction of noise enables one to see a low-light image which would be buried in the noise of a regular CCD camera.

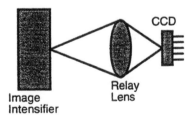

Fig. 1 Intensified CCD.

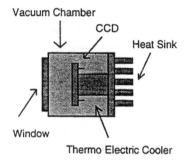

Fig. 2   Cooled CCD.

Even though the main goal of these types of cameras is higher sensitivity, they each have different characteristics which may be seen as both advantages and disadvantages when they are used in low-light applications. We are often asked which camera (ICCD or CCCD) is better; there is no simple answer. It depends on the application and what kind of characteristic is most important. This chapter describes the characteristics of each detector. This will enable you to determine which is most suitable for your application.

II. Sensitivity

Sensitivity is one of the most important characteristics for low-light cameras. We use the word "sensitivity" to describe how sensitive cameras are for low-light imaging. However, there is some misunderstanding as to which factors generate "high sensitivity."

A. What Does "High Sensitivity" Mean?

1. Brighter Image on the Monitor Screen (?)

We naturally think that the high-sensitivity camera generates a brighter image compared with a lower-sensitivity camera. Brightness of the image on a monitor screen is an important factor for the human eye to recognize the image; however, it does not always help to improve the sensitivity. For example, if the image is buried in noise the human eye cannot recognize the image, regardless of how bright it is.

Better Signal-to-Noise Ratio

If we consider that an image consists of signal and noise, we cannot recognize a signal smaller than the noise. Suppose we have two cameras to compare sensitivity; one generates higher signal than noise and the other generates lower signal than noise. In this case we can say that the former camera is more sensitive than the latter. In other words, the ratio of signal and noise at certain light levels indicates "sensitivity."

B. Signal-to-Noise Ratio (SNR)

Signal-to-noise ratio (SNR) is expressed by a very simple formula where S (electron) is the signal detected by the detector: and N (electron) is the total noise:

$$\text{SNR} = \frac{S}{N}. \tag{1}$$

1. Signal

Signal (S) (electron) is calculated from the input light level: I (photon/sec), the quantum efficiency: QE (electron/photon) of the detector and the integration time: t (sec).

$$S = I \cdot \text{QE} \cdot T \tag{2}$$

2. Noise

The noise is not only generated by the detectors but also by the signal itself. The combination of both types of noise becomes total noise (N).

a. Signal Shot Noise (N_{shot})

The minimum unit of light is a photon and the distribution of photons is considered to be Poisson distributed. In this case, statistical noise, signal shot noise (electron) of the signal S (electron), is calculated using Eq. (3):

$$N_{shot} = \sqrt{S}. \tag{3}$$

b. Camera Noise (N_{camera})

There are two major sources of camera noise. One is camera read noise (N_{read}) (electron) which appears when the signal is read from a detector and converted to a voltage signal. The other is camera dark noise (N_{dark}) (electron) which is caused by dark current (D) (electron/sec) generated by environmental heat around the detector. The total dark current is a function of the integration time T (sec) and it can be subtracted; however, the statistical noise caused by the dark current remains:

$$N_{dark} = \sqrt{D \cdot T}. \tag{4}$$

The camera noise (N_{camera}) (electron) is calculated using Eq. (5):

$$N_{\text{camera}} = \sqrt{N_{\text{read}}^2 + N_{\text{dark}}^2}. \tag{5}$$

c. Noise (N)

Noise is the combination of signal shot noise N_{shot} (electron) and camera noise N_{camera} (electron). Noise is calculated using Eq. (6).

$$N = \sqrt{N_{\text{shot}}^2 + N_{\text{camera}}^2}. \tag{6}$$

1. Noise Versus Light Level

Figure 3 shows the camera noise, signal shot noise, and total noise based on light level. Camera noise stays the same; however, signal shot noise increases relative to the light level increase. At lower light levels, camera noise is higher than signal shot noise and total noise is determined by camera noise. At higher light levels, signal shot noise is higher than camera noise and total noise is determined by signal shot noise. In other words, the camera noise is an important factor to achieve higher SNR only in low-light situations.

C. Approaches for Achieving "High Sensitivity"

Based on the consideration that higher signal-to-noise ratio realizes higher sensitivity in low-light situations, increasing signal and/or lowering noise will lead to "high sensitivity."

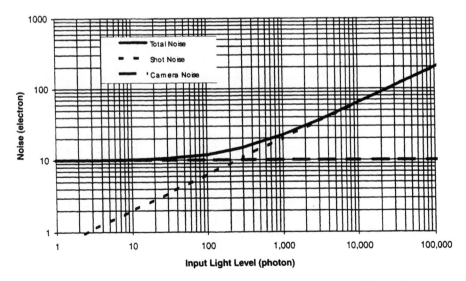

Fig. 3 Noise versus light level (camera noise = 10 electrons, QE = 0.4).

1. Increasing Signal

a. Higher QE (More Electrons)

Imaging detectors convert incoming photons into electrons. This conversion rate is known as quantum efficiency (QE). Higher QE generates more electrons and results in a higher signal. Figure 4 shows the signal-to-noise ratio for different QEs. In brighter light levels, the difference of SNR is not as significant; however, in lower light levels the difference of SNR is proportional to the difference of QE.

b. Signal Integration (Temporal/Spatial)

The signal can be integrated temporally and/or spatially on the detector or by using an image processor. This results in a higher signal at the expense of temporal and/or spatial resolution.

c. Larger Area of Detectors

What is the effect of increasing detector size on image brightness? One might expect that this would result in a decrease in intensity, since intensity falls off with increasing magnification. However, the opposite is the case. As shown in Fig. 5 for any given lens the larger the detector size the brighter the image. Since, in practice, the lens-to-detector faceplate distance is designed to be essentially constant, the object must be brought closer to the lens to form an image that just fills a larger detector. The lens collects a greater solid cone angle of light emanating from each object point and, consequently, point for point the image is brighter.

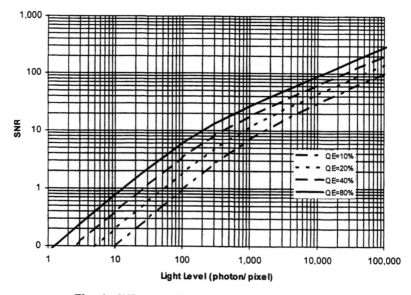

Fig. 4 SNR versus QE (camera noise = 10 electrons).

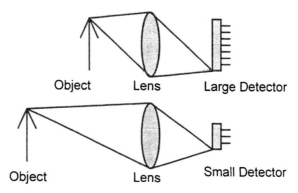

Fig. 5 Larger detector shortens the distance and results in higher light collection efficiency.

2. Reducing Noise

a. Intensification of Signal

Imaging detectors convert incoming photons into electrons. Those electrons can be intensified (multiplied) and the number of electrons are thusly increased. In this process, the signal shot noise also increases proportionally to the increase in the signal. Therefore, intensifying the signal is not equivalent to increasing the signal. However, intensification of the signal is relatively equivalent to reducing the camera noise since the camera read noise is not intensified. Figure 6

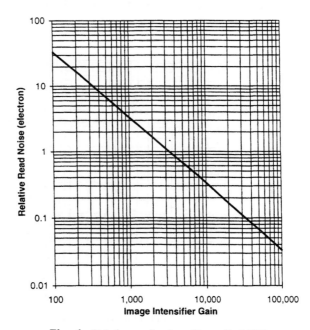

Fig. 6 Relative read noise of intensified CCD.

shows the relative read noise for various gains of the image intensifier based on the condition that the CCD camera noise is 150 electrons and the relay lens coupling efficiency is 0.045.

b. Lower Dark Noise

The dark current of the detector is primarily generated by environmental heat. Figure 7 shows the camera noise for various dark current. It is significant that the dark current is critical in low-light situations which require longer integration. By cooling the detector, dark current can be reduced. Normally, a 7 to 8°C reduction of the detector temperature reduces the dark current by one-half.

3. Intensified Camera (ICCD, SIT)

Standard CCD and tube cameras do not have enough sensitivity for low-light applications because of high camera read noise. The combination of an image intensifier and CCD or tube camera, such as the intensified CCD (ICCD) and silicon intensified target (SIT), makes the camera read noise relatively small and produces higher sensitivity.

4. Photon Counting Camera

A photon-counting camera is designed to have enough intensification for single photons to exceed the read noise of the camera. This enables us to discriminate camera read noise and produces a read-noise-free detector.

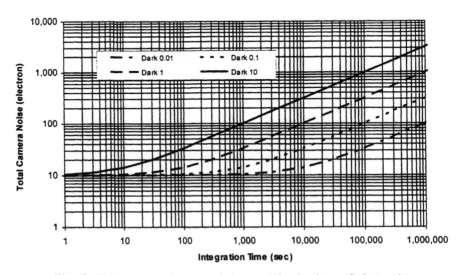

Fig. 7   Total camera noise versus dark current (read noise = 10 electrons).

Photon counting is a method to detect very weak light, such as that of luminescence. In high light levels, multiple photons come together and the light level can be measured as a function of signal intensity. In other words, higher signal means higher light and lower signal means lower light (see Figure 8).

In very low light levels, individual photons come separately. At this light level, signal intensity has no relation to light level. The signal only indicates the arrival of a photon. Therefore, light intensity can be measured as a function of the number of pulses which corresponds to the arrival of photons. In other words, a higher number of pulses within a certain time means a higher light, and a lower number of pulses means a lower light (see Figure 8).

A photon-counting detector generates pulses and the pulses are discriminated from camera noise and converted to binary data (1 or 0). This binary data is counted in memory and the result is displayed in gray levels. This method eliminates read noise completely and the only noise source will be dark noise. Normally the photon-counting detector is designed to have very low dark noise.

5. Cooled CCD

Standard CCD cameras do not have enough sensitivity for low-light applications because of high camera read noise. The cooled CCD has several features that result in higher sensitivity.

a. Slow Read-Out Reduces Read Noise

Read noise can be reduced by reading the signal from the CCD slowly. Normal CCD readout results in >100 electron read noise. Slow read-out reduces read noise to 5 to 10 electrons (Figure 9).

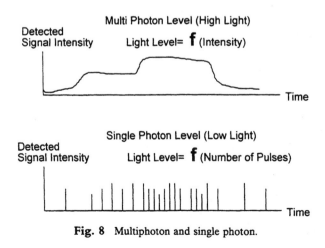

Fig. 8 Multiphoton and single photon.

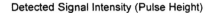

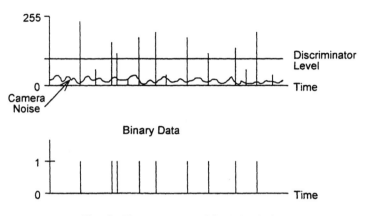

Fig. 9 Photon counting (discrimination).

b. CCD Has Higher QE

Standard full-frame CCDs have about 40 to 50% quantum efficiency (QE) at peak wavelength. Back-thin-type CCDs have about 80% QE at peak wavelength. The higher QE helps to increase the signal and results in higher sensitivity.

c. Cooling Makes Dark Noise Small

The cooling of a CCD reduces the dark noise of the CCD and enables us to integrate the signal for long periods without increasing the dark noise. This results in higher sensitivity at the expense of temporal resolution.

d. Pixel Binning Increases the Signal

The pixel of the CCD can be binned or averaged with adjacent pixels. This process adds the signal of the original pixel and the total signal can be read out at one time. This means the signal increases at the factor of the number of binned pixels without increasing the read noise. This produces higher sensitivity at the expense of spatial resolution.

D. Required Levels of Signal-to-Noise Ratio

Applications determine the level of signal-to-noise ratio required. Figures 10A–10E show 1000-sec photon counting taken with a photon-counting camera. Longer integration gives better SNR and results in higher resolution and higher dynamic range.

If the application requires one to recognize detail of the image or to measure the gray level of the image, such as the higher image quality (SNR) in Fig. 10E, 1000-sec Photon Counting is required. If the application requires simply

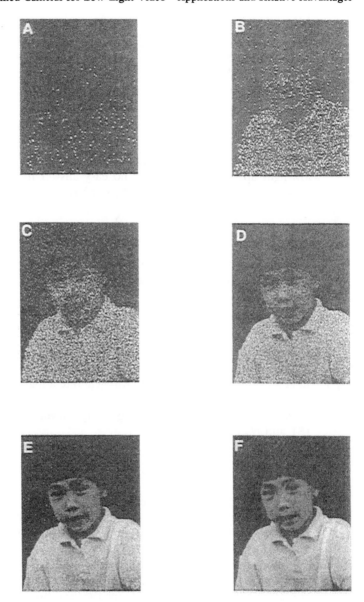

Fig. 10   (A) 1/30-sec photon counting; (B) 1-sec photon counting; (C) 10-sec photon counting; (D) 100-sec photon counting; (E) 1000-sec photon counting; and (F) high-light image.

recognizing the rough shape of an image, it does not require high image quality, 10-sec photon counting will be sufficient (Fig. 10C). If the application requires detection of an object, $\frac{1}{30}$-sec photon counting will be sufficient (Fig. 10A). It is important to determine the required image quality (SNR) for the application and choose the best detector to accomplish this.

E. Sensitivity Comparison

Intensified cameras, such as ICCD and SIT, and cooled CCD are used for low-light applications. There is no easy answer in determining which is more sensitive. From Eqs. (1), (3), and (6), SNR is

$$\text{SNR} = \frac{S}{\sqrt{N_{\text{shot}}^2 + N_{\text{camera}}^2}} = \frac{S}{\sqrt{S + N_{\text{camera}}^2}}. \tag{7}$$

In the situation of a very low light level, signal (S) is smaller than N_{camera}^2; and signal shot noise (N_{shot}) is negligible:

$$\text{SNR} \cong \frac{S}{\sqrt{N_{\text{camera}}^2}} = \frac{S}{N_{\text{camera}}}. \tag{8}$$

From Fig. 6, read noise of the intensified CCD (less than 0.1 electron at 100,000 gain) is much smaller than that of the cooled CCD (5 to 10 electrons). In this light level, the intensified CCD will generate a higher SNR because of its lower read noise.

In the situation of relatively higher light levels, N_{camera}^2 is smaller than signal (S), and N_{camera} is negligible:

$$\text{SNR} \cong \frac{S}{\sqrt{S}} = \sqrt{S}. \tag{9}$$

From Eq. (2), the signal is the function of quantum efficiency (QE), and higher QE generates higher SNR (higher sensitivity). Normally, a CCD has higher QE compared to the photocathode of an image intensifier; therefore, the cooled CCD will generate a higher SNR in this light level.

Figure 11 shows the SNR differences of intensified and cooled CCDs at various light levels. Cooled CCDs integrate the signal on the CCD chip and the signal is read out at one time. This means the read noise of the cooled CCD is fixed and it is not a function of the integration time. On the other hand, the signal of the ICCD is read out in real time and the signal, as well as the read noise, is integrated in an image processor. This means the read noise of the ICCD increases relative to the integration time. For shorter integration, the ICCD has an advantage over the cooled CCD because of smaller read noise; however, this advantage no longer exists for longer integration. Figure 12 shows the SNR differences for the intensified and cooled CCDs in various integration times.

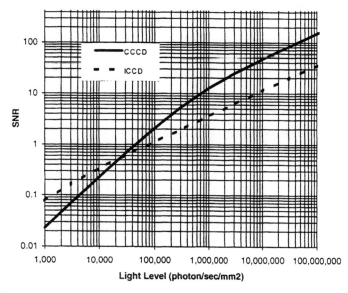

Fig. 11   Signal-to-noise ratio based on light level (integration time = 1 sec).

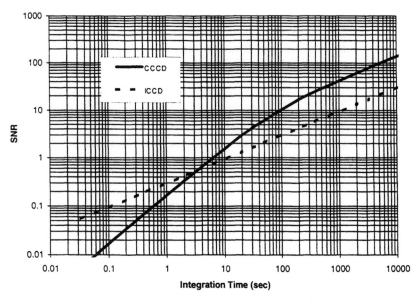

Fig. 12   Signal-to-noise ratio based on integration time (light level: 10,000 photon/sec/mm²).

III. Dynamic Range (DR)

Dynamic range (DR) is also one of the most important characteristics for low-light cameras. Dynamic range indicates the ratio of the brightest and the darkest signal which can be detected in the same scene.

A. Definition of DR for Video Camera

The dynamic range of a video camera is defined by the ratio of camera noise (N_{camera}), which is equivalent to the lowest signal detectable, and saturation (SAT), which is equivalent to the highest signal detectable:

$$DR_1 = \frac{SAT}{N_{camera}} \tag{10}$$

In case the signal is converted from analog to digital, the dynamic range will be limited by analog-to-digital conversion resolution which is the number of bits. For example, 8-bit conversion produces a dynamic range of up to 256 and 12-bit conversion yields up to 4096. However, a higher number of bits does not generate a dynamic range higher than the original analog dynamic range.

B. Dynamic Range Requirement from Applications

The dynamic range defined by Eq. (10) is useful when the application requires viewing both bright and dark objects in the same scene. However, if the application requires viewing a small signal on top of a bright background, Eq. (10) is not good enough to determine the smallest detectable signal. Since signal generates noise (signal shot noise), the noise often exceeds the camera noise (Eq. (9)). In this case the dynamic range is defined by the ratio of saturation, which is equivalent to the highest signal detectable, and the noise at saturation level:

$$DR_2 = \frac{SAT}{N_{SAT}} = \frac{SAT}{\sqrt{SAT}} = \sqrt{SAT} \tag{11}$$

The DR_2 is equivalent to the signal-to-noise ratio at saturation level. DR_2 can be calculated from the number of electrons at the saturation level. Here is one example:

For $N_{camera} = 10$ (electron) and SAT = 100,000 (electron):

$$DR_1 = \frac{SAT}{N_{camera}} = 10,000 \tag{12}$$

$$DR_2 = \sqrt{SAT} = 316. \tag{13}$$

C. Improving Dynamic Range

The dynamic range can be improved by lowering noise and/or increasing saturation level. This can be achieved by improving detectors. However, it is also possible to do this by integrating the signal temporally or spatially after digitization. The improvement of the dynamic range is the function of the integration time N:

$$DR_{1.N} = \frac{N \cdot SAT}{\sqrt{N} \cdot N_{camera}} = \sqrt{N} \cdot DR_1 \qquad (14)$$

$$DR_{2.N} = \sqrt{N \cdot SAT} = \sqrt{N} \cdot DR_2. \qquad (15)$$

The dynamic range of real-time video cameras such as ICCDs is normally limited to 256 which is 8 bit resolution. However, if the signal is integrated in a frame-grabber which has a higher number of bits (for example, 16 bits) the dynamic range will be improved. For example, when the signal is integrated for 256 frames, the dynamic range will be improved by a factor of 16, and this will be equivalent to 12 bits:

$$DR_1 \cong 256$$
$$DR_{1256} \cong \sqrt{256} \times 256 = 4096.$$

IV. Spatial Resolution

Spatial resolution of the detector is limited by the number of pixels and scanning lines. Another limitation to resolution is the amount of light (photons). In other words, the resolution of detectors may not be fully utilized in low-light situations.

A. Intensified Camera

An intensified camera such as the ICCD is made up of an image intensifier and video-rate CCD camera with relay optics. The resolution of an image intensifier is expressed as line pairs per millimeter (the number of line pairs, which are in black and white, in one millimeter of photocathode that can be resolved) and the resolution of the video-rate CCD is expressed as TV lines (the number of black-and-white lines in the picture size that can be resolved). Normally the resolution of the image intensifier exceeds that of a video-rate CCD and the resolution of the CCD limits that of the image intensifier.

Since the number of scanning lines is defined by TV standards such as RS-170 and CCIR, all video-rate CCD cameras have the same vertical resolution, which is limited by the number of scanning lines. In case of horizontal resolution, there is no standard to limit the number of pixels. As a result, the resolution in the horizontal direction exceeds that in the vertical direction.

Another limitation to resolution comes from the frame-grabbers. If the number of pixels of the frame-grabber is smaller than that of the CCD camera, the resolution will be lower than the original resolution of the CCD camera.

B. Cooled CCD

A cooled CCD normally has square pixels and gives the same resolution in both the vertical and horizontal directions. Since there is no standard to define the number of vertical lines in the cooled CCD, the resolution of the cooled CCD is expressed as the number of pixels in both the horizontal and vertical directions.

V. Temporal Resolution

Temporal resolution is limited by the frame rate of the detectors. The video-rate CCD camera of the ICCD has a 30 frames/sec frame rate (RS-170). The fastest frame rate for the cooled CCD is calculated from the total number of pixels and read-out speed. For example, the cooled CCD which has 1000 by 1000 pixels with 1 MHz read-out speed has a 1-frame/sec fastest frame rate. Since the full-frame CCD chip is commonly used in the cooled CCDs, additional exposure time and opening and closing time of a mechanical shutter is required. The frame rate will then be slower than the calculated result. For applications requiring higher frame rates, there are frame transfer type and interline-type cooled CCDs available.

In the case of an ICCD, there is another limitation of temporal resolution; decay lag of the phosphor screen of the image intensifier. Normally the decay lag of the phosphor screen is fast enough when the intensifier is used with a video-rate CCD (30 Hz). This concern has to be addressed when an image intensifier is used with higher-frame-rate cameras.

VI. Geometric Distortion

Intensified cameras have 2 to 3% geometric distortion, This is caused by the image intensifier itself and relay optics. Normally the inverter-type image intensifier has higher distortion compared to the proximity-type image intensifier. A cooled CCD has no geometric distortion since the signal of each pixel is handled independently.

VII. Shading

Intensified cameras have 10 to 20% shading. This is caused by the image intensifier itself and relay optics coupled to the CCD. Normally the center of view is brighter than the corners. A cooled CCD does not have shading like intensified cameras; however, there is a pixel sensitivity difference which can be a small percentage of the signal.

▬▬▬▬ VIII. Usability

It is important to understand basic limitations in practical use.

A. Intensified Camera

1. Standard Interface (RS–170, CCIR)

The standard video format makes it easy to use video peripherals such as video recorders, image processors, and image-processing software.

2. Fast frame rate (30 Hz for RS–170, 25 Hz for CCIR)

The frame rate is fast enough for most real-time applications and focusing.

3. Usable in Only Low Light Levels

The intensified camera can be used for high-light applications such as DIC and phase-contrast if the light source level is adjusted low enough for the detector. However, the image quality is not as good as that of high-light detectors. The image intensifier also has a risk for burn-in when excessive light is applied. The damage will be minimized with the tube protection circuit; however, special caution is required.

B. Cooled CCD

1. Low Light to High Light

Cooled CCDs can be used to obtain both low-light and high-light images. In both light levels, cooled CCDs generate a similar quality of image and there is no risk of detector burn in high light.

2. Pixel Manipulation (Binning, Subarray)

Binning combines the pixels and handles multiple pixels as one pixel. This increases sensitivity and reduces data size at the expense of spatial resolution. This is useful for very-low-light imaging and for applications which require a higher frame rate. Subarray scan allows one to read part of an image from the CCD. This maintains the original resolution and reduces data size. This is useful for applications which require a higher frame rate.

3. Nonstandard Interface

A cooled CCD generates digital video signal and each cooled CCD has its own interface to a computer. This makes it difficult to use standard video peripherals such as video recorders, image processors, and image-processing software. There is a movement to standardize interfaces, such as AIA (RS-422) and SCSI. So compatibility should be less problematic in the near future.

4. Cooling Required

Cooling of CCDs is normally done by a thermoelectric cooler; however, secondary cooling using water circulation is typically required.

5. Slow Frame Rate

Because of slow read-out and the existence of a mechanical shutter, the frame rate of a cooled CCD is slow. This limits the temporal resolution and makes focusing difficult under low-light conditions. Frame transfer-type CCDs and interline CCDs are available which generate faster frame rates.

IX. Advanced Technology

Imaging technology is improving every day and it is important to have up-to-date information in order to utilize advanced technology in research. The directions of improvements include higher sensitivity, higher spatial resolution, and higher temporal resolution.

A. High Sensitivity

The quantum efficiency (QE) of an image intensifier is lower than that of a cooled CCD. New photocathodes which have 30–40% QE are being developed and they will fill the gap between cooled CCDs and image intensifiers. On the other hand, a back-thin-type CCD is being developed which has about 80% QE at its peak and has good QE in the range of blue to ultraviolet.

Acknowledgments

The author greatly acknowledges Mr. T. Hayakawa and Mr. H. Iida of Hamamatsu Photonics K.K. Japan for their helpful discussions and advice.

References

Inoué, S. (1981). "Video Microscopy." Plenum, New York.
Wick, R. A., Iglesias, G., and Oshiro, M. (1993). "Bioluminescence and Chemiluminescence: Status Report: 47–59" Wiley, Chichester.

CHAPTER 4

Techniques for Optimizing Microscopy and Analysis through Digital Image Processing

Ted Inoué and Neal Gliksman

Universal Imaging Corporation, West Chester, Pennsylvania 19380

I. Fundamentals of Biological Image Processing

A. Introduction

The goal of this chapter is to become familiar with the fundamentals of video microscopy and digital image processing to make best use of the powerful computerized techniques now available. The use of digital image processors has greatly

METHODS IN CELL BIOLOGY, VOL. 56
Copyright © 1998 by Academic Press. All rights of reproduction in any form reserved.
0091-679X/98 $25.00

enhanced the utility of the optical microscope, permitting entire fields of study that were impossible until just a few years ago. By the end of this chapter, you will be exposed to the structure of a digital imaging microscope as well as to some of the fundamental techniques that will improve the quantitative as well as qualitative aspects of the using light microscopy. Finally, we will show examples of how and when to apply digital imaging so as to extract the most useful information from one's samples.

It was only 15 years ago when microscopy was a relatively "simple" art—one selected the optics and microscope stand one wanted from a relatively limited group of suppliers and then ordered the appropriate 35-mm camera. Once the system was set up, anybody could come in, install their film, set up the sample, and press a button to take a picture. The camera would then snap a picture of roughly the correct exposure and all that was left was to get the film processed.

Today, the typical research microscopy set-up often requires a trained technician just to turn on the power and display an image. Consider Table I, an abbreviated list of components used in a video imaging system used for fluorescence microscopy, many of which must be powered up in the right sequence to avoid damage.

With this many components, it is no wonder that most users find video microscopy systems to be very intimidating. The complexity has risen to such a degree that many users avoid such systems entirely, using simpler technologies that may be easier to use, but less optimal for the application.

Those who brave the complexities of video microscopy and thoroughly learn the fundamentals of the technology are acquiring amazing data that they could not gather using previous technologies. An example of one such application is 4D imaging. In 4D imaging the user rapidly acquires a set of images of a specimen taken at different focal planes. This is then repeated over time. The resulting image data set contains a representation of the three-dimensional structure of the specimen as it changes in time.

This chapter discusses the major components of the typical video microscopy system so that you are able to properly configure and utilize its features.

Table I
Components of a Typical Video Imaging System

Microscope components	Camera components	Imaging system	Other items
Microscope stand	Camera body	Computer	Video printer
Fluorescence illuminator	Image intensifier	Computer monitor	Video cassette recorder
Filter wheel	Camera controller	Camera digitizer	Laser printer
Light path shutters		Video display monitor	
Automated focus control			
Automated stage control			

B. Conventional Microscopy versus Digital Video Microscopy

Given the complexity and expense of the digital video microscopy system, what are the compelling benefits to the researcher? What are the advantages of digital video microscopy?

Consider the common steps in using a microscope to collect fluorescence images. Table II lists a sampling of these steps and some considerations relating to each.

Given the above considerations, it should be possible to make microscopy easier for the user. Digital video microscopy provides a solution for many of these considerations, making microscopy far more enjoyable and quantitative.

Starting with the microscope, how does a digital video microscopy system differ from a system designed for conventional visual light microscopy and photography? Overall, the digital video microscope, driven with the proper computer and software, is faster, more precise, and more sensitive than conventional, noncomputerized methods. The computerized system can automate many of the basic tasks of an experiment including changing filters, opening and closing shutters, changing the focus, and moving the stage reproducibly.

Table II
Considerations in the Use of a Conventional Light Microscope for Fluorescent Samples

Interaction with a microscope	Considerations
Look through the eyepieces at a specimen	Peering through eyepieces may produce eyestrain Eyes are not extremely sensitive at low light levels Eyes are not extremely sensitive to all wavelengths. Eyes are not extremely sensitive to seeing objects moving at high speeds.
Focus the microscope	Manual focus is difficult to set reproducibly. The focus may be different depending on viewer.
Move the stage	It may be difficult to relocate a sample if the stage is moved.
Select a filter set for fluorescence observation	The user must figure out how to change filters. Manual handling of filters make them susceptible to dust and fingerprints.
Take a photograph with a standard camera	Film lacks sensitivity to record fast events at low light levels. Quantitative capabilities of film are limited by nonlinear response. Long exposures can lead to photodamage of specimen. The color balance and contrast of film is highly dependent on processing conditions Film requires processing and printing. Pictures must be physically archived and protected.

Video microscopy systems do not come without a cost. The initial investment is far greater—in both price and time to learn. It can take months to understand the methods and terminology required to utilize a video microscope system effectively. However, once the investment has been made, the rewards can be tremendous.

C. Pixels and Voxels and How They Represent Images

1. How Do Images Get into a Video Microscope System?

In a video microscope system, a camera takes the place of your eyes as the light detector. On many microscopes there is a slider that is pulled out or the head of the microscope is rotated and the light that would have been sent to the eyepieces is sent to a camera port where a camera is attached.

There are two types of cameras used for video microscopy and image processing. The two types, analog and digital, are both discussed in Chapter 5 of this volume.

A digital camera scans the image and generates a numerical output representing the image. The numerical data is then sent to the image processor. The manner in which the image is scanned and the way in which the numerical data is transferred to the image processor is left intentionally vague, as there are several ways in which this procedure occurs.

Video cameras are the most common form of analog cameras used in biomedical science. Nevertheless, some confocal scanning microscopes use an analog detector that could be considered an analog camera for the sake of this discussion. In the analog system, the image is scanned from left to right and top to bottom in a pattern known as a raster scan, as shown in Fig. 1. The scan is shown as a solid arrow facing toward the right. After the device has scanned each line, the scan position returns to the left side of the next line. This continues until the entire image has been scanned.

As the video camera or confocal laser is scanned, a camera/detector detects the brightness at the spot being scanned and sends this as a voltage to the image

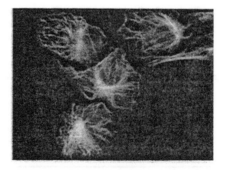

Fig. 1 Raster scanning of an image.

processor. The image processor will convert this voltage into a numeric value representing the brightness and work with it digitally (discussed later). After the image has been converted into a digital form, the image from a video camera or a digital camera can be treated identically and all subsequent discussions apply to both.

2. Pixel and Voxel Defined

Once an image from either a digital or analog camera has been received by the image processor and converted into a digital form (in the case of analog cameras) the images are fundamentally broken down into a finite grid of spots called "pixels," as shown in Fig. 2.

The pixel and the voxel are the smallest units that compose an image or a volume (a three-dimensional image). Just as an integer can only represent specific numbers within a range (for example, 2.5, 3.756, and 7.23 are not integers), a pixel and a voxel represent the sampled object at a specific location in space. Every pixel in an image has a unique X and Y coordinate and is assumed to have the same Z coordinate. Each voxel has X, Y, and Z coordinates. For historical reasons, pixels in an image are usually measured from the upper left corner of the image, starting with position 0,0. Likewise, voxels are measured as a distance away from the viewer, with the upper left corner in the top plane

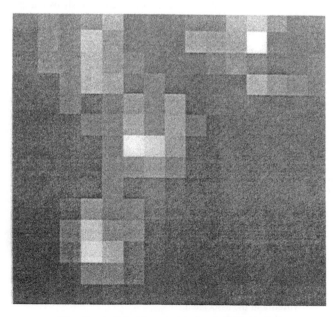

Fig. 2 A digital image showing individual pixels.

defining the origin: position 0,0,0. Figure 3 shows samples of a single plane and a volume, each with the associated coordinate system.

A digital image is somewhat analogous to a black-and-white photograph. Every tiny spot in a photograph is made up of a silver grain. Each silver grain in a photograph could be represented by a location and a value (the brightness of that grain). Likewise, each pixel has its own location and a value—its *gray value*. Typically, a pixel's value is an 8- or 16-bit number that the computer can use to represent 256 or 65,536 distinct brightness levels. Pixels in color digital images typically have a red, green, and blue value. Color images will not be discussed in this chapter.

3. Sampling of Images and Volumes

Pixels and voxels are evenly distributed throughout the image or throughout the volume: each is sampled at specific locations, as shown in Fig. 3. Since a pixel has a distinct size, you can think of the pixel value as the average of the values within the area covered by that pixel. A similar analogy is true for a voxel and the volume encompassed by the voxel. This concept critically effects the interpretation of a digital image.

Pixels and voxels are not infinitely small. This can cause artifacts when dealing with images of small objects. Suppose the imaging system is being used to look at a white line on a black background. The sampled image might have the values for black (usually 0) in the background and white (usually 255) in the foreground. This will not always be true. Figure 4 shows a portion of a greatly magnified

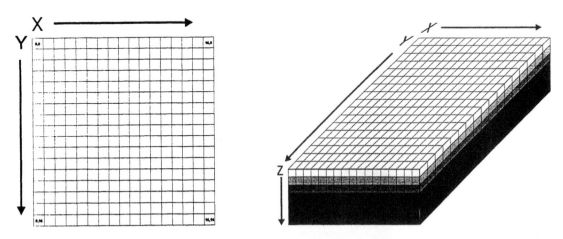

Fig. 3 (A) A sample pixel grid showing the location of the origin and the direction of increasing pixel X and Y coordinates. Corner pixel coordinates are shown. (B) A stack of images showing voxels and the associated.

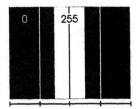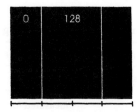

Fig. 4 (A) Expanded view of the original image (nondigitized) consisting of a single-pixel-wide white line on a black background. The thin lines represent the boundaries between pixels being sampled by the digitizer. (B) Expanded view of the digitized sample image in the image processor.

image with the white line on a black background. The marks below the diagram show the positions of the pixels. The white line is straddling two columns of pixels. Because of the way that the image processor broke the image down into pixels, two pixels cover half background and half white line and the digitized image contains the values 0 and 128 instead of the expected values of 0 and 255. In addition, note that the single line in the real sample, that is only a single pixel wide, appears in the digitized image as a two-pixel wide line (the right panel of Fig. 4). Depending on the exact alignment of the image with the way the digitizer breaks down the pixels, the resulting image could have pixel values anywhere from 0 to 255.

This artifact is minimized in most video digitizing systems by a filter in the digitizer that reduces its resolution. The filter causes all single-pixel-wide lines, regardless of alignment, to blur across multiple pixels and hence reduce in intensity. This is not true of direct digital CCD cameras which contain no such filter and will manifest the artifact if image details are on the scale of single pixels.

4. Tips for Visualizing Small Objects

Table III lists some tips for using digital imaging in regard to the size the objects being represented versus the size of a pixel.

Imagine what would happen in the previous example if the line's width increased relative to the size of the pixels. If the line being imaged crossed a path 10 pixels wide, then the digital representation of the line would be more accurate; the interior pixels would be white and only the edge pixels would have a value that is some fraction of white depending on how much of each pixel was covered by the line. As noted in Table III, the user's first recourse is typically to boost the magnification of the image to the camera to provide more pixels across the finest details in the image.

Table III
Methods to Improve Results when Imaging Small Objects

Application	Side effect of imaging	Ways to improve the accuracy of results
All imaging applications	Details smaller than one pixel will be represented inaccurately—the brightness measured for such objects will be altered and fine structures lost.	Choose a magnification such that the smallest details in the image are as large as possible relative to the pixel size. Use a higher resolution camera with smaller pixels.
Measurement of brightness	Objects or portions of objects which cover fractions of pixels will have sampled gray values which are different than those contained the actual image. Sampled values may be lower or higher depending on the brightness of the detail relative to the areas of the image around the detail.	Do not measure objects or details which contain only a few pixels. Measure the total brightness within an area when measuring the brightness of fluorescent objects.
Measurement of size	The boundaries of objects will blur across pixels resulting in an ambiguous determination of exactly where the boundary lies. This may result in measurements that are either too large or too small.	Characterize object boundaries at very high magnifications to determine the exact boundary shape. Use this to define how the boundary is characterized at lower magnifications.

II. Analog and Digital Processing in Image Processing and Analysis

A. Background

1. Analog and Digital Defined

This section discusses the relative merits of analog and digital cameras and analog and digital processing techniques. For years, the debate has raged about whether analog or digital is better, in fields from high-end audio to imaging. Now that computers are firmly entrenched in our lives, it would seem that digital is better. But is this the case? Both analog and digital cameras and analog and digital processing techniques have strengths and weaknesses.

Analog process: a process that functions over a continuum of values—analog methods represent and manipulate arbitrary values without regard to numerical precision.

Digital process: a process that stores or manipulates values with a finite precision and range. For example, in the case of the 8-bit pixel, gray levels may only take on integer values in the range of 0 to 255.

B. Practical Considerations

1. Comparison of Analog and Digital Devices and Processes

Table IV compares some of the advantages and disadvantages of digital and analog cameras. Table V compares some of the advantages and disadvantages of digital and analog image processing.

2. Matching a Signal's Dynamic Range to the Digitizer's Capacity

The choice between analog and digital processing can make the difference between solid data and useless images. For example, at even the most basic level of image acquisition, the appropriate choice of contrast control can make a critical difference. Figure 5 shows the voltages measured from a video signal from a line scanned through a high-contrast sample. The signal can be manipulated using only digital methods, but the part of the signal above the analog-to-digital converter's (A/D) saturation level (dotted line) is lost, leading to clipping

Table IV
The Relative Merits of Analog and Digital Cameras in Imaging

	Analog (video) cameras	Digital cameras

Advantages	Several standard interfaces
	  RS-170, NTSC, PAL
	Relatively high speed
	  The most common temporal resolution in the U.S. is 30 frames/sec at roughly 640 × 480 pixel resolution.
	Ease of use
	  Conventional analog video cameras stream voltages out a coaxial cable. The signal is easily manipulated with analog amplifiers to change the contrast and brightness.
		Potential for high speed performance
		  These cameras can acquire images at higher rates than conventional video.
		Higher potential resolution
		  Digital cameras exist with resolutions of ~4000 × 4000 pixels and larger arrays will be created.
		Image quality
		  Images are highly immune from electrical noise interference thus are easier to maintain digitally during transfer from the camera to an image processor.

Disadvantages	Flexibility
	  Standard interfaces limit the flexibility and performance of most video cameras.
	Resolution
	  Attributes such as vertical resolution and acquisition speed are intrinsically limited by the standard interfaces.
	Image quality
	  Video signals are subject to electrical noise and interference from sources such as fluorescent lighting.
		Expense
		  Instrumentation grade digital cameras tend to be more expensive than conventional video cameras.
		Ease of use
		  Digital cameras are moderately complex devices with many options. While they may be used in a "point and click" mode, current models are intrinsically more complicated than video cameras.
		Speed
		  Although images can be acquired faster than analog cameras, it often takes much longer to transfer the image into the image processor.

Table V
The Relative Merits of Analog and Digital Image Processing Techniques

	Analog signal manipulation	Digital number crunching

Advantages	Speed
	    Analog processing of signals, such as adjusting contrast, can be done at the speed of the signal transmission, limited only by the bandwidth of the electronics.
		Flexible
		    Digital systems are usually programmable, allowing the algorithms to change to suit the application.

Disadvantages	Inflexible
	    Most imaging systems manipulate analog signals through hardwired circuitry.
		Speed
		    Digital systems typically manipulate each piece of information using a single processor that serves as a bottleneck.

(Fig. 5B). The camera's analog contrast control can be used to reduce the voltage range of the signal so that it falls into the useable range of the A/D converter (Fig. 5C) and then the image can be digitized within the useful range of gray values accepted by the imaging system (Fig. 5D).

C. Some Caveats Regarding Video Camera Controllers

Video cameras were originally intended for purely visual purposes, and many cameras include controls or circuitry that allow one to "enhance" the image in

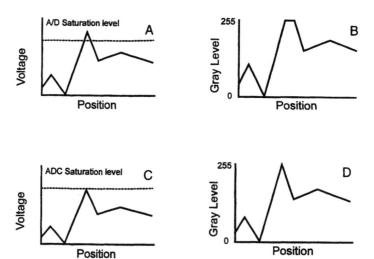

Fig. 5 (A) Line scan through an image having a dynamic range greater than that of the A/D converter. (B) Line scan of the digitized image showing clipping due to A/D saturation. (C) Video signal after analog signal reduction to bring signal within dynamic range of the A/D converter. (D) The digitized line scan of the reduced video signal now fits within normal 8-bit digitizer range.

ways that make it unsuitable for subsequent quantitation. Table VI lists some of these controls and how they operate. These options should be disabled in all cases where you may wish to measure the brightness within the image.

As is shown by the previous examples, it is important to know how images are processed as they flow from the camera and through the imaging system. Making adjustments to the digital image when it would be more appropriate to modify it in analog form can lead to poor or incorrect results. The next section covers more of the organization of the image processor, further clarifying the types of processes that should be done while an image is in an analog or digital form.

III. Under the Hood—How an Image Processor Works

The key to video microscopy is the image processor. An image processor is a conceptually simple device that takes in images, converts them into numbers, manipulates and stores the numbers, and converts those altered numbers back into video format for display.

The purpose of this section is to provide a working understanding of digital image processing hardware. It details each component of the image processor and provides information to guide you through determining the best solution for your imaging needs.

Many questions arise when discussing the details of image processors: How does the computer convert a video signal into numbers? How does it manipulate images? What features do you need? This section begins with the video signal coming from a video camera. The details would be almost identical for a digital camera except that the image processor would not contain an A/D converter.

Table VI
Camera Controls That Alter Gray Scale Response

Control name	Effect on video data (desired setting)
Gamma	Makes the camera's gray scale response nonlinear. Gamma should be set to 1.0 for any brightness measurements.
Autogain Autoblack	Alters the camera's gain and black level based on the brightness of the image. This feature should be disabled for virtually any use of the camera with an image processor.
Shading correction	Alters the response of the camera in different parts of the image. This feature should be disabled or set to a flat response for quantitation.
Intensifier protection	Reduces the light amplification of the intensifier when presented with bright objects. The protection can take effect too early, modifying the data. If possible, this feature should be disabled while making measurements.[a]

[a] Care must be taken not to overilluminate and damage the image intensifier after the circuits have been disabled.

A. The Components of a Digital Image Processor

Figure 6 shows a simplified diagram of a typical video image processor. This type of image processor is also called a pipeline image processor because the image flows through the system as if it was in a pipeline. The left side of the figure shows where the video image is dealt with in the analog domain, while the right side shows those portions that manipulate the image digitally.

1. The Analog-to-Digital Converter—Getting the Video Signal into the Computer

How does the output from a video camera—a time variant voltage signal representing the image—get into an image processor as a digital image?

The image processor contains a device called an analog-to-digital converter (A/D converter) which assigns a numerical value to each voltage level within a given range. (In video, it is most common to have an 8-bit A/D converter providing 256 values over a range of 0 to 0.7 volts.) The video signal comes in on a row-by-row basis (see Fig. 1 for a description of the pattern). The A/D converter samples the video signal at a set frequency that results in the desired number of samples (pixels) per row of the image. If you were to sample at high frequencies, you would obtain more pixels per row of the image which would result in higher resolution horizontally. Many manufacturers use A/D converters that will produce images with a 640 × 480 resolution. It also happens that an image with 640 columns and 480 lines results in pixels that can be considered square. Once the A/D has converted the video signal into numbers, the data pass into another device called an arithmetic and logic unit (ALU).

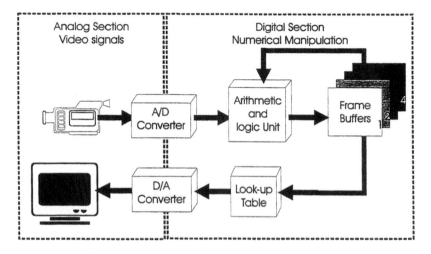

Fig. 6 Video image processor components and the data path.

2. The Arithmetic and Logic Unit—Processing Images at Video Rate

As noted previously, the speed at which a video camera generates images is called video rate. In the United States and other countries that have adopted a 60-Hertz powerline frequency, video rate is approximately 30 images/sec. In other countries that have adopted a 50-Hertz powerline frequency, video rate is 25 images/sec. For a typical 640 × 480-pixel resolution image containing 307,200 pixels, this translates to 9,216,000 pixel/sec passing through the image processor.

Even with today's fast computers, 9 million pixels is quite a few to process every second. In addition, the computer may experience bottlenecks which slow down the rate at which it can access and process pixels held in the image processor. For this reason, many modern, video rate image processors utilize specialized circuitry that is designed specifically for processing video images in real-time (video rate).

ALUs can often perform a number of other operations on and between images. When this occurs, the ALU is a pipeline component—it combines one pixel from each source image to compute a result. Typical computations of an ALU include: addition, subtraction, Logical OR, AND, and XOR. Most ALUs also have the capability to multiply or divide an image by a constant.

Many ALUs have two inputs so that the imaging system can combine the newly digitized video image with a stored image. One example of a stored image that is discussed later is the background reference image that is used to correct for background noise.

3. Frame Buffers—Holding the Images for Processing and Display

A frame buffer is memory within the image processor used to store one or more images that have been acquired, processed, or are awaiting subsequent display. Some image processing systems do not use frame buffers. Image processors without frame buffers grab images from the video camera and store them in the memory built into the computers that control them. However, most of the higher-speed image processors have their own frame buffers built into the image processing hardware because access to the computer's memory can be much slower than access to a frame buffer.

Besides access speed, there are other differences between frame buffer memory and the memory in a computer because the demands placed on it are different. For example, in the typical background subtraction application, background reference images must be read from one frame buffer at the same time as result images are written in to another frame buffer. In addition, if images are being averaged, there may be other frame buffers being read and written simultaneously. This type of memory is often referred to as multiport memory because of this unique capability.

4. Look-Up Tables—Converting Image Data On-The-Fly

After processing and storage, the digitized image usually passes through a look-up table (LUT). An LUT is simply a table of values, one per gray level in the image, that converts an image value to some other value. LUTs are particularly useful for rapidly modifying the contrast of the image while you view it on the monitor. Table VII demonstrates the use of a LUT to "stretch" the image contrast by a factor of two.

Why are LUTs used? Why not just modify the brightness values of pixels using the computer? Imagine a typical image with dimensions of 640×480 pixels and 256 gray levels. The image is composed of over a quarter of a million pixels. If the computer had to manually change pixel values as demonstrated in Table VII, it would require over a quarter of a million multiplications. Since there are only 256 starting gray levels, the computer could calculate the 256 multiplications to fill an LUT. The image processor could then reference the LUT for each pixel value (a fast operation) instead of doing a multiplication (a much slower operation).

LUTs are even more important when performing more complex operations on the pixel intensities. The calculations could be quite complicated and they could easily reach a point at which the computer could no longer compute the resultant image in a reasonable amount of time if it had to compute values for every pixel. Using an LUT, the computer only performs the computation 256 times, allowing the image processor to apply even the most complex functions as fast as the camera could generate images.

5. The Digital-to-Analog Converter—Displaying the Image Data on a Monitor

The converted values which pass out of the LUT usually enter the digital-to-analog converter (D/A converter). This device accepts numerical values and converts them into an analog video signal. The signal can be used by a video monitor to show the resultant image or by a VCR to record the resultant image.

An item that has been omitted from the above diagram is the computer. Often personal computers (PCs) control image processors, instructing them to change the function in the ALU, change the LUT, or change data in the frame buffers. For this discussion, you should imagine that the computer is connected to all parts of the digital image processor.

Table VII
Example of a Look-Up Table Used to
Double the Image Contrast

Image gray value	Table output value
0	0
1	2
2	4
3	6

B. Putting the Pieces Together—How Does an Image Processor Work?

1. Introduction

The previous section described the hardware components in a typical image processor. This section covers how the image processing software puts these components together to perform certain common imaging applications. This chapter addresses the fundamental operations that will handle most of your imaging needs. In addition, the section also covers some less commonly utilized operations that may be very useful.

2. Contrast Enhancement

Contrast enhancement is perhaps the most common application of image processing in microscopy. Inoué (1981) reported "The new combination provides clear images with outstanding contrast, resolution, diffraction image quality, and sensitivity." These remarkable results were made possible by a simple box that allowed a high level of contrast manipulation combined with optimal usage of the light microscope.

Contrast enhancement allows you to use the light microscope at the limits of its theoretical performance. Conventionally, microscopists were taught to close the condenser diaphragm to increase image contrast. Unfortunately, closing the condenser diaphragm reduces the resolution of the microscope considerably, making expensive 1.4-na oil objectives perform no better than 0.75-na objectives. Contrast enhancement allows the microscopist to use the microscope with its condenser diaphragm open so that the microscope has the highest image resolution and thinnest optical section. Clearly, the combination of appropriate microscope adjustments and the effective use of contrast enhancement can be critical to your imaging success.

a. What Is Contrast?

Contrast is the relative difference in brightness of two patches of light. Perceptually, the human visual system compares light intensities in multiplicative manner—two patches of light, one twice as bright as the other, are perceived as the same contrast regardless of the absolute brightness of the light. Mathematically, this implies that your eyes will perceive increased contrast in an image simply by subtracting a value from it. An example of this is shown in Table VIII. Note how in each example, the second brightness is 50 units less bright than the first. Regardless, the contrast value, defined as the ratio of the two brightnesses, changes dramatically.

b. Computing Values to Enhance Contrast

Contrast can often be improved by multiplying an image by a value greater than 1, then subtracting off a value so that the image does not saturate. Both the multiplication and the subtraction are necessary to improve contrast. If you

Table VIII
Contrast Values for Samples with the Same
Absolute Brightness Difference

First brightness	Second brightness	Contrast value
200	150	1.5
100	50	2
75	25	3
51	1	51

multiply the entire image by a value, then the relative brightness remains the same. In addition, if you multiplied by a value greater than 1, you may have increased the pixel values beyond the limited dynamic range of the system and the image will be saturated at the maximum value. Table IX shows an example of how this mathematical procedure could be used by multiplying pixel intensities by 5 and subtracting 100 from the results. Figure 7 shows the results of this operation.

c. How the Image Processor Modifies Contrast

Enhancing contrast with an image processor typically involves using an LUT assigned to the display. All of the data from the frame buffers goes through these LUTs at video rate before being displayed. The advantage of using a LUT for enhancing contrast is that it is fast and it is nondestructive. That is, it does not alter the image data stored in the frame buffers.

The image processor uses the LUT to modify the image contrast as shown in Table IX. The program computes the values for each grayscale element (the numbers 0 through 255) and places those values in the LUT. When the data come through the image processing pipeline, the value from the image enters the LUT and the converted value comes out of the LUT. The result is an image displayed on the screen that has a modified contrast.

LUTs can be very useful for a variety of contrast manipulations. For example, to reverse the contrast of an image the LUT would contain the values from

Table IX
Enhanced Contrast Values

Original values	Enhanced values ($n \cdot 5-500$)
100	0
110	50
120	100
130	150
140	200
150	250

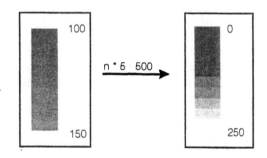

Fig. 7 Gray scale before and after contrast enhancement.

white-to-black (255 to 0), instead of black-to-white (0 to 255). Overall, LUTs are easy to program, operate in real-time, and can perform arbitrary transformations on image data.

3. Image Subtraction

Image subtraction serves a number of purposes in image processing and analysis. These uses range from enhancing images by removing noise from stray light (increasing the accuracy of quantitative brightness measurements), showing the movement of objects within the images over time, and showing the growth or shortening of objects within the images over time (Fig. 8).

a. How Does an Image Processor Subtract a Background Reference Image?
Subtraction of a background reference image from a video image can be used to remove shading and increase the detection of fine structures. Initially,

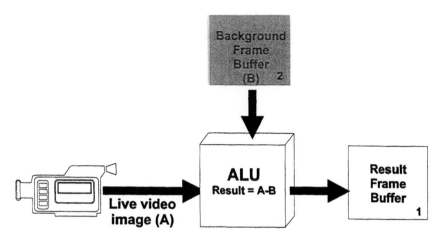

Fig. 8 Flow of data through a pipeline processor configured for background subtraction.

a background reference image is acquired, digitized by the A/D converter, and stored in a frame buffer. The live video image is taken in through the A/D converter and sent to the ALU at the same time that the frame buffer sends the reference image to the ALU. The ALU "sees" the image as a series of numbers and subtracts the reference image from the live video image. The image processor then stores the result in another frame buffer and or sends it through an LUT to enhance the image contrast. Finally, the D/A converter would change the numerical data back into a video image for display on a monitor or recording by a VCR.

In this example, note that two frame buffers are used—one containing the background or reference image and a second for storing the resultant image. The ALU subtracts the background image from the live video, pixel by pixel, and stores the subtracted pixels in the resultant frame buffer at the appropriate location.

b. Enhancing Motion with Subtraction

Image subtraction is an extremely sensitive motion detector which can greatly assist you in determining the presence or absence of any sort of change in a sample. Suppose that you are observing a very complex sample which contains very subtle motion—for example, a single cell moving from one area to another.

Figure 9 shows an image that has been modified to demonstrate what you might see in an image that changes. In this example, one blood cell has moved in the interval from time $t = 1$ to time $t = 2$. Without further processing, it would be very difficult to determine which cell has moved and by how much. However, subtracting the image at $t = 2$ from the image taken at $t = 1$ reveals movement in one cell very clearly (third panel).

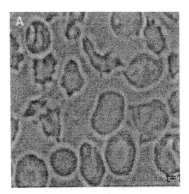 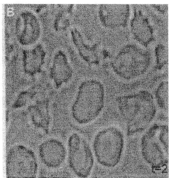 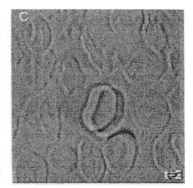

Fig. 9 Sample of images changing over time. (C) Image represents subtraction of the previous two images with an offset of 128.

4. Image Averaging

Image averaging is used to reduce or eliminate random noise from the image. Random noise has no pattern in the image and is usually due to photon statistics or noisy electronics. A pixel affected by random noise may be greater in intensity than the actual specimen; however, that same pixel is not likely to be affected by the noise in many other images. Averaging image frames should not affect the image of the specimen; however, averaging should reduce the intensity of a pixel that has been affected by random noise. Averaging more frames should reduce the random noise even further until it is so minuscule it cannot be distinguished from the specimen.

There are situations where averaging more image frames can hurt the image. Averaging more image frames exposes the sample to more light and can result in increased photobleaching or phototoxicity. Averaging more image frames can cause the image to blur if the sample is moving or has movement in it. The proper amount of image averaging should be a function of the amount of random noise present, the light stability of the sample, and the motility of the sample.

5. Image Integration

Image integration is used when the signal from the sample is extremely weak and close to the level of background noise. This situation often occurs with low-light-level fluorescence or luminescence microscopy. The image is integrated by leaving the shutter open for a longer period of time on the camera so that more photons can accumulate. As with image averaging, as the camera collects more photons, the signal from the sample builds up faster than the random noise in the camera detector and the signal-to-noise ratio increases.

This technique is more powerful than image averaging. For example, if the sample gives off a photon every 0.1 sec and electronic noise within the camera detector is the equivalent of several photons per 0.1 sec, the signal from the sample will never be seen even with image averaging if the acquisition time per frame is 0.1 sec or less (remember the standard video frame rate in the U.S. is every 0.033 sec). If the image is integrated on the camera for several seconds, the camera will record a signal from the sample. There will be a background signal, but because the background is random, it will be less than the signal from the sample.

Image integration is limited by several factors. First is the dynamic range of the camera. The image should not be integrated beyond the point that it saturates the camera. This is a function of the design of the camera detector and the dynamic range. Some cameras accept 8 bits of data—up to 256 gray levels. Other, more expensive cameras can acquire up to 14 bits of data—16,384 gray levels. Still others can collect further—up to 16 bits of data.

Another consideration for integrating images is time. The longer you integrate an image the more light your sample will be exposed to. This may cause photobleaching or phototoxicity. Longer times will blur any objects in motion just as image averaging would. These considerations should be kept in mind when using this feature.

6. Enhancing Details

Fine details within an image are usually low in contrast and easily obscured by out-of-focus features or low-frequency noise. A variety of techniques have been developed to enhance the details of an image. Often the goal of enhancing details is to increase the contrast of your details of interest or to reduce the contrast of everything but your details of interest. All of these techniques change pixel brightness depending upon the brightness of the surrounding pixels so care must be used in quantitating intensity information after using these techniques.

One of the most common enhancements is the removal of low-frequency information from an image. In this technique the image is treated as a two-dimensional signal. Brighter pixels are higher signals and darker pixels are lower signals. Fine details in an image would appear as high-spatial-frequency information. Out-of-focus features and shading in the optical system all contribute to low-spatial-frequency information—changes are much more gradual. Two common techniques for this operation are the fast Fourier transform and the unsharp mask. In general the unsharp mask may require less computation than the fast Fourier transform, but it has problems dealing with the edge of an image. An example of the unsharp masking technique is shown in Fig. 10B.

Another particular type of enhancement is edge enhancement. Edges of objects are usually seen as a pattern of increasing or decreasing brightness within the

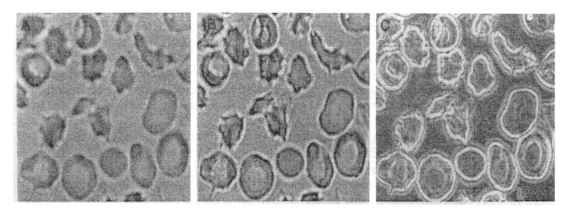

Fig. 10   A sample of detail enhancement. (A) A very detailed image of red blood cells. Many fine details are obscured. (B) Image after unsharp masking using a low-pass filter kernel size of 4 × 4 and scaling factor of 0.75 to decrease the intensity of the low pass image. (C) Image after Prewitt edge detection algorithm. Cell edges and some edges internal to the cells are enhanced.

image. If you are interested in size, shape, or other morphometric information, the edges of objects can be the most important part of the image. Several algorithms have been developed to enhance edges by making the brightness of a pixel a function of its immediate neighbors. This type of methodology is called a convolution and can be executed fairly rapidly by the ALU. An example of the Prewitt edge-detection technique is shown in Fig. 10A.

IV. Acquiring and Analyzing Images—Photography Goes Digital

A. Introduction

The first sections of this chapter covered the principles behind the operation of the digital image processor. This section provides an overview of some techniques and considerations when analyzing images.

B. How Well Does an Image Represent Reality?

How does the image captured by the image processor compare with physical reality? In order to answer this question, we must start at the light source and examine how the light interacts with the specimen, the microscope optics, the camera, and, finally, the image processor.

Ideally, you would like a method to directly extract the desired numerical data from a specimen. For example, if you wanted to measure the calcium in a cell, you would like to be able to directly probe the cell or better yet, somehow produce an image that directly shows the calcium levels. Unfortunately, under most circumstances, such direct measurements are too invasive, too difficult to make, or simply not possible.

Suppose some optical characteristic of the sample correlated with a parameter that you wished to measure. For example, if the amount of dye uptake per unit volume of a cell correlated with the desired parameter. You should be able to measure the optical density or fluorescence brightness in order to directly measure that parameter. Unfortunately, it is not that simple. This section covers some of the most important factors to be considered for accurately characterizing cellular parameters using optical imaging.

Returning to the above example, if we assume that the parameter of interest has an affect on the fraction of light that gets through the sample, we can mathematically work backward to calculate and characterize the parameter of interest. A common equation is listed below for dealing with parameters that affect the fraction of light that gets through a sample. We assume that the fraction of light is linearly proportional to the cell thickness, some parameter of interest, and a calibration factor, ω, characterizing the optical characteristics of the cell:

$$a = 1/(t \cdot p \cdot \omega),$$

where a is the fraction of light that gets through the sample (the amount light that gets through the sample/the amount of light illuminating the sample); t is the cell thickness in microns; p is the parameter of interest; and ω is the calibration factor.

Solving for p gives:

$$p = 1/(t \cdot \omega \cdot a).$$

Given this equation, image gray levels could be converted to the parameter p if the exact number of photons impinging on a spot in the sample and the corresponding number of photons coming out of that spot could be measured. Unfortunately, such direct measurements are not practical. The closest measurement is to detect the relative light intensity at a spot using a photodetector; a video camera in this case.

Why isn't the camera image a direct measure of the number of photons coming from the sample? In order for such measurements to be meaningful, you must understand and compensate for the factors that contribute to the image. The values measured at the camera plane represent some function of the number of photons striking and coming from the sample, the amount of stray light in the room, the transmission characteristics of the optics, the spectral and temporal response of the camera, the amplifiers in the camera and camera control box, and finally, the electronics in the image processor. Some of the most significant factors are shown in Fig. 11 where the light path to the detector (camera) is diagrammed. If you are aware of these corrupting factors then you may be able to quantify an image in a meaningful fashion.

Table X describes many of the factors that contribute to the gray value measured by an image processor.

Most of the factors listed in Table X can be written into a formula that represents the brightness that the imaging system reports at any given spot in the image. The equation for a simple case of a light source, a sample, lenses (objective and condenser), a video camera, and a digitizer is shown below:

$$R \sim (((I \cdot a \cdot L) + S_c) \cdot c_g + c_o + E) \cdot d_g + d_o,$$

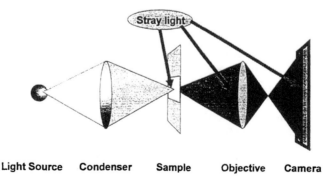

Light Source   Condenser   Sample   Objective   Camera

Fig. 11 A diagram of the light path from the source to the image detected at the camera plane.

Table X
Factors Contributing to the Gray Value Measured for an Image

Item	Description of contribution
Light source	Uniformity of illumination at the specimen is a measure of the relative number of photons that strike a given location in the specimen with another. This parameter may change over time.
Condenser and objective	Lenses absorb some light in a color-dependent manner. Lenses contain chromatic aberration which causes them to shift focus depending on color. Lenses can only accept a certain percentage of the light coming from a light source. Lenses typically have some loss towards the edge of the field. Dust on the lenses can scatter light and cause unevenness.
Sample	The thickness of the sample may vary. The amount of dye taken up by a specimen may vary depending on the location within the specimen. The specimen has intrinsic spectral characteristics. The slide which carries the specimen has its own optical characteristics.
Camera	The camera's response depends on color. The camera's sensitivity may vary across the image. Gray levels measured depend on the gain and black level settings of the camera. The linearity of response depends on the sensor and the electronics. Over long periods of time, some cameras will detect cosmic rays.
Stray light	Room light and light leaks can enter the light path. Internal reflections from optical components can also play a role in light transmission.
Electronic noise	Random electronic noise can occur inside the camera or during transfer of the image into the image processor. Nonrandom electronic noise can occur due to interference within the electrical system.

where R is the resultant brightness in the acquired image; I is the intensity of light from the light source; a is the fractional transmission of the sample; L is the transmission characteristic of the lens at this wavelength; S_c is the stray light coming in to the system and making its way to the camera; c_g and c_o are camera gain and offset (black level); E is electronic noise; and d_g and d_o are digitizer gain and offset.

Because of the mix of additive and multiplicative terms in the equation, you cannot adequately characterize the imaging system with random stray light entering at various places in the optical path. Recall that this stray light may be due to external sources (room light) and internal ones (reflections in the optical system). Problems with external sources of stray light can often be solved by blocking the light leaks or turning off the room lights.

Internal reflections are not easily compensated for in the computer and can be a significant source of error in measurements. The best solution for this problem is to track down the offending optical elements and set up light baffles or to replace the components with units with better antireflective properties.

Electronic noise can indicate a problem with your camera that the manufacturer may be able to fix. Noise can also come from your video cables and can often be reduced by using properly shielded cables. Nonrandom noise may be the results of electrical interference with other electrical equipment in the building. Try to turn various pieces of equipment off to find the culprit. Working in the evening when less electrical equipment is in use can help. Image integration or image averaging will reduce the effects of random electrical noise.

As is shown by the above equation, the *raw* image from the microscope and video system is a poor representation of the physical reality of the specimen. Nevertheless, with adequate precautions these factors can be eliminated or minimized.

C. Preparing the Video Image for Analysis

This section continues the analysis detailed in Section B to demonstrate how you can accurately analyze the brightness of objects in a video image.

Assuming that you have eliminated the problems of stray light and electrical noise, the previous formula characterizing the brightness reported by the imaging system can be reduced to a simpler one:

$$R \sim (I \cdot a \cdot L \cdot c_g + c_o) \cdot d_g + d_o.$$

You can simplify the formula even further by adjusting the camera offset to zero. To do this, follow these instructions:

1. Set the digitizer gain and offset to neutral settings.
2. Block all light to the camera.
3. Adjust the camera offset to produce a digitizer reading of 0.

Note that the default settings for the analog offset contained in the digitizing hardware may not allow a digitizer reading of zero with a given camera. Therefore it is often necessary to adjust both the camera and digitizer offset in order to achieve a digitizer reading of zero.

It is tempting to use digital image processing to compensate for the camera's offset. This can be a mistake. Recall that the digitizer has a limited *dynamic range*. If the camera or digitizer offset is not near zero, then the darkest black in your image will have values greater than zero. For example, if the camera and digitizer offset produce a black level with a gray value of 60 and if you are limited to 256 gray levels (8 bits), the dynamic range of the system has been reduced to only 196 gray levels.

Once the camera and digitizer offset terms have been removed, the formula is reduced further to:

$$R \sim I \cdot a \cdot L \cdot c_g \cdot d_g.$$

The illumination intensity, the light absorption of the lens, the camera gain, and the digitizer gain can all be characterized and kept relatively constant during an experiment. This simplifies the formula even further:

$$R \sim a \cdot \beta.$$

The key is to determine β for the imaging system.

Table XI summarizes the steps necessary before measuring meaningful image brightness data using a digital image processing system.

1. Characterizing the Optical System

Using the methodology in Table XI, it is quite simple to characterize an optical system in a manner that allows accurate video image analysis. Recall that β

Table XI
Methods for Increasing the Accuracy of Image Brightness Measurements

Step	Method
Remove external contributions from stray light.	Turn out the room lights or shield the microscope from light.
Remove internal contributions from stray light.	Determine sources of internal reflections and insert light baffles or replace the offending component.

Reduce or eliminate electrical noise from the image.	Use well-shielded cables.
	Turn off all unnecessary electronics and move other electronics as far from your system as possible.
	If using a video camera
	  Use image averaging to reduce the time variant noise contained in the illumination system, the camera electronics, and the digitizing system. Ideally, you want to average for as long as is practical to achieve the best signal-to-noise without blurring your images or causing photobleaching or phototoxicity. In practice 32 to 64 frames of averaging works quite well.
	If using an integrating camera
	  Integrate as long as possible without saturating the detector. This will reduce time variant noise as above.

Set camera and digitizer offsets to zero.	Block all light to the camera and adjust the camera offset until the digitizer produces a gray level zero. It may be necessary to adjust both the camera and digitizer offsets to achieve this condition.
Minimize variations in the illumination intensity.	Use a regulated or battery driven power supply for the light source. (Voltage fluctuations in the power lines can significantly degrade the reliability of measurements.)

represents the illumination intensity, camera gain, digitizer gain, and absorption of light by lenses and does not represent the sample. It is fairly easy to calculate β for the system. First, set the camera and digitizer offsets to produce a gray value of approximately zero. Set the illumination, camera and digitizer gains, and focal plane appropriately for the sample. Move the stage to a blank section of the slide *without changing any of the conditions* (illumination, camera and digitizer settings, and focal plane) and collect an image. The blank section of the slide represents an image in which the fractional transmission of the sample (a) should be constant. The collected image (R) contains a value of β for every point in the image and is often referred to as the *shading reference* or *white reference* image. A more accurate term might be "optical system characterization image."

Once the β for the image has been determined, it can be used to calculate the desired value for the fractional transmission (a) of a real sample:

$$a \sim R/\beta.$$

thus, to produce an image that contains the corrected fractional transmission of the sample at each pixel, divide the acquired image by the optical system characterization image.

2. Using Shading Correction

Figure 12 shows a dramatic example of this type of image correction used to extract a quantifiable image from a source image that is highly corrupted by shading, dirty optics, and poor camera design. Starting with the original image in the left panel, if analog enhancement is used to raise the contrast of the image,

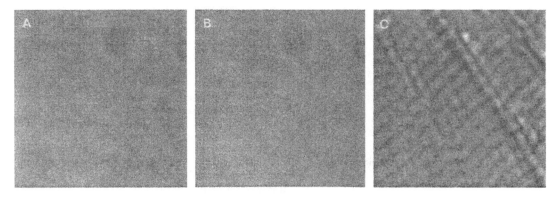

Fig. 12 (A) A very low contrast image of a thin muscle section. Virtually all image details are lost behind camera defects (repeating pattern), shading in the illuminator, and dirt on the optics. (B) Image of a blank section of the slide, used as the optical system characterization image. (C) Corrected image after digital contrast enhancement.

the image brightness quickly saturates the digitizer because the dynamic range of the shading pattern is larger than that of the specimen. Although some video camera controllers include a shading compensator that can help to reduce the effects of large, uniform shading, shading compensators would not be able to remove the nonuniform shading in this image caused by dust in the optical system. The best way to remove the effects of shading involves digital shading correction methods using an optical system characterization image. The optical system characterization image (Fig. 12B) was nearly identical to the original image containing the specimen (Fig. 12A). The resulting shading corrected image was extremely flat (not shown). Digital contrast enhancement was then applied to increase the dynamic range of the image (Fig. 12A). Table XII contains a list of steps we suggest for collecting and correcting images before measuring and image brightness information.

Figure 12 demonstrates the power of proper shading correction and image processing in increasing image accuracy. Note how clearly the muscle banding pattern appears. On first inspection, you may think that this represents a large change in the sample's brightness. However, the line scan in Fig. 13 shows the relative transmission of the sample magnified a 1000-fold. The full scale of the graph represents a tiny 2.5% variation in brightness! Small perturbations in the graph show evidence of brightness changes of less than one-half of a percentage, which is roughly one digitizer gray level at near-saturating brightness.

D. Analyzing Brightness with an Image Processor

Once the optical system has been characterized through shading correction, the brightness of various image features can be measured. Interpretation of the measurements remains the user's responsibility; nevertheless, if you follow the steps listed previously you can state with a fair degree of certainty that an object with a gray value of 100 is twice the brightness as an object with a gray value

Table XII
Steps for Preprocessing Images before Making Image Brightness Measurements

Recommended preprocessing tips
1. Set camera and digitizer black levels to zero.
2. Set the digitizer gain and camera gain controls to default settings.
3. Examine line scan of the video signal.
4. Adjust the signal amplitude to nearly fill the full range of the digitizer using the camera controls.
5. Use the analog controls of the digitizer to stretch the signal more if needed.
6. Collect an optical system characterization image and apply shading correction to reduce the effects of shading.
7. Adjust the contrast of the digital image to achieve optimal contrast for visualization and quantitation.

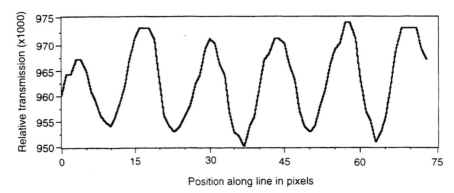

Fig. 13 Line scan showing relative transmission of the corrected image in Fig. 1. Note entire scale represents a relative contrast difference of 2.5%.

of 50. This is a fundamental requirement for a variety of applications including densitometric analysis of electrophoretic gels, quantification of fluorescence to measure protein concentration, quantification of luminescence from aequorin or luciferase assays, and so on.

The level of accuracy you can expect to see in you own measurements depends on a number of factors: light levels (the number of photons reaching the camera), how efficiently and accurately the camera detects that light, the electrical noise present in the digitizer and computer system, and the number of samples you can measure to improve the signal relative to random noise. Many other factors contribute to the overall accuracy of your measurements and a thorough discussion of sampling statistics is beyond the scope of this chapter. For more information, we encourage you to refer to the other references listed for this chapter.

Suggested References

Agard, D. A., Hiraoka, Y., Shaw, P., and Sedat, J. W. (1989). Fluorescence microscopy in three dimensions. *In* "Methods in Cell Biology," Vol. 30, pp. 353–377. Academic Press, San Diego.

Allen, R. D., and Allen, N. S. (1983). Video-enhanced microscopy with a computer frame memory. *J. Microscopy* **129,** 3.

Castleman, K. R. (1979). "Digital Image Processing." Prentice Hall, New Jersey.

Gonzalez, R. C., and Wintz, P. (1987). "Digital Image Processing," 2nd edition. Addison Wesley, Reading, MA.

Inoué, S. (1981). Video image processing greatly enhances contrast, quality, and speed in polarization-based microscopy. *J. Cell Biol.* **89,** 346–356.

Inoué, S. (1986). "Video Microscopy." Plenum, New York.

Russ, J. C. (1995). "The Image Processing Handbook," 2nd edition. CRC Press, Boca Raton, FL.

CHAPTER 5

Introduction to Image Processing

Richard A. Cardullo and Eric J. Alm
Department of Biology
The University of California
Riverside, California 92521

I. Introduction

Recent technological advances in computing and interfaces between video components and computers have allowed scientists to use processing routines to enhance or quantify particular characteristics of images. However, the use of image processing is not limited to scientists, and the introduction of digital cameras, video cards, and scanners has made image processing available to nearly every user of personal computers. With a minimum investment, one can readily enhance contrast, detect edges, quantify intensity, and apply a variety of mathematical operations to images. Although these techniques can be extremely powerful, the average user often digitally manipulates images with abandon, seldom understanding the basic principles behind the simplest image processing routines. Although this may be acceptable to some individuals, it often leads to an image which is significantly degraded and does not achieve the results that would be possible with some knowledge of the basic operation of an image processing system.

The theoretical basis of image processing along with its applications is an extensive topic which cannot be adequately covered here but has been presented in a number of texts dedicated exclusively to this topic (see References). In this chapter, we will outline the basic principles of image processing used routinely by microscopists. Since image processing allows the investigator to convert the microscope/camera system into a quantitative device we will focus on three basic problems: (1) reducing "noise," (2) enhancing contrast, and (3) quantifying intensity of an image. These techniques can then be applied to a number of different methodologies such as video-enhanced differential interference microscopy (VEDIC, Chapter 10), nanovid microscopy, fluorescence recovery after photobleaching, and fluorescence ratio imaging (see Chapters 11 and 12). In all cases, knowledge of the basic principles of microscopy, image formation, and image processing routines is absolutely required to convert the microscope into a device capable of pushing the limits of resolution and contrast.

II. Digitization of Images

An image must first be digitized before an arithmetic operation can be performed on it. For this discussion, a digital image is a discrete representation of light intensity in space (Fig. 1). A particular scene can be viewed as being continuous in both space and light intensity and the process of digitization converts these to discrete values. The discrete representations of intensity are commonly referred to as gray values whereas the discrete representation of position is given as picture elements (or pixels). Therefore, each pixel has a corresponding gray value which is related to light intensity (e.g., at each coordinate (x,y) there is a corresponding gray value designated as GV (x,y)). The key to digitizing an image is to provide enough pixels and grayscale values to adequately describe the original image.

Clearly, the fidelity of reproduction between the true image and the digitized image depends on both the spacing between pixels (e.g., the number of bins that map the image) and the number of gray values used to describe the intensity of the image. Figure 1B shows a theoretical one-dimensional scan across a portion of an image. Note that the more pixels used to describe, or sample, an image the better the digitized image reflects the true nature of the original. Conversely, as the number of pixels is progressively reduced, the true nature of the original image is lost. When choosing the digitizing array, such as a CCD camera, for a microscope, particular attention must be paid to matching the resolution limit of the microscope ($\sim$0.2 μm, see Chapter 1) to the resolution limit of the digitizer. A digitizing array which has an effective separation of 0.05 μm per pixel is, at best, using four pixels to describe resolvable objects in a microscope resulting

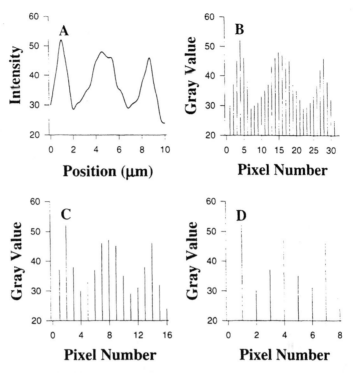

Fig. 1 (A) A densitometric line scan through a microscopic image is described by intensity values on the *y*-axis and its position along the *x*-axis. (B) A 6-bit digitized representation (64 gray values) of the object in A, with 32 bins used to describe the position across 10 microns. This digital representation captures the major details of the original object but some finer detail is lost. Note that the image is degraded further when the position is described by only 16 bins (C) or 8 bins (D).

in a highly digitized representation of the original image (Note: this is most clearly seen when using the digitized zoom feature of many image processors which results in a "boxy" image representation). In contrast, a digitizer which has the pixel elements separated by 1 μm effectively averages gray values five times above the resolution limit of the microscope resulting in a degraded representation of the original image.

In addition to describing the number of pixels in an image, it is also important to know the number of gray values needed to faithfully represent the intensity of that image. In Figure 1B, we have chosen to digitize the original image 6 bits deep (6 bits = 2^6 = 64 gray values from 0 to 63). The image could be better described by more gray levels (e.g., 8 bits = 256 gray levels) but would be poorly described by less gray levels (e.g., 2 bits = 4 gray levels).

The decision on how many pixels and gray levels are needed to describe an image is dictated by the properties of the original image. Figure 1 represents a low-contrast, high-resolution image which needs many gray scales and pixels to describe it. However, some images are by their very nature high contrast and low resolution and require less pixels and gray values to describe it (e.g., a line drawing may require only one bit of gray level depth, black or white). Ultimately, the trade-off is one in contrast, resolution, and speed of processing. The more descriptors used to represent an image, the slower the processing routines will be performed. In general, an image should be described by as few pixels and gray values as needed so that speed of processing can be optimized.

III. Using Gray Values to Quantify Intensity in the Microscope

A useful feature shared by all image processors is that they allow the microscopist a way to quantify image intensity values into some meaningful parameter. In standard light microscopy, the light intensity, and therefore the digitized gray values, are related to optical density which is proportional to the log of the relative light intensity. For dilute solutions, (i.e., in the absence of significant light scattering) the optical density may be proportional to the concentration of absorbers, C, the molar absorptivity, ε, and the pathlength, l, through the vessel containing the absorbers. In such a situation the optical density (O.D.) is related to these parameters through Beer's Law:

$$O.D. = \log(I_0/I) = \varepsilon Cl$$

where I is the intensity of light in the presence of an absorber and I_0 is the intensity of light in the absence of an absorber. Within dilute solutions, it therefore might be possible to equate a change in O.D. with either changes in molar absorptivity, pathlength, or concentration. However, with objects as complex as cells, all three parameters can vary tremendously and the utility of using O.D. to measure a change in any one parameter is difficult.

Although difficult to interpret in cells, measuring changes in digitized gray values in an optical density step wedge offers the investigator a good way to calibrate an entire microscope system coupled to an image processor. Figure 2 shows such a calibration using an upright brighfield microscope (Zeiss), a SIT camera (DAGE), an image processor (Universal Imaging), and an optical density wedge which had 0.15 O.D. increments. The camera/image processor unit was digitized to 8 bits (0 to 255) and the median gray value was recorded for a 100×100-pixel array in each step of the wedge. In this calibration, the black level of the camera was adjusted so that the highest optical density corresponded to a gray level of 5. At the other end of the scale (the lowest optical density used), the relative intensity value was normalized so that I/I_0 was equal to 1 and the corresponding gray value was ~95% of the maximum gray value (~243). As seen in Fig. 2, as the step wedge is moved through the microscope, the median value of the GV increased as the log of I/I_0. In addition to acting as a useful

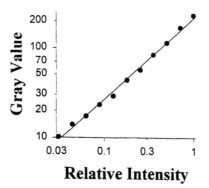

Relative Intensity

Fig. 2 Calibration of a SIT camera using an image processor. The light intensity was varied incrementally using an optical density step wedge (0.15 increments), and the gray value was plotted as a function of the normalized intensity. In this instance the camera/image processor system was able to quantify differences in light intensity over a 40-fold range.

calibration, this figure shows that an 8-bit processor can reliably quantify changes in light intensity over two orders of magnitude.

IV. Noise Reduction

The previous sections have assumed that the object being imaged is relatively free of noise and is of sufficient contrast to generate a usable image. Although this may be true in some instances, the ultimate challenge in many applications is to obtain reliable quantitative information from objects which produce a low-contrast, noisy signal. This is particularly true in cell physiological measurements using specialized modes of microscopy such as video-enhanced differential interference contrast (see Chapter 10), fluorescence ratio imaging (see Chapter 12), nanovid microscopy, and so on. There are different ways to reduce noise and the methods of noise reduction chosen depend on many different factors including the source of the noise, the type of camera employed for a particular application (i.e., SIT, CCD, cooled CCD, etc.), and the contrast of the specimen. For the purposes of this chapter we shall distinguish between temporal techniques and spatial techniques to increase the signal-to-noise ratio of an image.

A. Temporal Averaging

In most low-light-level applications there is a considerable amount of shot noise associated with the signal. If quantitation is needed, it is often necessary to reduce the amount of shot noise in order to improve the signal-to-noise ratio (SNR). Because this type of noise reduction requires averaging over a number of frames (≥ 2 frames) this method can only be used when static objects are used. Clearly, temporal averaging would not be appropriate if one were interested in optimizing contrast for dynamic processes such as cell movement, detecting rapid changes in intracellular ion concentrations over time, quantifying molecular motions using fluorescence recovery after photobleaching or single particle tracking, and so on.

Assume that at any given time, t, within a given pixel, i, a signal, $S_i(t)$, represents both the true image, I, and some source of noise, $N_i(t)$. Because the noise is stochastic in nature, $N_i(t)$ will vary in time, taking on both positive and negative values, and the total signal, $S(t)$, will vary about some mean value. For each frame, the signal is therefore just:

$$S_i(t) = I + N_i(t).$$

As the signal is averaged over M frames, an average value for $S_i(t)$ and $N_i(t)$ is obtained:

$$\langle S_i \rangle_M = I + \langle N_i \rangle_M$$

where $\langle S_i \rangle_M$ and $\langle N_i \rangle_M$ represent the average value of $S_i(t)$ and $N_i(t)$ over M frames. As the number of frames, M, goes to infinity, the average value of N_i goes to zero and therefore:

$$\langle S_i \rangle_{M \Rightarrow \infty} = I$$

The question facing the microscopist is how large M should be so that the SNR is acceptable. This is determined by a number of factors including the magnitude of the original signal, the amount of noise, and the degree of precision required by the particular quantitative measurement. A quantitative measure of noise reduction can be obtained by looking at the standard deviation of the noise which decreases inversely as the square root of the number of frames ($\sigma_M = \sigma_0/\sqrt{M}$). Therefore, averaging a field for 4 frames will give a 2-fold improvement in the SNR, averaging for 16 frames yields a 4-fold improvement, while averaging for 256 frames yields a 16-fold improvement. At some point the user obviously reaches a point of diminishing returns where the noise level is below the resolution of the digitizer and any improvement in the SNR is minimal (Fig. 3).

Although frame-averaging techniques are not appropriate for moving objects, it is possible to apply a running average where the resulting image is a weighted sum of all previous frames. Because the image is constantly updated on a frame-by-frame basis, these types of recursive techniques are useful for following moving objects but improvement in the SNR is always less than that obtained with the simple averaging technique outlined in the previous paragraph (Erasmus, 1982). Additional recursive filters are possible which optimize the SNR, but these typically are not available on commercially available image processors.

B. Spatial Methods

A number of spatial techniques are available which allow the user to reduce noise on a pixel-by-pixel basis. The simplest of these techniques generally use

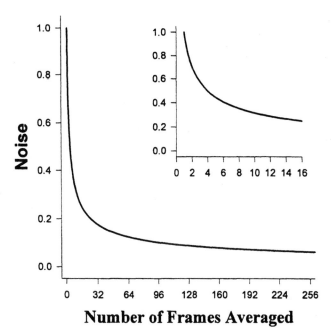

Number of Frames Averaged

Fig. 3 Reduction in noise as a function of the numbers of frames averaged. The noise is reduced inversely as the square root of the number of frames averaged. In this instance, the noise was normalized to the average value obtained for a single frame. The major gain in noise reduction is obtained after averaging very few frames (inset) and averaging for more than 64 frames leads to only minor gains in the signal-to-noise ratio.

simple arithmetic operations within a single frame or, alternatively, between two different frames. In general, these types of routines involve either image subtraction or division from a background image or calculate a mean or median value around the neighborhood of a particular pixel. More sophisticated methods use groups of pixels (known as masks, kernels, or filters) which perform higher-order functions to extract particular features from an image. These types of techniques will be discussed separately (see Section VI).

1. Arithmetic Operations between an Object and a Background Image

If one has an image which has a constant noise component in a given pixel in each frame, the noise component can be removed by performing an arithmetic operation (i.e., either subtraction or division) that would remove the noise and therefore optimize the SNR. These types of routines may substantially increase the SNR, but they also significantly decrease the dynamic range of the detector. Since this type of "noise" is generally caused by dirt on the optics or the camera or by incorrectly setting the black level or gain of the camera, these problems are best corrected by first attending to the microscope and camera system. Only after the dynamic range of the camera/image processor system has been optimized should these types of arithmetic manipulations be attempted.

With the camera/image processor system optimized, any constant noise component can then be removed arithmetically. In general, it is always best to subtract the noise component out from a uniform background image (Fig. 4). In contrast, image division greatly diminishes the signal while assigning a value of one to identical gray values in identical pixels. Thus, if a pixel within an image has a gray value of say 242, with the background having a gray value of 22 in that same pixel, then a simple subtraction would yield a resultant value of 220 (242 − 22) while a division would yield a resultant value of just 11 (242/22). Image subtraction therefore preserves the majority of the signal, and the sub-

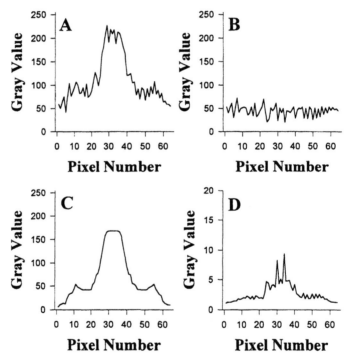

Fig. 4 Comparison of image subtraction with image division to reduce noise. (A) Line scan across an object and the surrounding background. (B) Line scan across the background alone reveals variations in intensity which may be due to uneven light intensity across the field, camera defects, dirt on the optics, and so on. (C) Image in A subtracted from image in B. The result is a "cleaner" image with high signal-to-noise ration retained in the processed image. (D) Image in A divided by image in B. In general, image division is sensitive to minor variations in both the image and the background and may lead to additional spikes in the processed image. In addition, image division results in a significant decrease in the signal and is therefore not routinely used to eliminate a constant noise source from the background.

tracted image can then be processed further using other routines. In order to reduce temporal noise, both images can be first averaged as described in Section IV,A.

2. Concept of a Digital Mask

A number of mathematical manipulations of images involve using an array (or a digital mask) around a neighborhood of a particular pixel. These digital masks can be used either to select particular pixels from the neighborhood (as in averaging or median filtering discussed in the next section) or alternatively can be used to apply some mathematical weighting function to an image on a pixel-by-pixel basis to extract particular features from that image (discussed in detail in Section VI). When the mask is overlaid upon an image, the particular mathematical operation is performed and the resultant value is placed into the same array position in a different frame buffer. The mask is then moved to the next pixel position and the operation is performed repeatedly until the entire image has been transformed. Although a digital mask can take on any shape, the most common masks are square with the center pixel being the particular pixel be operated on at any given time (Fig. 5). The most common masks are 3×3 or 5×5 arrays so that only the nearest neighbors will have an effect on the pixel being operated upon. Additionally, larger arrays will greatly increase the number of computations that need to be performed which can significantly slow down the rate of processing a particular image.

$W_{i-1, j-1}$	$W_{i-1, j}$	$W_{i-1, j+1}$
$W_{i, j-1}$	$w_{i, j}$	$W_{i, j+1}$
$W_{i+1, j-1}$	$W_{i+1, j}$	$W_{i+1, j+1}$

Fig. 5 Digital mask used for computing means, averages, and higher-order mathematical operations, especially convolutions. In the case of median and average filters, the mask is overlaid over each pixel in the image and the resultant value is calculated and placed into the identical pixel location in a new image buffer.

3. Averaging versus Median Filters

When an image contains random and infrequent intensity spikes within particular pixels a digital mask can be used around each pixel to remove them. Two common ways to remove these intensity spikes is to calculate either the average value or the median value within the neighborhood and assign that value to the center pixel in the processed image (Fig. 6). The choice of filter used will depend on the type of processed image that is desired. Although both types of filters will degrade an image, the median filter preserves edges better than averaging

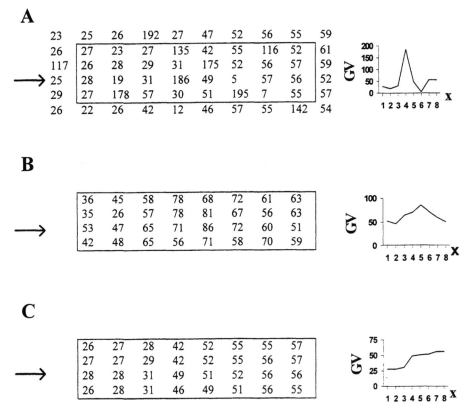

Fig. 6 Comparison of 3 × 3 averaging and median filters to reduce noise. (A) Digital representation of an image displaying gray values at different pixel locations. In general, the object possesses a boundary which is detected as a line scan from left to right. However, the image has a number of intensity spikes which significantly masks the true boundary. A line scan across a particular row (row denoted by arrow and scan is on right-hand side) reveals both high- and low-intensity values which greatly distort the image. (B) When a 3 × 3 averaging filter is applied to the image, the extreme intensity values are significantly reduced but the image is significantly smoothed in the vicinity of the boundary. (C) In contrast, a 3 × 3 median filter both removes the extreme intensity values while preserving the true nature of the boundary.

filter since all values within the digital mask are used to compute the mean. Further, averaging filters are seldom used to remove intensity spikes because the spikes themselves contribute to the new intensity value in the processed image and therefore the resultant image is blurred.

The median filter is more desirable for removing infrequent intensity spikes from an image since those intensity values are always removed from the processed image once the median is computed. In this case, any spike is replaced with the median value within the digital mask which gives a more uniform appearance to the processed image. Hence a uniform background which contains infrequent intensity spikes will look absolutely uniform in the processed image. Since the median filter preserves edges (a sharpening filter), it is often used for high-contrast images.

V. Contrast Enhancement

One of the most common uses of image processing is to digitally enhance the contrast of the image using a number of different methods. In brightfield modes such as phase contrast or differential interference contrast the addition of a camera and an image processor can significantly enhance the contrast so that specimens with inherent low contrast can be observed. Additionally, contrast routines can be used to enhance an image in a particular region which may allow the investigator to quantify structures or events not possible with the microscope alone. This is the basis for video-enhanced DIC which allows, for example, the motion of low-contrast specimens such as microtubules or chromosomes to be quantified (Chapter 10).

In order to optimize contrast enhancement digitally, it is imperative that the microscope optics and the camera be adjusted so that the full dynamic range of the system is utilized. The gray values of the image and background can then be displayed as a histogram (Fig. 7) and the user is then able to adjust the brightness and contrast within a particular region of the image. Within a particular gray value range, the user can then stretch the histogram so that values within that range are spread out over a different range in the processed image. Although this type of contrast enhancement is artificial, it allows the user to discriminate features which otherwise may not have been detectable by eye in the original image.

Stretching gray values over a particular range in an image is one type of mathematical manipulation which can be performed on a pixel-by-pixel basis. In general, any digital image can be mathematically manipulated to produce an image with different gray values. The user-defined function that transforms the original image is known as the image transfer function (ITF), which specifies the value and the mathematical operation that will be performed on the original image. This type of operation is a point operation which means that the output gray value of the ITF is dependent only on the gray value of the input gray value

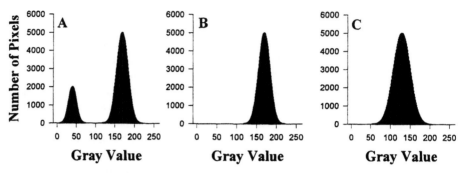

Fig. 7 Histogram representation of gray values for an entire image. (A) The image contains two distributions of intensity over the entire gray value range (0–255). (B) The lower distribution can be removed either through subtraction (if lower values are due to a uniform background) or by applying the appropriate image transfer function which assigns a value of 0 to all input pixels having a gray value less than 100. The resulting distribution contains only information from input pixels with a value greater than 100 (C) The histogram of the higher distribution can then be stretched to fill the lower gray values resulting in a lower mean value from the original.

on a pixel-by-pixel basis. The gray values of the processed image, I_2, are therefore transformed at every pixel location relative to the original image using the same ITF. Hence, every gray value in the processed image is transformed according to the generalized relationship:

$$GV_2 = f(GV_1),$$

where GV_2 is the gray value at every pixel location in the processed image, GV_1 is the is the input gray value of the original image, and $f(GV_1)$ is the ITF acting on the original image.

The simplest type of ITF is a linear equation of slope m and intercept b:

$$GV_2 = mGV_1 + b$$

In this case, the digital contrast of the processed image is linearly transformed with the brightness and contrast determined by both the value of the slope and intercept chosen. In the most trivial case, choosing values of $m = 1$ and $b = 0$ would leave all gray values of the processed image identical to the original image (Fig. 8). Raising the value of the intercept while leaving the slope unchanged would have the effect of increasing all gray values by some fixed value (identical to increasing the DC or black value control on a camera). Similarly, decreasing the value of the intercept will produce a darker image than the original. The value of the slope is known as the *contrast enhancement factor* and changes in the value of m will have significant effects on how the gray values are distributed in an image. A value of $m > 1$ will have the effect of spreading out the gray values over a wider range in the processed image relative to the original image.

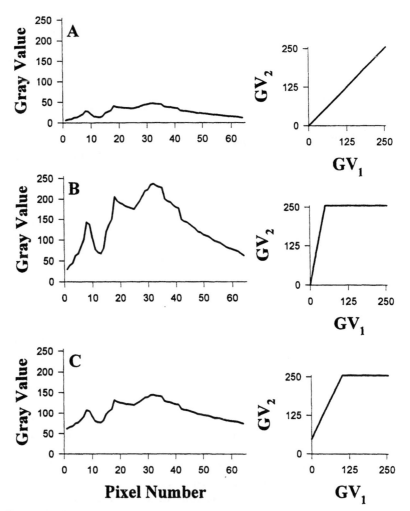

Fig. 8 Application of different linear image transfer functions to a low-intensity, low-contrast image. (A) Intensity line scan through an object which is described by few gray values. Applying a linear ITF with $m = 1$ and $b = 0$ (right) will result in no change from the initial image. (B) Applying a linear ITF with $m = 5$ and $b = 0$ (right) leads to significant improvement in contrast. (C) Applying a linear ITF with $m = 2$ and $b = 50$ (right) slightly improves contrast and increases the brightness of the entire image.

Conversely, values of $m < 1$ will reduce the number of gray values used to describe a processed image relative to the original (Fig. 8). As noted by Inoue (1986), although linear ITFs can be useful, the same effects can be best achieved by properly adjusting the video camera's black level and gain controls. However, this may not always be practical if conditions under the microscope are constantly changing or if this type of contrast enhancement is needed after the original images are stored.

The ITF is obviously not restricted to linear functions and nonlinear ITFs can be extremely useful for enhancing particular features of an image while eliminating or reducing others (Fig. 9). Nonlinear ITFs are also useful for correcting sources of nonlinear response in an optical system or to calibrate the light response of an optical system (Inoue, 1986). The actual form of the ITF, being linear or nonlinear, is generally application dependent and user defined. For example, nonlinear ITFs which are sigmoidal in shape are useful for enhancing images which compress the contrast in the center of the histogram and increase contrast in the tail regions of the histogram. This type of enhancement would be useful for images where most of the information about the image is in the tails of the histogram while the central portion of the histogram contains mostly background information. One type of nonlinear ITF, which is sigmoidal in shape and will enhance an 8-bit image of this type is given by the equation:

$$GV_2 = \frac{128}{(b - c)^a} ((b - c)^a - (b - GV_1)^a + (GV_1 + c)^a)$$

where b and c are the maximum and minimum gray values for the input image respectively and a is an arbitrary contrast enhancement factor (Inoue, 1986). For values of $a = 1$, this normally sigmoidal ITF becomes linear with a slope of $256/(b - c)$. As a increases beyond 1, the ITF becomes more sigmoidal in nature with greater compression occurring at the middle gray values.

In practice, ITFs are generally calculated in memory using a look-up table or LUT. A LUT represents the transformation that is performed on each pixel on the basis of that intensity value (Fig. 10). In addition to LUTs which perform particular ITFs, LUTs are also useful for pseudocoloring images where particular user-defined colors represent gray values in particular ranges. This is particularly useful in techniques such as ratio imaging where various color LUTs are used to represent concentrations of Ca^{2+} or pH when various indicator dyes are employed within cells.

VI. Transforms, Convolutions, and Further Uses for Digital Masks

In the previous sections we have briefly described the most frequently used methods for enhancing contrast and reducing noise using temporal methods, simple arithmetic operations, and LUTs. However, more advanced methods are often needed to extract particular features from an image which may not be possible using these simple methods. In this section, we will briefly introduce some of the concepts and applications associated with transforms and convolutions.

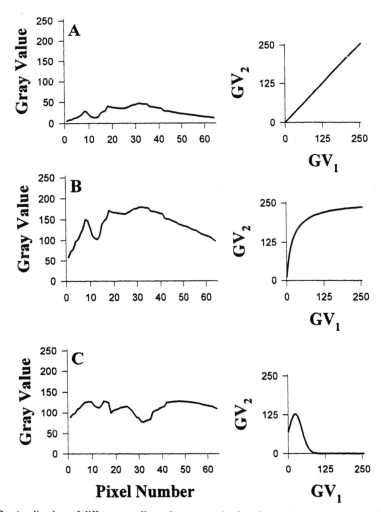

Fig. 9 Application of different nonlinear image transfer functions to the same low-intensity, low-contrast image in Figure 8. (A) Initial image and image transfer function (right) resulting in no change. (B) Application of a hyperbolic ITF (right) to the image results in amplification of low values and only slightly increases the gray values for higher input values. (C) Application of a Gaussian ITF (right) to the image results in amplification of low values, with an offset, and minimizes input values beyond 100.

A. Transforms

Transforms take an image from one space to another. Probably the most used transform is the Fourier transform which takes one from coordinate space to frequency space. In general, a transform of a function in one dimension has the form:

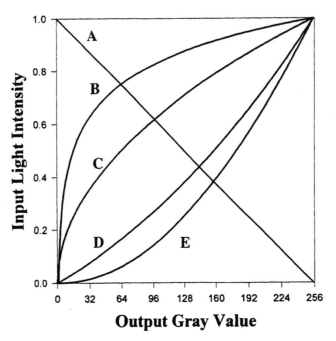

Fig. 10 Some different gray value LUTs used to alter contrast in images. (A) Inverse LUT;
(B) logarithmic LUT; (C) square root LUT; (D) square LUT; and (E) exponential LUT. Pseudocolor
LUTs would assign different colors instead of gray values.

$$T(u) = \sum_{x} f(x)g(x,u),$$

where $T(u)$ is the transform of $f(x)$ and $g(x, u)$ is known as the forward transformation kernel. Similarly, the inverse transform is given by the relation:

$$f(x) = \sum_{u} T(u)h(x,u),$$

where $h(x,u)$ is the inverse transformation kernel. In two dimensions, these
transformation pairs simply become:

$$T(u,v) = \sum_{v} \sum_{u} f(x,y)g(x,y,u,v)$$
$$f(x,y) = \sum_{v} \sum_{u} T(u,v)h(x,y,u,v).$$

It is the kernel functions that provide the link which brings a function from
one space to another. The discrete forms shown above suggest that these operations can be performed on a pixel-by-pixel basis and many transforms in image
processing are computed in this manner. In the case of the Fourier transform,

the computation of the discrete Fourier transform (DFT) can be speeded up considerably using different algorithms to yield a fast Fourier transform (FFT).

In the Fourier transform the forward transformation kernel is:

$$g(x,u) = \frac{1}{N} e^{-2\pi iux}$$

and the reverse transformation kernel is:

$$f(x,u) = \frac{1}{N} e^{+2\pi iux}.$$

Hence a Fourier transform is achieved by multiplying a digitized image pixel-by-pixel whose gray level is given by $f(x,y)$ by the forward transformation kernel given above. Transforms, and in particular Fourier transforms, can make certain mathematical manipulations of images considerably easier.

One example where conversion to frequency space using an FFT is useful is in identifying both high- and low-frequency components of an image that allows one to make quantitative choices about information which can be either used or discarded. Sharp edges and many types of noise will contribute to the high-frequency content of an image's Fourier transform. Image smoothing and noise removal can therefore be achieved by attenuating a range of high-frequency components in the transform image. In this case, a filter function, $F(u,v)$, is selected that eliminates the high-frequency components of that transformed image, $I(u, v)$. The ideal filter would simply cut off all frequencies above some threshold value, I_0 (known as the cutoff frequency):

$$F(u,v) = 1 \text{ if } |I(u,v)| \leq I_0$$
$$F(u,v) = 0 \text{ if } |I(u,v)| > I_0.$$

The absolute value brackets refer to the fact that these are zero-phase shift filters because they do not change the phase of the transform. A graphical representation of an ideal lowpass filter is shown in Fig. 11. Just as an image can be blurred by attenuating high-frequency components using a lowpass filter, so can they be sharpened by attenuating low-frequency components (Fig. 11). In analogy to the lowpass filter, an ideal highpass filter has the following characteristics:

$$F(u,v) = 0 \text{ if } |I(u,v)| \leq I_0$$
$$F(u,v) = 1 \text{ if } |I(u,v)| > I_0$$

Although useful, Fourier transforms are computationally intensive and are not routinely used in most microscopic applications of image processing. A mathematically related technique known as convolution, which utilizes digital masks to select particular features of an image, is the preferred method of microscopists since many of these operations can be performed at faster rates and perform the mathematical operation in coordinate space instead of frequency space. These operations are outlined in the following section.

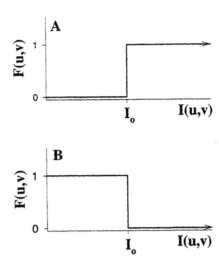

Fig. 11 Frequency domain cutoff filters. The filter function in frequency space, $F(u, v)$, is used to cutoff all intensities above or below some cutoff frequency, I_0. (A) A high-pass filter attenuates all frequencies below I_0 leading to a sharpening of the image. (B) A low-pass filter attenuates all frequencies above I_0 which eliminates high-frequency noise but leads to smoothing or blurring of the image.

B. Convolution

The convolution of two functions, $f(x)$ and $g(x)$, is given mathematically by:

$$\int_{-\infty}^{\infty} f(\alpha)g(x - \alpha)d\alpha,$$

where α is a dummy variable of integration. It is easiest to visualize the mechanics of convolution graphically as demonstrated in Fig. 12 which, for simplicity, shows the convolution for two square pulses. The operation can be broken down into a few simple steps:

1. Before carrying out the integration, reflect $g(\alpha)$ about the origin, yielding $g(-\alpha)$ and then displace it by some distance x to give $g(x - \alpha)$.

2. For all values of x, multiply $f(\alpha)$ by $g(x - \alpha)$. The product will be nonzero at all points where the functions overlap.

3. Integrating the product yields the convolution between $f(x)$ and $g(x)$.

Hence, the properties of the convolution are determined by the independent function $f(x)$ and a function $g(x)$ that selects for certain desired details in the function $f(x)$. The selecting function $g(x)$ is therefore analogous to the forward transformation kernel in frequency space except that it selects for features in coordinate space instead of frequency space. This clearly makes convolution an

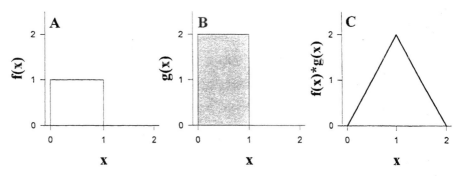

Fig. 12 Graphical representation of one-dimensional convolution. (A) In this simple example, the function, $f(x)$, to be convolved is a square pulse of equal height and width. (B) The convolving function, $g(x)$, is a rectangular pulse which is twice as high as it is wide. The convolving function is then reflected and is then moved from $-\infty$ to $+\infty$. (C) In all areas here there is no overlap; the product of $f(x)$ and $g(x)$ is zero. However, $g(x)$ overlaps $f(x)$ in different amounts from $x = 0$ to $x = 2$ with maximum overlap occurring at $x = 1$. The operation therefore detects the trailing edge of $f(x)$ at $x = 0$, and the convolution results in a triangle which increases from 0 to 2 for $0 < x \leq 1$ and decreases from 2 to 0 for $1 \leq x < 2$.

important image processing technique for microscopists who are interested in feature extraction.

One simple application of convolutions is the convolution of a function with an impulse function (commonly known as a delta function), $\delta(x - x_0)$:

$$\int_{-\infty}^{\infty} f(x)\delta(x - x_0)\, dx = f(x_0).$$

For our purposes, $\delta(x - x_0)$ is located at $x = x_0$ and the intensity of the impulse is determined by the value $f(x)$ at $x = x_0$ and is 0 everywhere else. In this example we will let the kernel $g(x)$ represent three impulse functions separated by a period, τ:

$$g(x) = \delta(x + \tau) + \delta(x) + \delta(x - \tau)$$

As shown in Fig. 13, the convolution of the square pulse $f(x)$ with these three impulses results in a copying of $f(x)$ at the impulse points.

As with Fourier transforms, the actual mechanics of convolution can rapidly become computationally intensive for a large number of points. Fortunately, many complex procedures can be adequately performed using a variety of digital masks as shown in the following section.

C. Digital Masks as Convolution Filters

For most purposes, the appropriate digital mask can be used to extract features from images. The convolution filter, acting as the selection function $g(x)$, can be

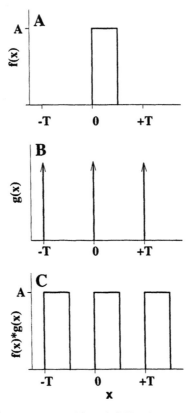

Fig. 13 Using a convolution to copy an object. (A) The function $f(x)$ is a rectangular pulse of amplitude, A, with its leading edge at $x = 0$. (B) The convolving functions $g(x)$ are three delta functions at $x = -\tau$, $x = 0$, and $x = +\tau$. (C) The convolution operation $f(x)*g(x)$ results in a copying of the three rectangular pulses at $x = -\tau$, $x = 0$, and $x = +\tau$.

used to modify images in a particular fashion. Convolution filters reassign intensities by multiplying the gray value of each pixel in the image by the corresponding values in the digital mask and then summing all the values; the resultant is then assigned to the center pixel of the new image and the operation is then repeated for every pixel in the image (Fig. 14). Convolution filters can vary in size (i.e., $3 \times 3, 5 \times 5, 7 \times 7$, etc.) depending on the type of filter chosen and the relative weight that is required from neighboring values from the center pixel.

For example, consider a simple 3×3 convolution filter which has the form:

$$
\begin{array}{ccc}
1/9 & 1/9 & 1/9 \\
1/9 & 1/9 & 1/9 \\
1/9 & 1/9 & 1/9
\end{array}
$$

applied to a pixel with an intensity of 128 and surrounded by other intensity values as follows:

Original Image

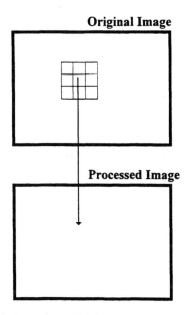

Processed Image

Fig. 14 Performing convolutions using a digital mask. The convolution mask is applied to each pixel in the image. The value assigned to the central pixel results from multiplying each element in the mask by the gray value in the corresponding image, summing the result, and assigning the value to the corresponding central pixel in a new image buffer. The operation is repeated for every pixel resulting in the processed image. For different operations, a scalar multiplier and/or offset may be needed.

$$
\begin{array}{ccc}
123 & 62 & 97 \\
237 & 128 & 6 \\
19 & 23 & 124
\end{array}
$$

The gray value in the processed image at that pixel therefore would have a new value of: $1/9 \times (123 + 62 + 97 + 237 + 128 + 6 + 19 + 23 + 124) = 819/9 = 91$. Note that this convolution filter is simply an averaging filter identical to the operation described in Section IV,B,3 (in contrast, a median filter would have returned a value of 128). A 5×5 averaging filter would simply be a mask which contains 1/25 in each pixel whereas an 7×7 averaging filter would consist of 1/49 in each pixel. Since the speed of processing decreases with the size of the digital mask, the most, frequently used filters are 3×3 masks.

In practice, the values found in the digital masks tend to be integer values with a divisor which can vary depending on the desired operation. In addition, because many operations can lead to resultant values which are negative (since the values in the convolution filter can be negative), offset values are often used to prevent this from occurring. In the example of the averaging filter, the values in the kernel would be:

$$
\begin{array}{ccc}
1 & 1 & 1 \\
1 & 1 & 1 \\
1 & 1 & 1
\end{array}
$$

with a divisor value of 9 and an offset of zero. In general, for an 8-bit image, divisors and offsets are chosen so that all processed values following the convolution fall between 0 and 255.

Understanding the nature of convolution filters is absolutely necessary when using the microscope as a quantitative tool. User defined convolution filters can be used to extract information specific for a particular application. When beginning to use these filters it is important to have a set of standards which the filters can be applied to in order to see if the desired effect has been achieved. We have found that simple geometric objects (circles, squares, isosceles and equilateral triangles, hexagons, etc.) are useful test objects for convolution filters. Many commercially available graphics packages provide such test objects in a variety of graphics formats. Examples of some widely used convolution masks are given in the following sections.

1. Point Detection in a Uniform Field

Assume that an image consists of a series of grains on a constant background (e.g., a dark field image of a cellular autoradiogram). The following 3×3 mask is designed to detect these points:

$$
\begin{array}{ccc}
-1 & -1 & -1 \\
-1 & 8 & -1 \\
-1 & -1 & -1
\end{array}
$$

When this mask encounters a uniform background, then the gray values in the processed center pixel will be zero. If, on the other hand a value above the constant background is encountered, then its value will be amplified above that background and a high contrast image will result.

2. Line Detection in a Uniform Field

Similar to the point mask in the previous example, a number of line masks can be used to detect sharp, orthogonal edges in an image. These line masks can be used alone or in tandem to detect horizontal, vertical, or diagonal edges in an image. Horizontal and vertical line masks are represented as:

$$
\begin{array}{ccc}
-1 & -1 & -1 \\
2 & 2 & 2 \\
-1 & -1 & -1
\end{array}
\qquad
\begin{array}{ccc}
-1 & 2 & -1 \\
-1 & 2 & -1 \\
-1 & 2 & -1
\end{array}
$$

whereas diagonal line masks are given as:

$$\begin{array}{ccc} 2 & -1 & -1 \\ -1 & 2 & -1 \\ -1 & -1 & 2 \end{array} \qquad \begin{array}{ccc} -1 & -1 & 2 \\ -1 & 2 & -1 \\ 2 & -1 & -1 \end{array}$$

In any line mask, the directions of nonpositive values used reflect the direction of the line detected. When choosing the type of line mask to be utilized, the user must *a priori* know the directions of the edges to be enhanced.

3. Edge Detection Computing Gradients

Of course, lines and points are seldom encountered in nature and another method for detecting edges would be desirable. By far, the most useful edge detection procedure is one that picks up any inflection point in intensity. This is best achieved by using gradient operators which take the first derivative of light intensity in both the x- and y-directions. One type of gradient convolution filters which are often used are the Sobel filters. An example of a Sobel filter which calculates horizontal edges is the Sobel North filter expressed as the following 3×3 kernel:

$$\begin{array}{ccc} 1 & 2 & 1 \\ 0 & 0 & 0 \\ -1 & -2 & -1 \end{array}$$

This filter is generally not used alone, but is instead used along with the Sobel East filter, which is used to detect vertical edges in an image. The 3×3 kernel for this filter is:

$$\begin{array}{ccc} -1 & 0 & 1 \\ -2 & 0 & 2 \\ -1 & 0 & 1 \end{array}$$

These two Sobel filters can be used to calculate both the angle of edges in an image and the relative steepness of intensity (i.e., the derivative in intensity with respect to position) of that image. The so-called Sobel Angle filter returns the arctangent of the ratio of the resultant Sobel North filtered pixel value to the Sobel East filtered pixel value while the Sobel Magnitude filter calculates a resultant value from the square root of the sum of the squares of the Sobel North and Sobel East values.

In addition to Sobel filters, a number of different gradient filters can be used (specifically Prewitt or Roberts gradient filters) depending on the specific application. Figure 15 shows the design and outlines the basic properties of these filters.

4. Laplacian Filters

Laplacian operators calculate the second derivative of intensity with respect to position and are useful for determining whether a pixel is on the dark side or light side of an edge. Specifically, the Laplace-4 convolution filter, given as:

Name   Kernels   Uses

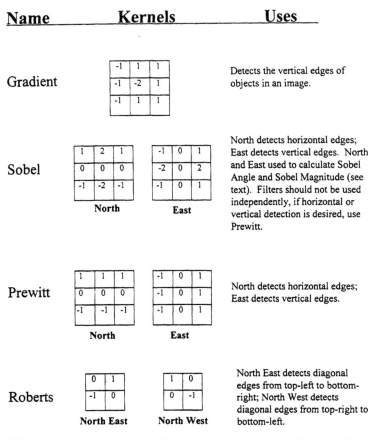

Gradient — Detects the vertical edges of objects in an image.

Sobel — North detects horizontal edges; East detects vertical edges. North and East used to calculate Sobel Angle and Sobel Magnitude (see text). Filters should not be used independently, if horizontal or vertical detection is desired, use Prewitt.

Prewitt — North detects horizontal edges; East detects vertical edges.

Roberts — North East detects diagonal edges from top-left to bottom-right; North West detects diagonal edges from top-right to bottom-left.

Fig. 15 Different 3 × 3 gradient filters used in imaging. Shown are four different gradient operators and their common uses in microscopy and imaging.

$$\begin{array}{ccc} 0 & -1 & 0 \\ -1 & 4 & -1 \\ 0 & -1 & 0 \end{array}$$

detects the light and dark sides of an edge in an image. Because of its sensitivity to noise, this type of convolution mask is seldom used by itself as an edge detector. In order to keep all values of the processed image within 8 bits and positive, a divisor value of 8 and an offset value of 128 is often employed.

Another type of Laplacian filter is the Laplace-8 filter which is used to detect isolated pixels:

$$\begin{array}{ccc} -1 & -1 & -1 \\ -1 & 8 & -1 \\ -1 & -1 & -1 \end{array}$$

This filter most often uses a divisor value of 16 and an offset value of 128. Unlike the Laplace-4 filter which only enhances edges, this filter enhances both the edge and other features of the object.

VII. Conclusions

The judicious choice of image processing routines can greatly enhance an image and can extract features which are not otherwise possible. When applying digital manipulations to an image, it is imperative to understand the routines that are being employed and to make use of well-designed standards when testing them out. In the future, more sensitive detectors and even faster computing power will enable investigators to further expand the capabilities of the modern light microscope.

Acknowledgments

We acknowledge a number of individuals who have been extremely helpful in the AQLM and in the preparation of this manuscript. Douglas Benson from Inovision Corporation as well as Ted Inoue and Dan Green from Universal Imaging Corporation have provided us with assistance and insights on image processing over the past 8 years. In addition, a number of academic faculty members from DAGE Inc., Photometrics Ltd., Hamamatsu Corp., and Videoscope have routinely provided us with cameras and "video toasters." Frederick Miller, Christine Thompson, and Elizabeth Thompson have been invaluable course assistants at the AQLM course and have allowed us to continually try out new exercises on unsuspecting students every year. Finally, we thank the other members of our laboratory, particularly Dr. Catherine Thaler, Qin Chen, and Scott Herric for stimulating conversations and sharing new ways to use image processing in their research.

References

Andrews, H. C., and Hunt, B. R. (1977). "Digital Image Restoration." Prentice-Hall, Englewood Cliffs, NJ.

Bates, R. H. T., and McDonnell, M. J. (1986). "Image Restoration and Construction." Oxford University Press, New York.

Castleman, K. R. (1979). "Digital Image Processing." Prentice-Hall, Inc., Englewood Cliffs, NJ.

Chellappa, R., and Sawchuck, A. A. (1985). "Digital Image Processing and Analysis." IEEE Press, New York, NY.

Erasmus, S. J. (1982). Reduction of noise in a TV rate electron microscope image by digital filtering. *J. Microsc.* **127,** 29–37.

Gonzalez, R. C., and Wintz, P. (1987). "Digital Image Processing." Addison-Wesley, Reading, MA.

Green, W. B. (1989). "Digital Image Processing: A Systems Approach." Van Nostrand Reinhold, New York, NY.

Inoue, S. (1986). "Video Microscopy." Plenum, New York.

Jahne, B. (1991). "Digital Image Processing." Springer-Verlag, New York.

Pratt, W. K. (1978). "Digital Image Processing." Wiley, New York.

Russ, J. C. (1990). "Computer-Assisted Microscopy. The Measurement and Analysis of Images." Plenum, New York.

Russ, J. C. (1994). "The Image Processing Handbook." CRC Press, Ann Arbor, MI.

Shotton, D. (1993). "Electronic Light Microscopy: Techniques in Modern Biomedical Microscopy." Wiley-Liss, New York.

CHAPTER 6

Quantitative Video Microscopy

David E. Wolf

Department of Physiology
Biomedical Imaging Group
University of Massachusetts Medical School
Worcester, Massachusetts 01655

XXXIII
There was the Door to which I found no Key:
There was the Veil through which I could not see:
Some little talk awhile of Me and Thee
There was—and then no more of Thee and Me.

XXXIV
Then to the rolling Heav'n itself I cried,
Asking, "What Lamp had Destiny to guide
Her little Children stumbling in the Dark?"
And—"A blind Understanding!" Heav'n replied.
The Rubyat of Omar Kayham, Charles Fitzgerald, 1859

I. What Is an Image?

Several years ago I was presenting a lecture on the "Physical Optics of Image Formation" and one of the wise sages of microscopy asked me "What is my definition of an image"? This question has an answer which is at the core of the subject of quantitative video microscopy. My response was to go to the blackboard and draw something like what is shown in Fig. 1. "This is essentially the old engineering problem of the black box. The microscope and all of our fancy video instrumentation form a black box. Into the black box we put a signal which is reality. The black box transforms reality and out comes our image. So the image is the transformation of reality. If we call reality $R(x,y)$ with x and y the coordinates in object space, the transformation T, and the image $I(x',y')$ with x' and y' the coordinates in image space, then

$$I(x',y') = T * R(x,y)."$$ (1)

Our job is to determine the inverse transform T^{-1} such that

$$R(x,y) = T^{-1} * I(x',y').$$ (2)

What I was saying is that the goal is to determine how the optics and electronics distort important aspects of the true object and produce the image. We need to know this so that we can digitally reverse the distortion and correctly define the properties of the object. When I say that we need to know how the distortion takes place, I do not mean that we need to follow mathematically the tortuous

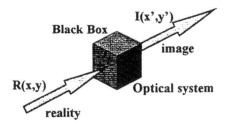

Fig. 1 Optical systems as block boxes. Optical systems can be treated as black boxes. Reality in this case is a two-dimensional set of intensities: $R(x,y)$ enters the optical system and is transformed or distorted to produce the image $I(x',y')$. The goal of the video microscopist is to understand the mathematical rules of the transformation so that one can go backward from $I(x',y')$ to $R(x,y)$.

route of light through the system. Rather we need to empirically define the transformation. As a simple example, suppose that you are concerned about the intensity of objects. If you put a signal through the system with a given intensity and then double the intensity does the intensity of the image also double?

You will also note that I have used a cryptic * in defining the process of transformation in Equation (1). This is because different properties transform in different ways. They are not always linear and as anyone who has meditated upon transcendental functions for any length of time knows, not always analytically reversible. However, as my quantum mechanics professor once pointed out to me: "This is the age of computers!" That after all is what we are talking about here, taking an image and converting it digitally by computer processing to reality.

II. What Kind of Quantitative Information Do You Want?

The particular kind of quantitative information which you want obviously depends upon the experiment you are doing. I have tried to list some of these out in Table I. Basically, these fall into two categories: those which require critical assessment of spatial relationships within an image and those which require critical assessment of intensity relationships.

III. Applications Requiring Spatial Corrections

A. Maximizing Resolution before You Start

Let's begin with the issue of obtaining accurate spatial information. In Table I we have distinguished between information above and below the classical resolution of the microscope. A good place to begin is to discuss how resolution is classically defined. A complete discussion of this subject is outside of the scope of this chapter and I refer the reader to two good texts describing the whys and

---

Table I
Categories of Quantitative Microscopy Applications

Applications requiring spatial corrections
1. I need to accurately know the position of objects within the resolution limit.
2. I need to accurately know the position of objects below the resolution limit.
3. I need to extract from my image the finest resolution possible.
4. I need to determine color coincidence within an image.

Applications requiring intensity corrections
1. I need to know the relative intensity between points in an image.
2. I need to know the absolute intensity at different points in an image.
3. I need to know the spatial distribution of intensities at two or more wavelengths.

wherefores of microscope resolution (Inoue, 1986; Born and Wolf, 1980). For our purposes, let us first follow our intuition which tells us that resolution, defined as the smallest distance, r, between which two self-luminous points would still be distinguishable, should be related to wavelength. We then define the Rayleigh diffraction limit r as:

$$r = .61 \ \lambda/na_{obj} \tag{3}$$

where na_{obj} is the numerical aperture of your objective (you will find it written on the barrel of the objective). For a bright field application the numerical aperture of the condenser, na_{cond} also factors in and equation (3) becomes:

$$r = 1.22 \ \lambda/(na_{obj} + na_{cond}). \tag{4}$$

What this tells you is that for 500 nm light and for an objective with an na = 1.4 objective at say 500 nm the resolution is going to be around .2 μm.

A striking point is that resolution does not involve any eyepiece or camera projection lens. These contribute what is called "empty magnification." They make images bigger but don't contribute any added detail.

Empty magnification can, in fact, be really important in video applications. Video detectors can be thought of as arrays of individual pixel detectors. These are separated by some distance. Recognize that at the detector surface the resolution limit r in the object plane maps to a distance R where

$$R = r \times \text{(objective magnification)} \times \text{(empty magnification)}. \tag{5}$$

It is critical that this R be significantly greater than the interpixel distance. If it isn't, your camera is going to reduce the resolution of your microscope.

What do we mean by the phrase "significantly greater"? This is actually a difficult question to answer. If resolution is the most critical thing in your application then I might suggest that R be at least 10 times the interpixel distance. However, there are other considerations. First, image intensity scales as the inverse of the magnification squared. If you start with light in a single pixel and then spread it out over a square 10 pixels on a side then each pixel is going to be 1/100th as bright. Second, if .2 μm is spread out over 10 pixels and you have a 1000-by-1000-pixel array then the image is going to only cover a 20 μm $\times$ 20 μm area of your sample. This might not suit your application. The moral is that you have to make compromises between the need for greater intensity and the need for resolution and the need to see your object.

B. Aspect Ratio

Before going any further I want to warn you about aspect ratio. You are probably going to want to measure the magnification of your system. Take a stage reticule and image it through your system. Then determine the number of pixels between two points of known distance on the reticule's rulings. This tells you the interpixel distance projected back down to object space. Now do the same calibration vertically. Do you get the same number? Sometimes you don't. This usually relates to whether your camera and video frameboard (This is the computer board where your image processor stores an image) are assuming the same aspect ratio. For instance, trouble arises if your camera has square pixels and the frameboard assumes they are rectangular. The image becomes elongated in one dimension. Circles become ellipses and squares become rectangles.

Most often this problem arises when you are transforming from one system to another. For instance, your confocal microscope might assume square pixels and portray geometrically correct images on the computer screen display. But try to move this data digitally into a video printer, which assumes it is working with the standard rectangular aspect ratio of conventional video signals, and distortions arise.

Some of you will have a problem in performing the above calibration. You will find that horizontal in the microscope does not translate to horizontal in the video system. Moving along a horizontal line in the object plane also moves you vertically in the video plane. The simple solution to this is to rotate your camera until the video plane horizontal and vertical match perfectly with the object plane horizontal and vertical. If you can't manually rotate the system read on. I describe below how to digitally rotate images.

Recognize also that in some systems because the detector pixels are rectangular the interpixel distance will be smaller in one direction. This results in the resolution being greater in that direction. If your structures of interest line up in a particular direction then you can maximize resolution in that direction again by the simple expediency of rotating the camera.

C. Image Translation

So now that you have optimized resolution and been warned about aspect ratio you are ready to look at all of the other nasty things your optical system can do to your image. Let's start off with something simple. Suppose that you image a submicron fluorescent bead which you have positioned dead center in your microscope. Will it be at the center of your image? Probably not!

This may or may not matter to you. When it tends to be a real problem is when you are concerned with issues like color coincidence. Different colors tend to be shifted by different amounts and proper registration can only be achieved by making at least a relative correction for these shifts. Additionally, as already noted, this represents a first simple example which leads us into the more complicated issue of nonlinear distortions.

As a practical matter you might take a two-color bead and measure its position with one filter set and then with another. This will give you the shifts for each filter set. The issue will be whether the shifts are the same for the two colors. They are probably not. Armed with the relative shifts you are in a position to shift one image relative to the other and therefore to bring them back into coincidence.

In general terms (see Fig. 2) if a bead is in position (x,y) in object space it's location will shift off in going to image space and will settle in at position (x',y'). Usually these shifts are small and can be described by a simple pair of linear equations like:

$$x' = x + a \tag{6}$$

and

$$y' = y + b \tag{7}$$

where a and b are constants. As already noted, this kind of correction becomes really important in applications such as color registration. Typically the as and bs will be wavelength dependent, especially if you are changing wavelengths by shifting filter cubes. Additionally you cannot assume *a priori* that these shifts are independent of where in the object you are. The shifts may be more at the edges of the image than at the center.

Equations (6) and (7) are easily inverted to obtain x and y. That after all is our goal: we want to know how a position in the image (x',y') translates to a position in the plane of the object. We have that

$$x = x' - a \tag{8}$$

and

$$y = y' - b. \tag{9}$$

D. Image Rotation

Above I described the situation where the camera's horizontal and vertical did not match the object's horizontal and vertical. I suggested the simple solution

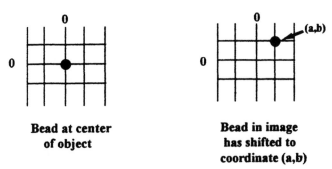

Bead at center
of object

Bead in image
has shifted to
coordinate (a,b)

Fig. 2 Linear transformation. One form of transformation in an optical system is linear translation. Here, we show a hypothetical bead at coordinate (0,0) in the reality or object plane. The optical system translates the bead to coordinate (a,b) in the image.

of carefully rotating your camera. This might not always be convenient. Alternatively you might have a specific need to rotate an image about the center of the field. This can be done using a very similar process to that described for point translation in Equations (4) to (7). We will again use the convention that (x,y) describe the position in the object space and (x',y') in the image space. (see Fig. 3) "if the image is rotated by an angle θ relative to the object we have that.."

$$x' = x \cos\theta + y \sin\theta \tag{10}$$
$$y' = y \cos\theta - x \sin\theta. \tag{11}$$

Upon inversion equations (10) and (11) become

$$x = x' \cos\theta - y' \sin\theta \tag{12}$$
$$y = x' \sin\theta + y' \cos\theta. \tag{13}$$

Of course, in a real world you are liable to find a combination of translation and rotation. So the inversion becomes slightly more involved.

E. Image Warping

We are now in a position to consider what happens when we image a rectilinear pattern like a hemacytometer grid through a microscope and video system. Typically rectilinear patterns are not transmitted true through the "black box" of your optical system. Figure 4 illustrates the two most common results, pincushion and barrel distortion. These distortions are not really surprising since despite the best efforts of microscope designers, optical systems tend to focus to curved rather than planar surfaces and the detector of your video camera is more or

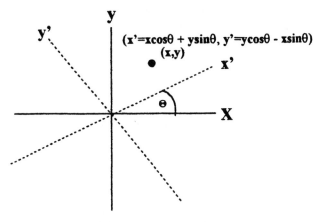

Fig. 3 Rotational translation. Optical systems can also rotate. Here, the optical system rotates the object through an angle of θ so that a point with coordinate (x,y) in the object is located at (x',y') in the image.

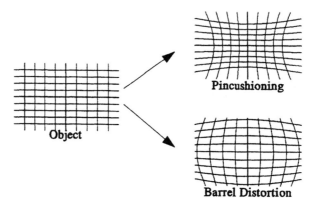

Fig. 4 Pincushioning and barrel distortion. Because optical systems tend to focus to curved rather than to planar surfaces, rectilinear patterns in the object plane tend to produce "pincushioned" or "barrel-distorted" images. These can be corrected for as described in the text.

less flat. You may also note that the edges of the image are slightly out of focus. My calculus teacher taught me that "a sphere is not isomorphic to a plane." That means that if you cut a rubber ball in half and try to flatten it out perfectly you will never succeed. So you can't flatten a curved image surface to a flat detector surface without distortion.

The implications of warping can be very disturbing. Implicit in warping is the fact that magnification changes with position in your image. Also if warping is different for different wavelengths then the magnification becomes wavelength or color dependent.

The bad news here is that we do not live in a linear world. The good news is that in a digital world we can easily deal with distortions like that of Fig. 4. So the next question is: How does one correct for this type of distortion?

1. Ignoring the Problem

First some practical advice, be aware that in some cases you can ignore this kind of distortion. Determine over what area of the center of the field this distortion remains less than a pixel. If you stay within that area you can safely ignore it for most purposes.

2. Dewarping Algorithms

Many image processing software packages come with "canned" software to dewarp an image. You should always make the effort to understand what the algorithm is doing. In the end you are the judge of the validity of a particular algorithm for your applications.

An example of a dewarping algorithm is polynomial dewarping. This is a logical step up from a linear transformation. The idea is to assume that the shape of the distorted pattern in image space is a generalized polynomial of the rectilinear pattern of the object. For most purposes a second-order polynomial will suffice. Again following our convention the image coordinates (x',y') that correspond to a point in the object plane are given by

$$x' = a + bx + cy + dx^2 + exy + fy^2 \tag{14}$$
$$y' = A + Bx + Cy + Dx^2 + Exy + Fy^2. \tag{15}$$

where a, b, c, d, e, f, A, B, C, D, E, and F are constants. Since each point in object space has two coordinates, these 12 constants are determined by choosing six points in object space and measuring the values of (x',y') in the image. Typically, these points are chosen to be major intersections of your rectilinear grid; Equations (14) and (15) then enable you to calculate the constants. For example, suppose you use the point you define as the origin in the object plane, that is $(x,y) = (0, 0)$, which maps to a point (x',y') in the image plane, then Equations (14) and (15) tell us that $a = x'$ and $A = y'$ respectively.

Once the constants are determined you are ready to dewarp your image. As in the cases of translation and rotation this requires inversion of Equations (14) and (15). This looks like an algebraic nightmare. This is because it is. However, the fact is that there are several readily available symbolic language software packages which can do this for you. Not only will they solve this for you, many of them will generate software in your favorite computer language to do the calculating for you. This is not meant to be facetious. Modern personal computers have gotten so fast that imaging calculations like this, which a few years ago required array processors to accomplish in reasonable time, can now be accomplished in the blink of an eye.

IV. Two-Camera and Two-Color Imaging

With the popularity of ratio imaging one is often confronted by the problem of processing images from two cameras or two colors. In both of these cases the images are not in perfect registration and there tend to be magnification differences. This is in fact a very similar situation to the warping problem. The same generalized polynomial dewarping algorithm can be used to map an image obtained with one camera onto that obtained with the other. This is in fact one of the most common uses of dewarping algorithms.

V. A Warning about Transformations—Don't Transform away What You Are Trying to Measure!

My goal so far has been to introduce the concept of digitally correcting your images. I have tried to focus on some of the more critical types of transformation.

For a much more in-depth discussion of image process I refer you to Chapters 4, 5, 11, and 12 in this volume and to several specialized texts (Baxes, 1984; Castleman, 1979; Pratt, 1991; Taylor and Wang, 1989; Wang and Taylor, 1989). Recognize that image processing represents a form of data processing. There are, of course, many other examples of data processing such as fitting your data to a theoretical curve. I cannot overemphasize the admonition I made above that you should always be aware of the consequences of what you are doing.

Suppose you want to accurately determine the area under some curve. The data is noisy. So you choose to smooth it out using one of several conventional approaches before analyzing it. This smoothing represents a transformation. Does your transformation preserve the area under the curve or does it systematically distort it in a particular direction? Obviously you want the area to be true. So you will want to choose your smoothing algorithm accordingly.

VI. The Point Spread Function

I have tried here to separate spatial from intensity distortions. However, the two come together when we consider how an optical system treats a point of light. At a practical level one can achieve this by using fluorescent beads of a diameter of .1 to .2 μm. Figure 5A shows a pseudocolor image of such a bead. What one observes is that the light from a point source is spread out over neighboring pixels. This pattern is referred to as the point spread function (PSF). It is not the result of bad optics. Indeed a microscope will never create a perfect point from a point because it cannot capture all of the high-order diffraction bands (Inoue, 1986; Born and Wolf, 1980). The point spread function represents what is referred to as a convolution. That is, all points in the object are distorted in the same way to create the image. Mathematically we can write

$$I(x',y') = \int_{-\infty}^{+\infty} \int_{-\infty}^{+\infty} \rho(z - x, t - y)R(x,y) \, dz \, dt. \tag{16}$$

A more complete discussion of convolutions may be found in Chapter 5. For our purposes here, let's suppose that you have a noisy image. You would like to smooth it out and make it less noisy. One way to proceed is to replace the intensity at each pixel by the average of itself and its eight nearest neighbors. Performing such an operation leads to a quieter image but a fuzzier one. As a next level of correction you might decide not to give equal weighting to all of the pixels. Maybe you would use the full intensity of the center pixel but only add in half of the intensity of the neighboring pixels in calculating your average. The point spread function is performing such a convolution on the intensity distribution of the object $R(x,y)$. Since it tails off as one gets farther and farther away from the center it is a weighted average.

Now, of course, as always, our task is the more daunting one of inverting Equation (16). Suppose, for a moment, that the PSF was the equal weighted averaging convolution described above. A reasonable inversion, referred to as deconvolution, could be achieved by replacing the intensity at each point in the image by itself less 1/8 of the intensity in each of the neighboring pixels. This is in fact, the philosophy behind some of the simple sharpening algorithms employed in image processing.

There are a number of ways of deconvolving real PSF's. One direct approach is the method of Fourier transforms. This is discussed in Chapter 15. Furthermore, it is important to recognize that the PSF is three dimensional. This is illustrated clearly in Fig. 5 where we look at the point spread function different angles relative to the optical axis (see the legend to the figure). One sees that the distortion of the point source along the axis is even more extreme than in the plane perpendicular to the optical axis. So there really should be a third axis in Equation (14). This is where the remarkable power of deconvolution lies. Three-dimensional deconvolution techniques enable one to remove signal from out-of-focus planes in the image. This is discussed in detail in Chapter 15.

VII. Positional Information beyond the Resolution Limit of the Microscope

One might assume that one cannot determine position in the microscope to a greater resolution than that defined by Equation (4). This is not really the case. The resolution defined by Equation (4) describes the ability of the observer to distinguish between one and two points of light. As the points get closer and closer they get lost in each other's PSFs. The question of how accurately you can determine position is quite a separate issue.

Recent work on the technique of single particle tracking has demonstrated that position may be determined to an accuracy of 10 or 20 nm (Gelles *et al.*, 1988). The key to this is to accurately determine the image of a particle. This is typically done using differential interference microscopy (see Chapter 9). This image is in a sense equivalent to a PSF and is referred to as the kernel. The analysis algorithms then involve convolving the kernel around the image and using statistical correlation techniques to find the "best" position.

A. Applications Involving Intensity Corrections

We next turn our attention to the issue of applications which involve or require intensity correction. What I am going to assume is that it is your goal to be able to accurately compare intensities from different points on the image and then to follow and monitor the intensity at these points over time.

B. Take Control of Your Camera

While there are no absolutes, it is probably the case that the first thing you are going to want to do is turn off your automatic gain control and your automatic black level adjust. This puts you in charge of your camera. You can then, for instance, calibrate the response of your camera at a particular gain setting and know how it will respond in an experiment conducted at those settings.

C. Dynamic Range—Don't Waste It

In my experience the most common error in making intensity measurements, an error by no means confined to imaging applications, is not paying attention to dynamic range. In a typical imaging application an analog signal (i.e., voltage) comes out of a video camera. This signal is then digitized using some kind of analog-to-digital (A/D) converter. The function of an A/D converter is shown schematically below in Fig. 6. In this example the A/D converter will digitize a 1-V signal into 10 parts. A signal such as that shown in Fig. 6A, which ranges from 0.0 V to about 0.7 V, transmitted reasonably well through the A/D converter. Here we have exaggerated the step nature of the digital output. Most real-life A/D converters divide the signal into more than 10 parts. The trouble arrises if all of your signal falls between 0 and .05 V as in Fig. 6B The A/D converter will treat it as all black. If all of your signal falls between 0 and .15 V the A/D converter will treat it as either black for $V \leq .1$ and gray for $V > .1$. The obvious solution is to multiply your signal voltage that is apply gain to your signal making the maximum 1.0 V so that you maximize the gray level resolution known as dynamic range. In Fig. 7 we demonstrate this for a video image. Figure 7 shows a properly adjusted dynamic range giving a full 8-bit (256) gray level image resolution. Successive images so the degradation associated with improper setting of dynamic range to 4, 2, 1, and 0 bits respectively. We see progressive loss of detail and information. The 1-bit image is binary regions are either black or white. In the 0-bit image all information is lost.

Typically, video cameras have a gain control which enables you to adjust the gain setting so that it totally fills the full range of the A/D converter. This will provide you with maximum dynamic range. In some systems this control is provided by a computer setting rather than a manual switch. For a further discussion of this issue see Chapter 1.

D. Correcting for Nonlinearity

As discussed by Berland and Jacobson, certain types of video cameras do not have linear gain. That is, that if you plot camera output vs light input the response is not a straight line (see Fig. 8). The example in the plot is hypothetical so as to exagerate nonlinearities which more typically are only a few percentages. To generate such a plot for your camera system place an intensity step wedge with a series of graded optical densities between the light source and the sample. In actuality you do not need a sample. You can just look at a blank slide and get a "homogeneous" image. You then systematically read the output from the camera as you change the intensity with the step wedge. This curve can then be

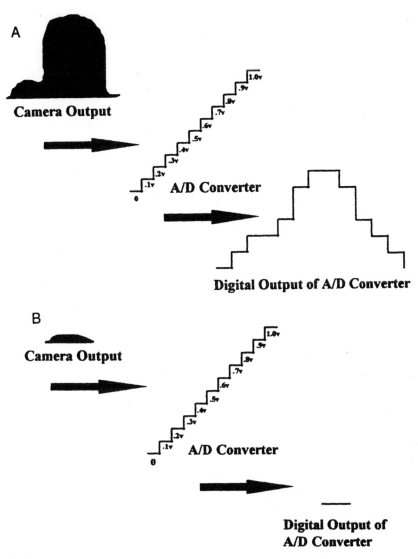

Fig. 6 The importance of dynamic range. One should always maximize the dynamic range in image processing. Ultimately, the signal passes through an A/D converter which divides the signal into a finite number of discrete steps. By setting the output gain on the camera one can adjust the voltage going into the converter. (A) We see a camera output gain reasonably well matched to the A/D converter. The digital output of the converter does a good job of detecting the shape of the original object. (B) All of the signal falls within the first step on the converter and no output above zero is obtained.

Fig. 7 The importance of dynamic range. Here, we illustrate the value of dynamic range in resolving image detail. (8) Shows an 8-bit image of a rat seminiferous tubule labeled with eosin. The same image is shown successively at 4-, 2-, 1-, and 0-bit resolution. Progressive image detail is lost.

used to digitally calibrate the camera's response. This is typically done using what is known as a look-up table or LUT. Suppose you have an 8-bit video system. So you get intensity values from 0 to 255. The LUT tells the system that a value of, say, 240 is underestimated and should be, say, 250. Every intensity value from 0 to 255 has a corresponding value in the LUT which corrects the intensities in the image.

A major complication here is that modern cameras have very broad dynamic range. You can change the sensitivity by changing the gain. Often there is more than one gain to change. For instance, a SIT camera will have the intensifier gain and the output gain. Linearity tends to depend upon these settings, so you need to do calibrations for all possible settings.

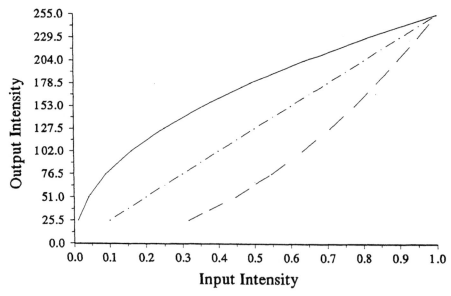

Fig. 8 Camera linearity: output intensity vs input intensity. With certain cameras one cannot assume that the output intensity is linear with the input intensity (figure courtesy of Ted Inoue courtesy of Universal Imaging Corporation, Westchester, PA).

E. Flat Field Correction

Now that your dynamic range is properly set and you have corrected for any nonlinearities in gain you are ready to analyze an image. Figure 9A shows a fluorescence image of a seminiferous tubule labeled with eosin. An obvious problem is going to be the background. For this experiment, we obtain a back-

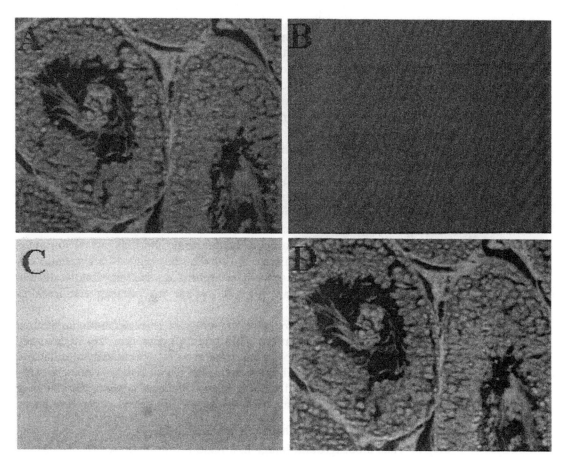

Fig. 9 Flatfield correction. In order to get a true representation of intensities over an image, it is necessary to perform flat field correction. (A) The original image of rat seminiferous tubules labeled with eosin; (B) the background, in this case an unlabeled no-cell region of the sample; (C) the image of a homogenous sample. One clearly sees that the illumination and/or detection is not homogeneous. (D) The flat field corrected image which is $(A - B)/(C - B)$. One additional point to note is the dirt spots in the optics clearly visible in both B and C. While one can, by this flat field approach, correct for such blemishes, one should be aware that one may be seriously compromising dynamic range in those regions of the sample. It is always important to keep one's optics as clean as possible (images courtesy of Scott Blakely of Universal Imaging Corporation, Westchester, PA).

ground by moving to a region of the sample where there is no cell and take a background image (Fig. 9B). We then pixel by pixel subtract the background image from the image of the tubule. *An obvious warning here is watch your dynamic range.* This is where you can lose it. If you examine Fig. 9A closely you'll see a couple of disturbing points: first, there are what appear to be dirt blotches on the image, and second, there appears to be a general tendency of the intensity to fall off as one moves away from the center of the image. Background subtraction does not completely solve these problems.

In order to correct for these problems one needs to perform a flat field correction which corrects for inhomogeneities of illumination and detection. You create a sample which is a homogeneous fluorescent surface. You don't want this to be a thick solution. Some kind of a film is best. What we tend to do is to take solution of a protein labeled with the same fluorophore being used in the experiment and let a drop of this sit on a coverslip for 20 to 30 min. We do not allow this drop to dry. Upon gently washing away the drop you will find that a film of fluorophore remains on the coverslip. You search out a region on the coverslip where the film is essentially homogeneous and then defocus very slightly to correct for any minor imperfections. We then take an image of this field (9C). We then subtract a background image (again a region of no fluorescence). We then measure the intensity of the center of homogeneous image field. What we tend to do is to take the average intensity of the center nine pixels. We then divide the background corrected homogeneous image by this value. This will create a matrix which in general is one at the center and falls off as you move away from the center. The final step is to divide the original background subtracted image by this matrix in point-by-point fashion, creating what is referred to as a flat field corrected image (Fig. 9D) where the intensity relationships are true.

In my description I am assuming that you are doing this calculation in floating point rather than integer arithmetic. With modern computers this is generally doable and preferable. I do not assume that you can necessarily store such a large floating point matrix. I will now repeat the sequence of events as a step-by-step algorithm in more mathematical terms. Since in general the flat field image will be prestored I will start with the acquisition of this corrected image:

Step 1. Take an image of a homogeneous field: $H(x,y)$

Step 2. Take a background image in a region of no fluorescence: $B_H(x,y)$

Step 3. Background correct homogeneous field image: $H'(x,y) = H(x,y) - B_H(x,y)$

Step 4. Calculate average intensity of central nine pixels in $H'(x,y)$: H_o

Step 5. Take image of your sample: $I(x,y)$

Step 6. Take background image in a region of no fluorescence: $B_I(x,y)$

Step 7. Background correct your image: $I'(x,y) = I(x,y) - B_I(x,y)$

Step 9. Perform that flat field correction: $F(x,y) = I'(x,y) H_o/H'(x,y)$

VIII. Intensity Changes with Time

Taking a temporal series of images causes a few additional concerns. The two most common concerns are whether the intensity of the illuminator changes with time and whether the image bleaches with time. Source fluctuations can be corrected for by monitoring what should be a constant point in the image. You could, for instance, add some fluorescent beads to your sample. Alternatively you could add a photodiode to monitor lamp intensity and divide your images by the diode intensity. Photobleaching can also be dealt with using a constant region of your sample. If your experimental conditions are reproducible you can do a precalibration for photobleaching.

A bigger problem would be if you determine that lamp flicker or wander are causing variations in the spatial intensity distribution of the source light in your image. Fortunately this is usually not a problem. Kohler illumination is designed to optimally confuse light coming from different regions of the source. Once again we stress the importance of proper Kohler illumination of the microscope.

IX. Summary

Obviously there are many variations and embellishments on the topic of quantitating your image. I have tried here to offer you a very practical guide which highlights some of the critical issues paying particular attention to problems which can prevent or compromise your success. These include: errors due to aspect ratio, use of automatic camera settings, improper setting of dynamic range, and use of integer arithmetic. Further information can be found in several other chapters in this volume as well as in the references below. Additionally you shouldn't overlook the technical manuals which come with your video systems. The field of video microscopy has evolved as a cooperative effort between academia and industry. As a result you will find that many of the technical support personnel from microscope, video, and image processing companies are well versed on the issue of video imaging in biology and are more than willing to assist you.

Acknowledgments

The author thanks Christine A. Thompson and Scott Blakely for help in the preparation of figures for this manuscript.

References

Baxes, G. (1984). "Digital Image Processing." Prentice Hall, Englewood Cliffs, NJ.
Born, M., and Wolf, E. (1980). "Principles of Optics." Pergamon, NY.
Castleman, K. R. (1979). "Digital Image Processing." Prentice Hall, Englewood, NJ.

Gelles, J., Schnapp, B. J., and Sheetz, M. P. (1988). Tracking kinesin-driven movements with nanome-
ter scale precision. *Nature* **331,** 450–453.

Inoue, S. (1986). "Video Microscopy." Plenum, NY.

Pratt, W. (1991). Digital Image Processing, Wiley, NY.

Taylor, D. L., and Wang, Y.-L. (1989). "Fluorescence Microscopy of Living Cells in Culture," Part
A, Vol. 29. Academic Press, San Diego, CA.

Wang, Y.-L., and Taylor, D. L. (1989). "Fluorescence Microscopy of Living Cells in Culture," Part
B, Vol. 30. Academic Press, San Diego, CA.

CHAPTER 7

Proper Alignment of the Microscope

H. Ernst Keller

Carl Zeiss, Inc.,
Thornwood, New York 10594

The light microscope, in many of its configurations, is a somewhat complex tool with many adjustable components. Good alignment is essential for good image quality, especially so for quantitative studies. In this chapter we will try to provide a few simple guidelines for the best alignment of all those components of the light microscope which can be focused and/or centered. A better understanding of the function of these components and how their control influences the image has become even more critical for electronic imaging. While analog or digital image processing can, to a small extent, compensate for poor mechanical and optical alignment, the best end result, free of artifacts, is derived from the best possible optical image. When all the microscope's controls are routinely set correctly, the video image will be at its best.

METHODS IN CELL BIOLOGY, VOL. 56

═══ I. Key Components of Every Light Microscope

Before we discuss the alignment of the microscope, let us follow the light path through the microscope from source to detector, look at the main components, learn the terminology, and understand their functions. Those familiar with the light microscope can proceed directly to the section of this chapter that discusses Koehler illumination (Fig. 1).

First a few words about the basic stand: Whether the microscope is upright or inverted, the stand is designed to provide the stability essential for high-resolution microscopy and to rigidly hold all components such that they can be ergonomically controlled. The stand incorporates functions like coaxial and bilateral focus for stage, substage, or sometimes nosepiece; control of diaphragms; and often switching of filters and diffusers. Furthermore, many stands provide receptacles for neutral density, color or polarizing filters, waveplates or so-called compensators, and special sliders for contrast enhancement.

A. Light Source

The lamp is either mounted inside the stand or in an external lamp house for better heat dissipation, particularly for higher-intensity sources. Most modern microscopes are equipped with a low-voltage tungsten-halogen illuminator, either 6 V or 12 V, ranging in power consumption from 10 W to 100 W. Either an adjustable lamp socket or a lamp house control provides for precise centration and focus; sometimes the bulb has its own precentered socket. The shape of the tightly woven flat filament of the bulb can be rectangular or square and this in turn mandates the best lamp alignment, taking into account a reflector behind the bulb, which is part of most lamp housings. Frequently in modern microscopes the power supply for the illumination is integrated into the stand.

More demanding is the mounting and alignment of gas discharge sources such as mercury or xenon lamps, commonly used for fluorescence microscopy or when an exceptionally bright source is required for specimens which absorb large amounts of light. Be sure to follow the manufacturer's instructions and warnings closely when installing and aligning high-pressure discharge lamps! Because many of these lamps are under high pressure wearing safety goggles is essential. Since these lamps emit large amounts of UV radiation, UV-protective eyewear must be worn in the presence of an unshielded arc lamp. Discharge lamps usually require special lamp collectors to optimally feed the small, bright arc source image to condenser or objectives.

Tungsten-halogen sources have a continuous spectral emission over the visible and near IR ranges with a color temperature of between 2800° and 3200°K. Xenon lamps are a somewhat continuous spectral emitter with a color temperature of 5500°K; mercury lamps have very discrete emission peaks from the UV through the visible spectral ranges and cannot provide good color balance.

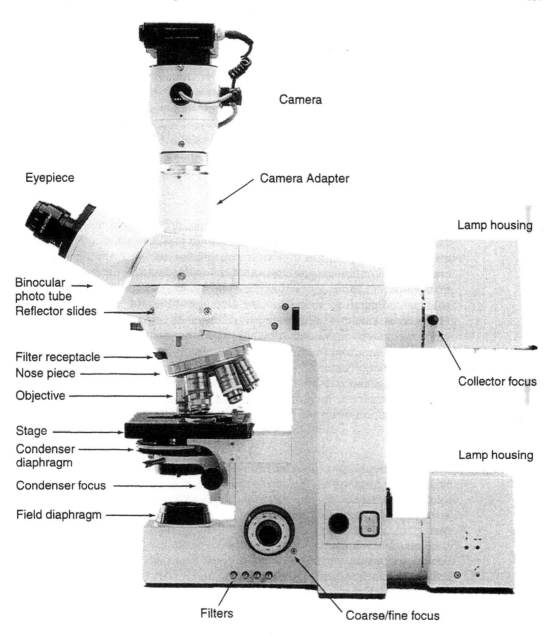

Camera

Eyepiece

Camera Adapter

Lamp housing

Binocular photo tube

Reflector slides

Collector focus

Filter receptacle

Nose piece

Objective

Stage

Condenser diaphragm

Lamp housing

Condenser focus

Field diaphragm

Filters

Coarse/fine focus

B. Lamp Collector

As the name implies, the collector captures as much light as possible from the source and relays this light to the condenser, which is the objective in incident light applications such as fluorescence microscopy. Either the collector is focusable or the lamp can be axially moved to assure focus of the source in the front focal plane (aperture) of the condenser or the focal plane of the objective for even, uniform illumination. The focal length, the aperture, and the degree of the optical correction of the collector lens directly influence the light flux and the uniformity of illumination at the specimen. Special quartz collectors are available and recommended for UV microscopy or for fluorescence excitation in UV.

C. Diffusers and Filters

In transmitted or incident light illumination, diffusers are often used to more evenly distribute the nonuniform illumination of a filament or arc. Special low-angle diffusers accomplish this with a negligible loss of overall light intensity; conventional ground glass diffusers are not recommended. Filters serve a variety of functions, including correcting the color balance, contrast enhancement, neutral light attenuation, aiding in the selection of specific spectral regions for fluorescence excitation and for improved chromatic correction in black and white video microscopy or photomicrography. The use of infrared (IR) blocking filters is especially important because objectives are not well corrected for these wavelengths yet video cameras are sensitive to this light. As a consequence, image quality is degraded if IR blocking filters are not used. The field diaphragm is an essential component for proper specimen examination and we will learn more about its function when we get to adjustment for Koehler illumination.

D. Condensers

In transmitted light, the condenser's important function is to provide even, bright illumination of the specimen field for a wide range of different magnifications. Its numerical aperture, or cone angle of illumination, directly influences resolution, contrast, and depth of field. Condensers come in a range of optical corrections, achromatic and aplanatic, offer different numerical apertures and specimen field diameters, and sometimes are fitted with turrets to provide for special contrasting techniques such as phase contrast, modulation contrast, or differential interference contrast.

As indicated earlier, in incident light applications the objective is also the condenser. Being one and the same, alignment between condenser and objective is always perfect, greatly facilitating the set-up of epi fluorescence or other incident light methods.

E. Condenser Carrier or Substage

On an upright microscope the condenser carrier is usually attached to the stage carrier and moves with the stage focus. The inverted microscope has a separate condenser carrier attached to the transmitted light illumination column. In both cases the condenser carrier not only holds the condenser but also allows for its focus and centration.

F. The Specimen Stage

The control knobs on the stage move the specimen in the x and y planes. Depending on the application, the specimen stage can come in a variety of configurations. Some stages offer rotation of the specimen.

G. The Objective

This component is most instrumental to a quality image. Its parameters and its degree of optical correction largely determine not only the magnification ratio but more importantly the resolution, the contrast, the image brightness, the depth of focus, the sharpness from center to edge, and the color rendition. From Achromats and Planapochromats through a variety of very specialized lenses, the microscope objective needs to be carefully selected according to budget and application. Keep in mind, however, that the best performance of an objective is only attainable if everything else is perfectly aligned.

H. Revolving Nosepiece

Precisely machined turrets with receptacles for four to seven objectives allow for fast changing between magnifications and help maintain focus (parfocality) and registration of specimen field (parcentration).

I. Infinity Space

Most modern microscopes have infinity-corrected objectives providing this space, which can readily accept a variety of accessories like reflectors, filters, and so on, which, as long as they are plane parallel, have no negative impact on focus or alignment. A tube lens at the far end of this infinity space focuses the infinite conjugate rays into the intermediate image plane, frequently the location where electronic detectors can be directly mounted.

A brief warning may be in order here: different manufacturers use different design concepts for the full correction of the intermediate image; also, the focal length of the tube lens varies and the parfocal distance (objective shoulders to specimen plane) may not be the standard 45 mm. This means that switching objectives between different manufacturers is not possible today as it was in the past with finite optics, where some standards prevailed.

J. Tube, Eyepiece, and Video Adapters

Comfortable viewing of the image with a fully relaxed eye is the function of tube and eyepiece with the latter enlarging the real intermediate image formed by the objective to such an extent (usually around 10X) that all detail resolved by the objective can also be resolved by the eye. Video and camera adapters serve the same purpose and must be carefully chosen depending on the pixel resolution of the detector (see Chapter 8).

II. Koehler Illumination

Koehler illumination is one of the oldest and today most universally used techniques to provide even illumination over the specimen field and allow for control over resolution, contrast, and depth of field. The technique was first described in 1893 by August Koehler, then a young assistant professor at the University in Giessen, Germany, who later joined the Zeiss Works in Jena to head up their microscopy development. It took almost 50 years for Koehler illumination to gain universal acceptance. Today, practically every microscope provides the necessary controls for its implementation.

Figure 2a shows the two intertwined geometric optical light paths in a microscope with Koehler illumination for transmitted light; to simplify things, the ray path is shown for an instrument with finite image conjugates and traces typical illumination (1) and imaging (2) rays. The same principles are directly applicable to the modern infinity corrected microscopes.

1. The *light source* is imaged by the collector into the front focal plane of the condenser, where we usually have an adjustable diaphragm, the *condenser* or *aperture* diaphragm. From here the source image is projected by the condenser into infinity, traversing the specimen as a parallel pencil of light for each source point imaged in the condenser aperture and uniformly distributing each source point's intensity over the full field. The objective receives these parallel pencils and forms a second image of the light source in its *back focal plane.* A third image of the source appears in the *exit pupil* of the eyepiece, also called eyepoint, usually approximately 15 mm above the eyepiece. If we include the source itself, we have four source conjugated planes in the microscope.

2. The imaging beam path conjugate planes begin with the *field diaphragm,* which with a properly focused condenser will be imaged into the *specimen plane.* From here the objective images specimen plus field diaphragm into the *real intermediate image,* which lies near the front focal plane of the eyepiece. The eyepiece projects the image to infinity and the lens of the eye focuses the *final image* on the retina. Again, four image conjugated planes can be found or three if the imaging detector receives the intermediate image directly.

In incident light Koehler illumination (Fig. 2b) there is one additional source conjugated plane to allow for aperture control of the illumination independently from the observation or objective aperture. All else is identical to transmitted light, only that condenser and objective are one and the same.

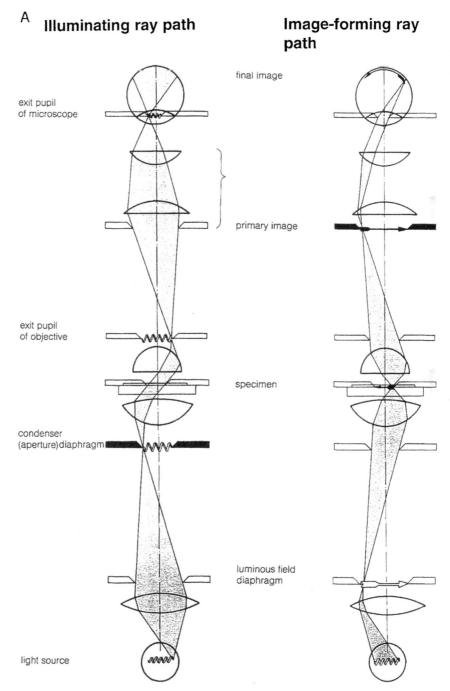

Fig. 2A   Light path in transmitted light Koehler illumination.

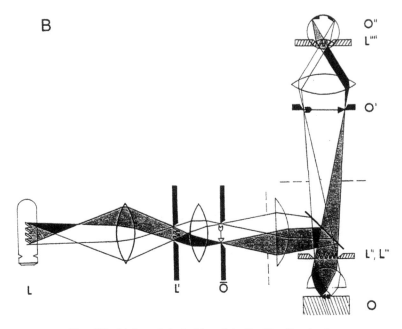

Fig. 2B Light path in incident light Koehler illumination.

Understanding illumination and imaging ray paths will greatly facilitate an optimal alignment and can be very useful in finding image artifacts or dust particles in the optical system, etc.

A. Aligning the Microscope for Koehler Illumination

Follow these simple steps:

1. Center and focus the light source (unless precentered) following the manufacturers instructions. With the condenser and diffuser removed, place a piece of paper into the light path at the level of the condenser carrier. An image of the lamp filament should be visible and controls on lamp housing or socket permit focus and centration of this filament image in the circular area representing the condenser aperture.

2. Insert condenser and diffuser (if present). Place a specimen on the stage, and use a medium power objective (10x, 20x or 40x) to focus on any specimen feature. If not enough light is available, open field and aperture diaphragms fully. Too much light may saturate the camera and may require rheostat control of the light source or the placement of neutral density filters in the light path. Do not use the condenser aperture as a means to control light intensity.

3. Once specimen focus is obtained, close the field diaphragm and focus (usually raise) the condenser until the field diaphragm comes into sharp focus with the specimen.

4. Center the condenser to align the image of the field diaphragm concentric with the circular field of view.

5. Open the field diaphragm just beyond the field of the objective used. For videomicroscopy do not open the diaphragm beyond the field of the sensor, which is always less than the visual field. Also, when using a video camera do not close the image of the field diaphragm into the field of the sensor for reasons outlined in the chapter by Sluder and Hinchcliffe.

6. Slowly close the condenser aperture diaphragm until the desired image contrast is achieved. Be sure to *never* use this diaphragm to control image brightness. Both contrast and resolution are directly affected. An open aperture stop usually gives poor contrast. Closing it partially enhances contrast but at the expense of resolution. Setting of the condenser aperture for best balance between resolution and contrast is what sets the expert microscopist apart from the novice and is very much a function of the specimen itself.

A centering telescope or Bertrand lens lets us focus on the condenser aperture diaphragm to verify its setting and centration. When using such a lens, set the condenser aperture to just impinge upon, or become visable, in the objective lens aperture.

Contrary to transmitted light Koehler illumination, in epi fluorescence microscopy the aperture diaphragm, if available, can be used to control the excitation intensity, often useful to obtain a flare free, high contrast image. Not all epi fluorescence systems provide an aperture iris, however, and light attenuation by neutral density filters or arc control can be used to reduce excitation intensity if desired. A good test specimen to set up Koehler illumination in epi-fluorescence is a H & E stained tissue section, which will fluoresce with almost any filter combination and display a minimum bleaching.

B. What Are the Benefits of Koehler Illumination?

By limiting the imaging ray bundles to the field covered by a given objective/eyepiece or detector combination, no stray light or internal "noise" is generated in the microscope. With the exceptional sensitivity of electronic detectors, "hot spots" may occur unless the field diaphragm is closed to the edges of the detector's field.

Even, uniform illumination of the specimen is achieved by distributing each source point's energy over the full field. Thus, minimal shading correction is required by the video electronics.

Full control over the illumination (condenser) aperture allows for the best compromise between resolution and contrast and for some effect on the depth of field.

C. Resolution and Contrast

Since the theory of image formation, the diffraction limits of resolution, and contrast generation have all been described in books and articles, it would go beyond the scope of this practical guide to repeat Abbe's theory and wave optical imaging. However, the reader is well advised to gain some understanding of these concepts in order to better appreciate good alignment and optimal setting of controls and diaphragms.

Here is just a very brief summary of the basic concepts, taking into account the electromagnetic wave nature of light, its wavelength, λ, and the interaction of wavefronts with the specimen's structures. Ernst Abbe was the pioneer who developed the theory of image formation through the microscope. He differentiated between two types of specimens: the self-luminous object (light sources, luminescence, fluorescence) which follow the Rayleigh criterion for resolution, which is based on diffraction in the objective. Here the minimum point-to-point spacing resolution (d) is given by:

$$d(\mu\mathrm{m}) = \frac{0.61\lambda}{\mathrm{NA\ (objective)}} .$$

In the case of an illuminated object, the Abbe criterion for resolution is:

$$d(\mu\mathrm{m}) = \frac{\lambda}{\mathrm{NA\ (objective)} + \mathrm{NA\ (condenser)}} ,$$

where λ is the wavelength of the emitted or illuminating light and NA is the numerical aperture of objective or condenser, given by the sine of half its collection angle times the refractive index of the medium between specimen and objective. Note that this formula provides the reason why one does not want to close the condenser diaphragm simply to control light intensity or to enhance contrast. The penalty is lost resolution. Also, note that increasing the condenser numerical aperture to a value greater than that of the objective, though in principle it may provide increased resolution, is not recommended for brightfield applications. The problem is that under such conditions one is adding a darkfield image to the brightfield image; consequently, image contrast is severely compromised.

With the self luminous object it is diffraction in the objective itself, which causes the smallest image point to expand into an "airy disk" (Fig. 3), the diameter of which becomes

$$D_{(\mu\mathrm{m})} = \frac{1.22\lambda}{\mathrm{NA\ (objective)}} .$$

In Fig. 3 (left) W' indicates a spherical surface in the optical path which is equidistant from the center of the image plane O'. All light rays from this surface follow equidistant paths to O' and arrive there in phase. This leads to a bright central maximum at O'. At other points along the image plane we can have either constructive or destructive interference or some combination of both. This results in the airy disk profile in the image plane (Fig. 3, right). The airy disk and point spread functions in general will be further discussed in Chapters 6 and 15.

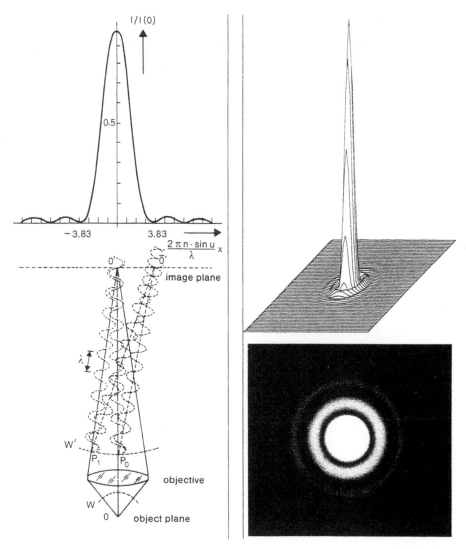

Fig. 3 Airy disc of a self luminous, infinitely small object point. The dark rings around the central disc are zones of destructive interference between wavefronts diffracted on the objective's internal aperture.

In the illuminated object, diffraction is generated by the specimen's structural features and their spacing. The diffraction angle (α) is determined by the wavelength (λ) and the spacing or distance (d) between features:

$$\sin \alpha = \frac{\lambda}{d}.$$

The numerical aperture of an objective is a direct measure for the diffraction angle it is capable of collecting. Abbe was able to prove that for a given structural spacing to be resolved at least two orders of diffracted light produced by this spacing need to participate in the image formation. It is interference between diffracted and nondiffracted wavefronts in the intermediate image plane which resolves structural detail and determines the contrast at which the image is rendered.

CHAPTER 8

Mating Cameras To Microscopes

Jan Hinsch

Leica, Inc.
Allendale, New Jersey 07401

I. Introduction

The compound microscope forms images in two stages. The objective (first stage) projects a real image of the specimen into the intermediate image plane (IIP) of the microscope at a ratio magnification (M). The resolution in the IIP is typically between 40 and 100 line pairs per millimeter (lp/mm). Since the resolution of the human eye is of the order of 5 lp/mm, additional, virtual magnification (V) by the eyepiece (second stage) is necessary to match the resolution of the eye to that of the microscope. The total microscope magnification (V_{total}) is the product of $M \times V$. The designation V_{total} is used because the eye views a virtual image in the microscope.

The diameter, in millimeters, of that part of the intermediate image that is utilized by the eyepiece is called the field of view number (FOV). Depending on the numerical aperture (nA) and magnification of the objective and the FOV number of the eyepiece, microscope images may contain up to five million image elements, referred to as pixels. Video cameras can resolve only a fraction of this available information due to the inherent resolution limits of the camera detectors. Thus, when mating video cameras to microscopes, choices need to be made whether to sacrifice some of the field of view in the interest of matched resolution or vice versa. This chapter discusses how best to mate video cameras to microscopes at magnifications which preserve specimen image resolution when necessary. This paper will also consider what needs to be done to maintain the optical corrections of objective and eyepiece in the video image.

METHODS IN CELL BIOLOGY, VOL. 56

The microscope's resolution (r) depends on the numerical aperture (nA) of the objective. It is commonly calculated at:

$$r = \frac{1.22 \times \text{lambda}}{\text{nA}_{\text{obj}} + \text{nA}_{\text{cond}}},$$

where r is the minimum spacing of specimen detail, lambda is the wavelength of the illuminating light, and nA_{obj} and nA_{cond} are the numerical apertures of objective and condenser, respectively. In the case of green light (lambda 0.55 μm) and a condenser numerical aperture equal to that of the objective one can state the resolution simply as:

$$r(\mu\text{m}) \simeq \frac{1}{3 \times \text{nA}}$$

or, converted to line pairs per millimeter (lp/mm):

$$r(\text{lp/mm}) = \text{nA} \times 3000.$$

Resolution usually refers to distances in the object plane. In the context of this chapter the term circle of confusion (Z) is sometimes used to signify resolution in planes other than the object plane; for example, the intermediate image plane, where a point source appears as a confused circle of radius $Z = M \times r$.

Magnification is a means of diluting the density of the image information to suit the limited resolution of the sensing structure of the receiver. The human eye, for example, typically has a resolution of 0.2 mm at a viewing distance of 25 cm, the so-called conventional viewing distance (CVD). At a magnification $V_{\text{total}} = \text{nA} \times 600$, $Z = V_{\text{total}} \times r$ and becomes 0.2 mm; thus, the instrument and the eye are matched in regards to resolution. A somewhat broader range of magnifications from $V_{\text{total}} = \text{nA} \times 500$ to $V_{\text{total}} = \text{nA} \times 1000$ are considered reasonable if the eye is to appreciate all the spatial information an objective produces. This is called the range of useful magnifications. At $V_{\text{total}} < \text{nA} \times 500$ there is information overload and the eye cannot detect the finest specimen detail resolved by the microscope. At $V_{\text{total}} > \text{nA} \times 1000$ no information is revealed that was not already apparent at lower magnifications; we have entered the zone of empty magnification $Z = r \times V_{\text{total}}$ (see Table I), where $r(\mu\text{m}) = 1/3$ nA, lp/mm $= \text{nA} \times 3000$ and $w =$ the angle subtended by two image points spaced by the distance Z ($\tan w = Z/\text{CVD}$)

$$\text{Total pixels} = \frac{\text{lp/mm} \times \text{FOV no.}}{(2 \times M \text{ IIP})} {}^{\wedge}2 \times \text{pi}.$$

In mating video cameras to microscopes the resolution limits of a video system need to be taken into account just as those of the human eye. In fact, we have to concern ourselves with both. First we have to determine at what magnification, expressed in multiples of the aperture, we need to project the image on to the faceplate of the video camera in order to resolve specimen detail on the monitor

Table I
Range of Useful Microscope Magnifications for the Human Eye

Z (μm) at CVD of 25 cm	Z (lp/mm)	Angle	Product (nA $\times V_{total}$)	Pixels in millions at FOV 25	
100	10.0	1'23"	300	4.91	↑
112	8.9	1'32"	336	3.91	Zone of
125	8.0	1'43"	375	3.14	lost
140	7.1	1'56"	420	2.50	information
160	6.3	2'12"	480	1.92	↓
175	5.7	2'24"	525	1.60	↑
200	5.0	2'45"	600	1.23	Range of useful
220	4.5	3'02"	660	1.01	magnifications
250	4.0	3'26"	750	0.79	(nA $\times$ 500 to
280	3.6	3'51"	840	0.63	nA $\times$ 1000)
320	3.1	4'24"	960	0.48	↓
350	2.9	4'49"	1050	0.40	↑
400	2.5	5'30"	1200	0.31	Zone of empty
440	2.3	6'30"	1320	0.25	magnification
					↓

screen. Next, we need to consider the maximum permissible viewing distance from the screen at which our eye can still fully appreciate all the displayed information.

The answer to the first part of the question can be worked out reliably in an empirical way: A test object close to the resolution limit of the lens is chosen, such as the diatom *Pleurosigma angulatum* in the case of an objective of 40× magnification with an nA of 0.65 (40/0.65). This diatom shows a periodic structure of dots spaced by 0.65 μm in hexagonal arrangement. Our objective will just comfortably resolve this structure. If the dots are clearly resolved on the screen, the video system takes full advantage of the resolution available from the microscope.

What one will find, typically, for an ordinary video camera of 2/3-inch tube diameter is a minimum magnification at the faceplate of the camera of 130:1 which translates, more generally, into nA $\times$ 200 to be necessary in order to match the resolution of the camera to the microscope. Thus, between the intermediate image plane and the faceplate of the video camera additional magnification of 3.25:1 is needed in the case of the 40/0.65 objective. How much of the field of view will be captured under these conditions? Division of the faceplate diagonal (11 mm) by the magnification occurring between the IIP and the camera faceplate (3.25:1) yields 3.4 mm. Of the 25-mm field of view, customary in today's microscopes, 98% of the area had to be sacrificed to fully match the resolving power of the objective lens. This is the worst-case scenario but even with the best video technology we would be lucky to keep the lost area to 80%.

To reconcile the demand for imaging large fields for surveying purposes at one time and small fields to preserve resolution at other times the use of a pancratic (zoom) type eyepiece is helpful. A zoom range of 1:6 will comfortably cope with both extremes and for many samples lesser zoom ranges may be perfectly adequate. The magnification changers built into many microscopes serve the same purpose. It is worth remembering that the light intensity changes inversely to the square of the magnification and sometimes higher intensity may have to be given priority over resolution, as, for example, in fluorescence applications.

Now to the second part of the question. A structure clearly resolved on the screen must be viewed from the proper distance. If we move too close to the screen the TV lines become annoyingly apparent. That tends to happen at viewing distances of less than five screen diagonals. If we move too far away the eye's resolution may become the limiting factor. To address this concern in a rational way we return to the earlier statement that the range of useful magnifications lies somewhere between the extremes of nA $\times$ 500 and nA $\times$ 1000. This rule holds for the TV screen as well. How do we calculate the magnification? The magnification of real images such as the intermediate image in the microscope or the image on the TV screen can be measured with a ruler or a scale if the dimensions of the object are known. This we call the ratio of magnification. To distinguish it from virtual magnification we say "ten to one" and write 10:1. In this chapter the letter M is used to signify the ratio of magnification.

Virtual images, on the other hand, are formed, for example, by magnifying glasses (simple compound microscopes). The image exists only in the plane of our retina and is thus inaccessible to measurement. We define the magnification of such images by relating the retinal image size to what it would have been if the object were viewed from a distance of 25 cm without any optical aids. This reference distance of 25 cm is, by agreement, the closest accommodation distance of the average human eye and called the conventional viewing distance (CVD). From a distance of 25 cm we see things at 1$\times$ magnification. A magnifying glass of f = 12.5 cm permits the infinity-accommodated eye to cut the conventional viewing distance effectively by half, therefore the magnification, being reciprocal, is called "two times" and written as 2$\times$.

The ratio of magnification on the monitor screen is the product of several factors:

$$M \text{ (objective)} \times M \text{ (magnification changer)} \times [M(\text{video adapter})$$
$$\text{or } V(\text{eyepiece})] \times V(\text{camera lens})(=f\text{lens}/25 \text{ cm})$$
$$\times \text{ Electronic magnification} = M \text{ screen}$$

"Electronic magnification" is the ratio of the screen diagonal over the camera faceplate diagonal. For a 15-inch screen and a 2/3-inch TV tube the electronic magnification is 35:1.

In the case of an objective 40/0.65, an eyepiece 10X, a camera lens 0.32X, and electronic magnification of 35:1 we have a total screen magnification of $M = 4480:1$. If viewed from a distances of 25 cm we also have a magnification of $V = 4480\times$. From viewing distance of 50 and 100 cm the magnification shrinks to $2240\times$ and $1120\times$, respectively, or generally:

$$V = \frac{M \text{ screen} \times 25 \text{ cm}}{d}, \text{ or } d = \frac{M \text{ screen} \times 25 \text{ cm}}{V},$$

where d is the distance in centimeters from the screen to the observer.

For V to be in the range of useful magnifications of 500 to 1000 times the objective nA, we calculate the minimum and maximum distances for an objective nA of 0.65:

$$d \text{ min} = \frac{M \text{ screen} \times 25 \text{ cm}}{nA \times 500} = \frac{112000 \text{ cm}}{325} = 345 \text{ cm}$$

$$d \text{ max} = d \text{ min} \times 2 = 690 \text{ cm}$$

Thus, within a range of viewing distances from 345 to 690 cm we will be able to appreciate the full resolution of the optics. Had we chosen a lesser M_{screen}, then d_{min} soon becomes less than five screen diagonals and some of the information literally falls through the cracks of the screen's line raster.

II. Optical Considerations

Classification of microscope objectives into Achromats, Fluorites, and Apochromats refers to their color and spherical corrections. In a perfect objective the image does not change its axial location with wavelength. It may, however, change in size. A lateral chromatic difference of magnification (CDM) of 1% across the extremes of the visible spectrum is not unusual. Color fringes, increasingly apparent the larger the FOV number, result from this condition if left uncorrected. For many decades CDM was successfully eliminated by the use of compensating eyepieces. Some of these eyepieces may additionally contribute to improved flatness of field.

C-Mount adapters connect cameras with standardized C-mounts to the photo tube of the microscope. The simplest kind of C-mount adapters just place the camera target into the IIP and then have $1/\times$ magnification. Theoretically, the image is marred by CDM yet in practice the shortcomings of the TV technology are even more severe and tend to mask the image defects, which are minor anyway across a typical FOV of 11 mm.

The shrinking size of camera chips requires proportionally reduced magnifications for the C-mount adapters. Magnifications of 0.63:1, 0.5:1, and so on demand appropriate internal optics with properties similar to the compensating eyepieces.

In recent years the chromatic aberration correction philosophy of most manufacturers has changed. Nikon and Olympus eliminate CDM entirely in the objec-

Table II
Matching Projection Magnification to Camera Faceplate Size

Class	Dimensions and diagonal of faceplate area (mm)	M projective[a]
1-inch TV tube	8.9 × 11.9 (d 14.9)	1.25 : 1
⅔-inch TV tube	6.6 × 8.8 (d 11)	1.00 : 1
½-inch TV tube	4.85 × 6.5 (d 8)	0.63 : 1
⅓-inch TV tube	3.6 × 4.8 (d 6)	0.50 : 1

[a] The magnification of projectives is stepped so that a similar FOV# of 11 to 12 is obtainable with all TV tubes.

tive. Leica and Zeiss prefer to assign the correction of CDM to the tube lens which is a necessary part of all infinity-corrected microscope optical systems which are now ubiquitous. Either way the intermediate image is free of CDM; thus the optics of the C-mount adapters need to be chromatically neutral. To use the same adapter for various makes of microscopes continues to be elusive because of mechanical constraints as well as different approaches used to achieve flatness of field. It is important therefore to use the proper adapter both in terms of vintage of the optical system and the manufacturer of the microscope (Table II).

A lens system designed to transfer the intermediate image onto the faceplate of a camera is sometimes called a projective. Standard C-mount adapters (with the exception of the lensless 1× adapter) are projectives. A different approach has been taken traditionally in microscope cameras made for use with films. Here a standard eyepiece is used and the rays are brought into focus by an optically neutral camera lens. Such an arrangement is not without merit for video microscopy, either. It assures proper correction of CDM and other lens aberrations; it permits the change of the final magnification by changing the eyepiece power. Furthermore, eyepieces which contain a reticle for length measurements are usable. If the eyepiece is of the pancratic type its magnification is adjustable to the size of the camera faceplate, even beyond the zoom range, by the proper choice of focal length for the camera lens.

It is perhaps worthwhile to add at this point that a system rather immune to vibration problems can be created if the video microscope is arranged in two units. One consists of the microscope including the photo eyepiece. The camera with its lens forms the second. No direct mechanical connection exists between the two and the camera is supported independently of the microscope, for example, by a copy stand. Any motion between the two, as long as it results only in a parallel displacement of the optical axis, will not affect the image at all. This strategy was indispensable in the era of microcinematography and may serve a useful purpose even today when heavy video cameras are used, the weight of which, when directly borne by the microscope stand, make it much more susceptible to externally generated vibrations.

CHAPTER 9

High-Resolution Video-Enhanced Differential Interference Contrast (VE-DIC) Light Microscopy

E. D. Salmon and Phong Tran

Department of Biology
University of North Carolina
Chapel Hill, North Carolina 27599

I. Introduction

Differential interference contrast (DIC) light microscopy was an immediate success after its introduction in the 1960s (Allen *et al.*, 1969) because it could produce high-contrast optical images of the edges of objects and fine structural detail within transparent specimens without the confusion induced by the interference fringes typical of phase-contrast microscopy. DIC contrast depends on gradients in optical path (OP = nd; refractive index, n, × thickness, d) and not absolute values of OP. As a result, DIC methods can produce clear optical sections of relatively thick transparent specimens. Resolution in DIC microscopy is also superior to phase contrast. In transmitted light microscopy, lateral resolution, r, depends on the wavelength of light, λ, and the numerical aperture, NA, of both the objective and condenser:

$$r = \lambda/(\mathrm{NA_{obj}} + \mathrm{NA_{cond}}). \qquad (1)$$

The best objectives ($\mathrm{NA_{obj}} = 1.4$) potentially can achieve a diffraction limited resolution in green light ($\lambda = 550$ nm) of about 200 nm when their aperture is fully and evenly illuminated ($\mathrm{NA_{cond}} = 1.4$). Vertical resolution is approximately two times the lateral resolution (Inoué, 1989; Inoué and Oldenbourg, 1993). In DIC, the NA of the condenser illumination can match the objective NA and the resolution limit is close to that predicted by Eq. 1 (Inoué and Oldenbourg, 1993). In phase contrast, the annular cone of condenser of illumination is usually only 60% of the objective NA, and the resolution that can be achieved is less than that by DIC.

When viewed by the eye or recorded by conventional photography, fine structural detail near the limit of resolution is often invisible because it has too little contrast. In the early 1980s, Allen, Inoué, and others (Allen *et al.*, 1981a, b; Inoué, 1981, 1989) discovered that the electronic contrast enhancement capabilities of video cameras could make visible the fine structural detail barely resolvable by the DIC microscope and structures such as 25-nm diameter microtubules (Fig. 1) which are nearly an order of magnitude smaller than the diffraction limit. Both cameras and digital image processors have advanced significantly since the early 1980s (Inoué, 1986; CLMIB Staff, 1995; Inoué and Spring, 1997) and now video-enhanced DIC (VE-DIC) methods are easily implemented (Schnapp, 1986; Salmon *et al.*, 1989; Weiss *et al.*, 1989; Walker *et al.*, 1990; Salmon, 1995).

VE-DIC has been important in many aspects of cell biology including: measuring the assembly dynamics of individual microtubules and force production by microtubule assembly/disassembly (reviewed in Inoué and Salmon, 1995); seeing the motility of bacterial flagella (Block *et al.*, 1991) or the assembly of sickle cell hemoglobin filaments (Samuel *et al.*, 1990); studies of intranuclear spindle dynamics and cell cycle regulation in genetic organisms such as the yeast *Saccharomyces cerevisiae* (Yeh *et al.*, 1995); providing functional assays for the discovery and function of microtubule motor proteins (reviewed in Salmon, 1995), including the measurement of the stall force and 8-nm step size for the microtubule motor protein kinesin in combination with an optical trap (Kuo and Scheetz, 1993; Svoboda *et al.*, 1993; Svoboda and Block, 1994); assays for membrane dynamics and mechanical properties in combination with optical traps (Kuo and Scheetz, 1992); laser microsurgery (Cole *et al.*, 1996); measurements of dynamic changes in cell volume, organelle transport, cell morphogenesis, and motility (reviewed in Scholey, 1993); and monitoring changes in cellular distributions and cell fate during embryogenesis (Thomas *et al.*, 1996).

The quality of images in VE-DIC microscopy depends critically on both the quality of the image projected onto the camera detector by the microscope as well as on the electronic contrast enhancement capabilities of the camera elec-

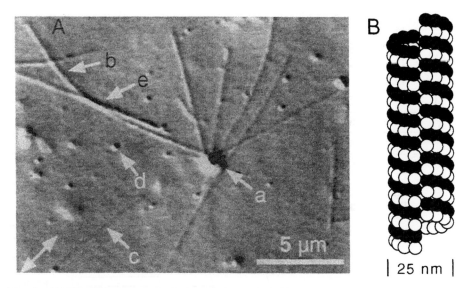

Fig. 1 (A) VE-DIC micrograph of microtubules nucleated from a centrosome (a) in cytoplasmic extracts of sea urchin eggs (Gliksman *et al.*, 1992). The double arrow indicates the DIC shear direction. Contrast is much greater for a microtubule oriented approximately perpendicular (b) to the shear direction in comparison to one oriented in a parallel direction (c). Points in the specimen are imaged in DIC as pairs of overlapping airy disks of opposite contrast as seen for the tiny particle at d. The separation between airy disks in a pair is the shear, which is usually chosen to be less than the resolution limit. The width of the airy disk pair expands the size of objects so that the 25-nm-diameter microtubules (B) appear as 400- to 500-nm diameter filaments in VE-DIC images (b). Overlapping microtubules are not resolved, but have about twice the contrast as individual microtubules (e). (B) Microtubules are 25-nm-diameter cylindrical polymers of α,β-tubulin dimers (see Inoué and Salmon, 1995).

tronics and the digital image processor (Fig. 2). In this article, we will briefly describe the basic concepts of DIC image formation and analog and digital contrast enhancement, the practical aspects of component selection, alignment and specimen chambers for VE-DIC, and several test specimens for measuring microscope and video performance. Further details about DIC, video signals, cameras, recorders, and display devices are beyond the scope of this article and can be found in other sources (Inoué, 1986; Pluta, 1989; Schotten, 1993; CLMIB, 1995; Inoué and Spring, 1997.

II. Basics of DIC Image Formation and Microscope Alignment

A. Microscope Design

A DIC microscope is a "dual beam interferometer" which uses a brightfield polarizing microscope containing polarizer, analyzer, and rotating stage (Figs. 2

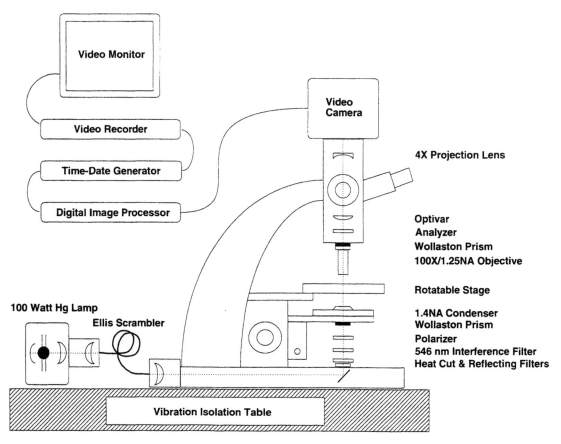

Fig. 2 Schematic diagram of a high-resolution, VE-DIC microscope system. A research polarized light microscope on a vibration isolation table is equipped with an illuminator having a bright 100-watt mercury light source and Ellis fiber optic light scrambler, heat cut, heat reflection and 546-nm interference filters, efficient polars, rotatable stage, and DIC optics. The Optivar contains a telescope lens for focusing on the objective back focal plane during microtubule alignment. The specimen image is projected at high magnification onto the faceplate of a video camera, enhanced using analog electronics in the camera, then passed through a digital image processor for further enhancement (background substraction, exponential averaging, and contrast improvement). This image is recorded on a videocassette recorder or memory optical disc recorder, and the final image is displayed on a video monitor. See text for details.

and 3). The polarizer is inserted in the light path beneath the condenser. The analyzer is inserted in the light path above the objective in a position where the light from the objective is mainly parallel to the microscope axis. The polarizer is usually aligned with its transmission vibration direction of the light electric field in the E–W direction as you face the microscope and the analyzer is aligned in a N–S direction to extinguish light from the polarizer (Fig. 3, left).

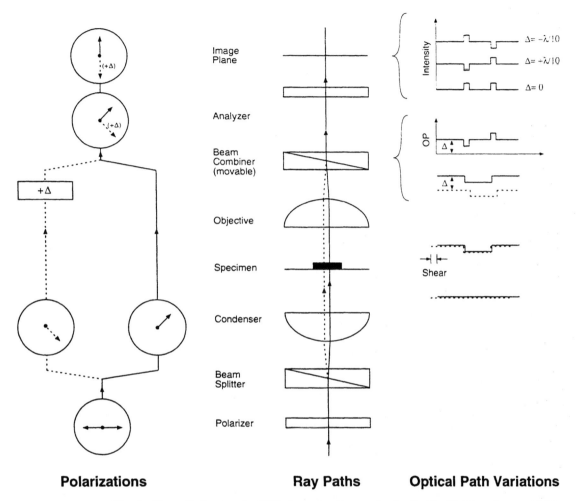

Fig. 3 The optical system for DIC microscopy. See text for details. Modified from Spencer (1982).

In the basic system a prism beam splitter is inserted near the condenser diaphragm plane with its wedge direction at 45° to the polarizer direction (Fig. 3). This birefringent prism splits the light coming from the polarizer into two divergent polarized light waves whose planes of polarization are orthogonal to each other and at ±45° to the initial plane of polarization of the polarizer (Fig. 3). The divergent beams are converted by the condenser into two wavefronts which pass through the specimen separated laterally from each other in the wedge direction (termed the "shear direction," Fig. 3) by a tiny amount which is less than the resolution limit for the objective condenser lens combination (Eq. 1). These two wavefronts are recombined by the objective lens and another prism.

In the original system divided by Smith (1955), the beam splitting prisms are Wollaston prisms, one mounted at the front focal plane of the condenser (the condenser diaphragm plane) and one at the back focal plane of the objective. For medium- and high-power objectives, the back focal plane is usually inside the lens system. Objectives made for Smith DIC contain the Wollaston prism at this position, much like a phase objective with its phase ring, making them unsuitable for other microscopy modes. To avoid this problem, Nomarski (Allen *et al.*, 1969) modified one wedge of the Wollaston prism so that it could be placed in front of the condenser diaphragm or above the objective (Pluta, 1989). These Nomarski prisms act like Wollaston prisms located at the focal planes of the condenser or objective lenses.

The DIC image contrast depends on the "compensation" or "bias retardation" (Δ_c) between the two wavefronts along the microscope axis (Fig. 3, right). When the objective prism is perfectly aligned with the condenser prism ($\Delta_c = 0$) and the background light is extinguished (Fig. 3, upper right), the edges of objects appear bright in the image much as seen in dark field images. When one wavefront is retarded relative to the other, then the background brightens. One edge of the object becomes brighter than the background while the other edge becomes darker than the background. This produces the "shadow-cast" appearance of DIC images (Fig. 1). Reversing the sign of retardation or compensation produces an image of the opposite contrast (Fig. 3, upper right).

Compensators in polarized light microscopes are devices that can add or subtract bias retardation between the two orthogonal beams. In the design in Fig. 3, the objective beam combiner also acts as a compensator when it is translated in the shear direction away from the extinction position. This is done by sliding the prism in the wedge direction (Fig. 3, right). Translation in one direction produces positive bias retardation while translation in the other direction produces negative bias retardation. Some instruments are designed with the objective and condenser prisms aligned and fixed in the extinction position. In these instruments (not shown in Fig. 3), bias retardation between the two wavefronts is typically achieved using a deSenarmount compensator. A deSenarmount compensator uses a quarter-wave retardation waveplate in the extinction position mounted on the prism side of a rotatable polarizer or analyzer. The combination of the quarter waveplate and rotation of the polar induces the bias retardation where $\Delta_c =$ (degrees rotation) $\lambda/180$ (Bennett, 1950).

B. Major Features of a VE-DIC image

There are four major features of a typical DIC image as illustrated by the high-magnification VE-DIC micrograph of the field of microtubules and other small cellular organelles in Fig. 1.

1. Notice that objects appear "shaded" with one side being brighter and the other side darker than the background light intensity. This is easily seen for the 1-μm diameter centrosome (Fig. 1a). Contrast is greater the steeper the gradient in optical path in the shear direction so that contrast occurs mainly at the edges of objects.

2. DIC image "contrast" is directional and maximum in the "shear" direction, the DIC prism wedge direction in the microscope, and this direction is indicated by the double-headed arrow in Fig. 1. Notice that a microtubule aligned perpendicular to the shear direction has the highest image contrast (Fig. 1b) while one in the direction of shear have the least (Fig. 1c). Because contrast is directional, a DIC microscope usually has a rotating stage so that structural detail of interest can be rotated to achieve maximum image contrast.

3. Each point in the specimen is represented in the image by a pair of overlapping "Airy disks." An Airy pattern is the "point spread function" produced by the diffraction of light through the objective aperture. Diffraction by the objective aperture spreads light from each point in the object into an Airy disk in the image. The objective beam-splitting prism in the DIC microscope splits the scattered light from a point in the specimen into two beams producing overlapping pairs of Airy disks for each point in the specimen. By compensation, one Airy disk image is adjusted to be brighter and the other to be darker than the background light. A line in the plane of the image through the centers of the Airy disk pairs is the shear direction while the distance between the centers of the Airy disks is termed the "shear distance." The overlapping pairs of Airy disks are best seen for tiny particles 50 nm or so in size (Fig. 1d). Images of linear objects such as 25-nm diameter microtubules are made from a linear series of Airy disk pairs, while objects larger than the resolution limit, such as the centrosome (Fig. 1a), are 2D and 3D arrays of overlapping Airy disk pairs. Their light dark contrast cancels out except at the edges of the object, so that one side of the object is bright and the other darker than the background.

4. Image size and resolution in VE-DIC images depends on the size of the Airy disk pair. For the images in Fig. 1, the $NA_{obj} = NA_{cond} = 1.25$ and $\lambda = 546$ nm. In this very high-contrast image, each Airy disk appears about 250 nm in diameter and the pair is about 450 nm wide. The diameter of the central peak intensity of the Airy disk depends on wavelength and inversely on NA and the smaller the diameter the better the resolution as given in Eq. 1. For objects much less in width than the resolution limit, such as the small particles (Fig. 1c) and microtubules (Fig. 1b), they appear in VE-DIC images as objects with a width expanded by the size of the Airy disk pair. This explains why the microtubules in Fig. 1 appear almost 10-fold wider than predicted by their 25-nm diameter and the magnification of the microscope. For larger objects than the resolution limit, the size of the Airy disk becomes less a factor and the object dimension becomes closer to the value predicted from the magnification of the microscope optical system.

A VE-DIC microscope cannot "resolve" a pair of adjacent microtubules since these are only 25 nm apart, tenfold less than the resolution limit for the microscope used to make the image in Fig. 1. However, pairs of microtubules have twice the mass and twice the contrast as a single microtubule as seen in Fig. 1e, where two microtubules cross each other. So as mass increases, contrast increases.

Although resolution is limited to about 200 nm in VE-DIC images, movements of objects can be tracked with subnanometer precision by tracking the centroid of the Airy disk image pair of point sources in the specimen (Gelles *et al.*, 1988; Svoboda *et al.*, 1993; Svoboda and Block, 1994).

C. How Much Bias Retardation Is Best

Intensity in DIC images depends on the difference in optical path between the two beams when they interfere with each other at the analyzer (Fig. 3, right). The intensity of light (Fig. 4) at an edge of a specimen, I_s, and a point in the background media, I_b, are given by:

$$I_s = (I_p - I_c)\sin^2[2\pi((\Delta_c + \Delta_s)/2)/\lambda] + I_c \qquad (2)$$

$$I_b = (I_p - I_c)\sin^2[2\pi(\Delta_c/2)/\lambda] + I_c \qquad (3)$$

where I_p is the intensity of light that would leave the analyzer if the analyzer axis was rotated parallel with the polarizer axis of vibration, Δ_c and Δ_s are the retardations induced by the compensator bias retardation and the specimen (Fig. 3, right), and I_c is the intensity of light with the specimen out of the light path when the analyzer and polarizer are crossed and Δ_c and Δ_s are both zero.

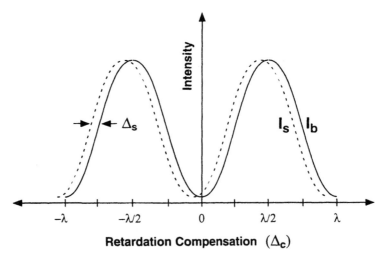

Fig. 4 Plots of intensity verses compensator retardation, Δ_c, for the intensity of the background light, I_b, and for the edge of a specimen, I_s, which exhibits retardation, Δ_s. See text for details.

I_c represents the stray light in the system induced by birefringence in the lenses, rotation of the plane of polarization by highly curved lens surfaces (Inoué, 1986; Inoué and Oldenbourg, 1993), and by imperfect polarization of light by the polarizer and analyzer (see below). I_c is usually about 1/100th of I_p in a high-resolution (NA = 1.4) DIC microscope system. If I_c is any more than this value it can be a significant noise factor interfering with visualization of fine structural detail or objects like microtubules (see below). As seen in Eq. (3) and Fig. 4, when the retardation is $\lambda/2$, the background light is maximum, and when the bias retardation is zero, the only light in the image is produced by the specimen structural detail and the stray light (Fig. 3, upper right).

In practice, the best optical contrast is achieved when bias compensation is adjusted to extinguish the light coming from one edge of the object of interest, the edge of a microtubule for example in Fig. 1. This occurs when $\Delta_c = -\Delta_s$. For the edges of organelles and cells, Δ_s corresponds to about 1/10th the wavelength of light or greater, but for microtubules and tiny organelles in cells, Δ_s is very small, less than 1/100th the wavelength of green light. For microtubules, we still use about 1/15th–1/20th the wavelength bias retardation so there is sufficient light to the camera. A similar result was found by Schnapp (1986), but Allen and others argue that redardations near 1/4 wavelength are best (Allen *et al.*, 1981; Weiss *et al.*, 1989). Near or at extinction, there is a diffraction anomaly in the Airy pattern produced by the rotation of the plane of polarization at the highly curved surfaces of high-NA lenses (Inoué, 1986; Inoué and Oldenbourg, 1993). This can be corrected by rectifiers, but they are only available for particular lenses. This anomaly disappears with increasing bias retardation (Allen *et al.*, 1981b) and is not apparent in our images (Fig. 1) obtained with about 1/15th wavelength bias retardation.

III. Basics of Video-Enhanced Contrast

A. Video Basics

In video microscopy the image produced by the light microscope is projected onto the detector of the video camera (Fig. 2). Electronics in the video camera scan the detector in a raster pattern, converting variations in light intensity into variations in voltage which are encoded into a video signal where white is 0.7 volts and black is about .03 volts. The video signal also includes negative horizontal and vertical sync pulses which guide the generation of the raster pattern in the image display (Inoué, 1986). Conventional video cameras produce images at 30 frames/sec for NTSC (the standard TV rate in America) and 25 frames/sec for PAL (the TV rate in Europe). A frame is made by successive scans of two interlaced fields; each field is scanned in 1/60 sec. An NTSC image contains 480 active TV lines displayed on the video screen. The video signal produced by the camera can be immediately displayed on a video monitor and/or sent to a time-date generator, digital image processor, or recording device (Fig. 2).

Resolution in video is measured by the number of black plus white lines visible in either the horizontal or vertical direction. Horizontal resolution is usually limited by the resolution of the detector, bandwidth of the analog electronics or the speed of a digitizer, and number of picture elements "pixels" in a frame-storage memory buffer. Vertical resolution is limited by the about 480 active TV lines in a video frame. A horizontal TV resolution of 800 lines means that 800 vertical lines, 400 black and 400 white alternating lines, are visible in the video image over width equal to the height of the active picture area.

If a detector or image processor has 800 pixels horizontally, it can't reliably detect 800-TV-line resolution because of insufficient pixels and the problem of "alising" which occurs when the lines are not centered on the pixel detectors, but centered on the junctions between pixels (Inoué, 1986). At this position, adjacent pixels get average intensities from black and white lines and they are not resolved. The same alising problem occurs in the vertical direction. In general, resolution is given as about $0.7 \times$ number of pixels in either the horizontal or vertical direction. For example, video images with 480 active horizontal lines, have $0.7 \times 480 = 336$ TV lines of vertical resolution. In contrast, horizontal resolution is often 800 TV lines, or almost threefold better than the vertical resolution.

B. Analog Contrast Enhancement

As shown in Fig. 5, a field of microtubules, invisible by eye in the DIC microscope (Fig. 5a), can be see with high contrast using analog and digital contrast enhancement methods. Most video contrast enhancement is produced by analog electronics in the video camera (Fig. 5b and the diagram in Fig. 6). Contrast enhancement is produced by subtracting voltage (black level or pedestal control) from the video signal and amplifying the difference (gain control) for each horizontal line scan of the video raster image.

Increasing the analog contrast enhancement to levels necessary to give distinct images of microtubules induces a new set of technical problems. Image quality

---

Fig. 5 An example of how individual microtubules can be seen by VE-DIC microscopy because of the enormous gains in contrast that can be achieved by analog and digital image processing. Taxol-stabilized microtubules assembled from pure brain tubulin were bound to the coverslip surface by *Xenopus* egg extracts treated with 5 m*M* AMPPNP (adenosine 5′-(β,γ-imino)triphosphate). The microtubules were viewed with the vertical optical bench VE-DIC microscope described by Walker *et al.* (1990). Video images were obtained with a Hamamatsu C2400 Newvicon camera and Argus 10 digital image processor (Hamamatsu Photonics Inc.), stored in a MetaMorph image processing system (Universal Imaging, Inc.), and printed on a Tektronics Phaser IISDX dye-sublimation printer. (a) A typical image when viewing by eye through the microscope. (b) The video image after analog contrast enhancement by the video camera electronics. Microtubules are visible, but the image is contaminated by shading, mottle, dirt on the camera detector, and random noise. (c) The background image made by defocusing the microscope. (d) The video image after background substraction from live video, the addition of a bias grey level, and further digital contrast enhancement. (e) An exponential average of four frames substantially reduces the random noise. Scale = 2.5 μm. From Salmon (1995).

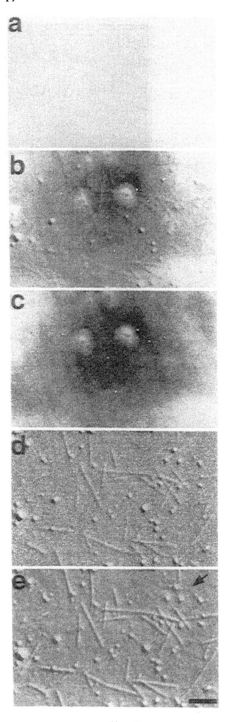

Fig. 5

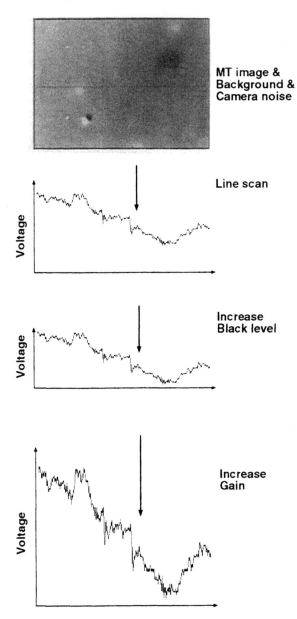

Fig. 6 Analog contrast enhancement. Video cameras with black level and gain controls can improve image contrast by first subtracting voltage (termed black level or pedestal) from the video signal and then amplifying the difference (using the gain control) to stretch the remaining information in each horizontal scan line. This process also increases the magnitude of uneven background intensity and random noise. Modified from Salmon *et al.* (1989).

becomes limited by noise from several sources which is amplified as gain is increased. There are two different types of noise. One is stocastic, "shot" or "random," noise which is produced by noise in the video camera and by variations in photon flux between successive frames. This noise makes the image "grainy." The second is "fixed pattern" noise that does not change between successive frames and can obscure weak-contrast objects like microtubules (Fig. 5b). "Shading" across the image is produced by uneven illumination, birefringence in the lenses, or variations in sensitivity across the detector faceplate. "Mottle" comes from out-of-focus dirt on the lens surfaces. "Dirt" on the faceplate of the detector makes distinct spots.

As gain is increased, the bright and dark regions of the image exceed the dynamic range of the camera and specimen detail is lost because of "clipping" (Fig. 6). Some high-performance video cameras (e.g., DAGE VE-1000 and Hamamatsu C2400) have analog electronic controls which can be used to flatten-out uneven background brightness, thus allowing higher contrast enhancement. These devices are very useful for correction the unidirectional gradient in background light intensity produced by imperfectly matched DIC prisms. However the mottle and dirt in the image cannot be removed by this method.

C. Digital Contrast Enhancement

Allen and Allen (1983) were the first to recognize that fixed pattern noise could be eliminated from the video image by subtracting a stored background image using a digital image processor operating at video rates. The background is obtained by defocusing the microscope until the microtubule image just disappears. Then a number (usually 128 or 256) of successive frames are averaged and stored in a "background" frame buffer in the image processor (Figs. 5c and 7). The microscope is refocused and the background is continuously subtracted from the live image. A bias gray level is added to the displayed image (usually 128 out of the total of 255 gray levels in an 8-bit digital image) to produce a uniform background gray level. After background subtraction, further contrast enhancement (four- to sixfold) can be achieved by using the output look-up tables of the digital image processor (Figs. 5d and 8).

This method removes much of the fixed pattern noise, but does not reduce the random noise in the image from camera noise or variations in photon flux. Averaging of successive frames with the real-time image processor solves this problem by sacrificing temporal resolution. The increase in signal-to-noise depends on the square root of the number of averaged frames (N). Too much averaging blurs images of moving objects. The image in Fig. 5e was obtained using a four-frame sliding or exponential average after background subtraction (Fig. 8).

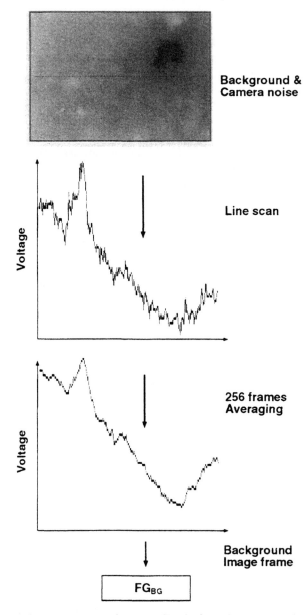

Fig. 7 Background image average and storage. To obtain an image of the fixed pattern noise, such as dirt particles and uneven background in the video image, the microtubule image is slightly defocused and 256 sequential video frames are acquired and averaged to produce a low-noise image, which is saved in a framestore board. Modified from Salmon *et al.* (1989).

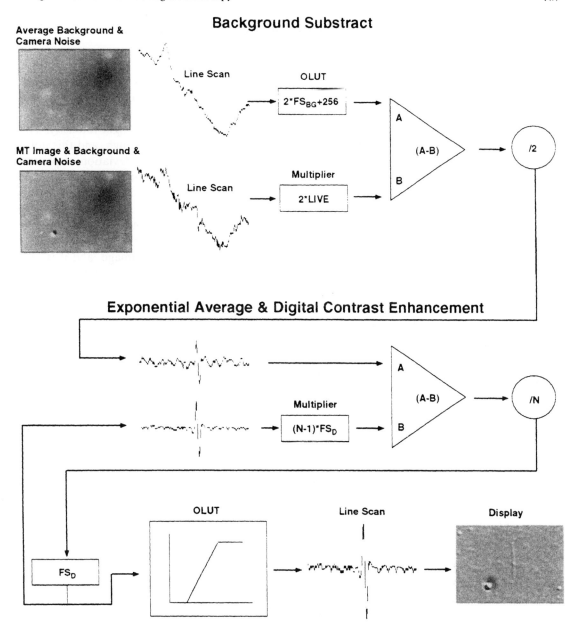

Fig. 8 Flow diagram of a real-time digital image processor. An arithmetic logic unit subtracts a background image (stored in a frame buffer) from each incoming live video frame (every 33 ms) and sets the average background intensity at 128. The resulting frame is then exponentially averaged with a fraction of the previously obtained frame and stored in a display frame buffer, FS_D. For digital contrast enhancement, the image in FS_D is passed through an output look-up table (OLUT), which converts the gray scale values of each pixel according to the function stored in the OLUT. To accomplish these operations in real-time, we use the Argus 10 or Argus 20 digital image processor from Hamamatsu Photonics, Inc. Modified from Salmon *et al.* (1989).

IV. Selection of Microscope and Video Components

A. General Considerations

Achieving high-resolution, high-contrast VE-DIC images requires the highest performance of both the optical and video components of the video microscope system.

One particularly important design consideration is image light intensity. The image will appear too "noisy" if the image is not bright enough, and this noise is amplified by video contrast enhancement. The standard deviation of photon noise in the image is the square root of the number of photons per second, per picture element, or "pixel." If only 100 photons are captured per pixel in a video frame, the noise will be equivalent to 10% of the brightness, and the image will appear grainy and resolution poor. If 1000 photons are captured, then the noise will be only 3.3% of the image brightness and a much higher-quality image is achieved. The video cameras also contribute noise to the image, and it is also important that the image light is sufficiently bright so that the video signal is well above the "noise floor" for the video camera.

A number of factors in the microscope multiply each other to reduce the image light intensity, I_{image}, at the camera detector. These include:

1. The brightness of the illumination.
2. Reduction of light intensity by the DIC optics as described by Eqs. (2) and (3), which can be 100-fold.
3. The need for near monochromatic illumination.
4. Transmission efficiency of the microscope optics including the polarizer, analyzer, filters, and lenses.
5. Magnification, M, from the specimen to the camera which reduces light intensity by $1/M^2$, or about $1/333^2 = 1/10,000$ for the $333\times$ magnification needed in our VE-DIC microscope system to match the optical resolution to the camera detector resolution.
6. Camera exposure time, which is 1/30th sec at video frame rates.

The reduction of light intensity by the DIC optics and magnification is a property of a VE-DIC microscope system properly aligned for high-resolution, high-contrast imaging. It is important not to have any further losses in light intensity by microscope components because of poor transmission efficiency and these should be selected carefully.

The brightness of the light source required depends on the exposure time of the camera. At video frame rates, the exposure time is short, 1/30th sec, and a very bright light source like a 100-watt mercury lamp with its bright peak at 546 nm is required. If time resolution can be sacrificed, then cameras which allow longer integration times on the detector can be used (Shaw *et al.*, 1995; Cole *et al.*, 1996). The longer the integration time, the less bright must be the light source for the same quality image (assuming motion is not a problem). For example, a

600-ms exposure on the detector requires about 1/20th the brightness from the light source as needed for video rates, and the standard 100-watt quartz halogen illuminator is suitable.

B. Microscope Components and Alignment

1. Illumination

The objective aperture must be fully and uniformly illuminated to achieve high-resolution and high-quality images (Inoué, 1989; Salmon *et al.*, 1989; Inoué and Oldenbourg, 1993). It is the high-NA rays of condenser illumination which are important for achieving the maximum resolution predicted by Eq. (2). Uneven illumination causes the image to move obliquely as one focuses through the specimen, reduces depth discrimination, and enhances out-of-focus mottle in the image. The highest-resolution objectives require oil immersion of both the objective and condenser. It is critical that there be no air bubbles in the oil and that sufficient oil is used, particularly for the condenser, so the objective aperture can be fully illuminated.

Finally, it is important the specimen field be illuminated uniformly, since uneven illumination limits the amount of contrast enhancement that can be achieved.

The Köhler method is the standard method in bright field microscopy to achieve uniform illumination of the specimen. This is the best method for VE-DIC microscopy and it is usually described in detail in the operating manual for the microscope. Briefly:

1. The image of the field diaphragm is adjusted by moving the condenser to be in-focus and centered on the specimen.

2. The field diaphragm is closed down until only the specimen of interest is illuminated to reduce scattered light from surrounding objects.

3. The condenser diaphragm controls the illumination NA, and the condenser diaphragm is typically opened so that it is just visible at the peripheral edges of the objective back focal plane.

4. The image of the light source is projected in-focus and centered onto the condenser diaphragm plane using adjustments of the lamp collection lens or projection optics along the illumination path.

Steps 3 and 4 are usually made using a telescope to view the objective back focal plane, where the image of the condenser diaphragm and light source should be in focus.

The light intensity in the image depends on the bias retardation as given in Eq. (3). For Δ_c near 1/20 the wavelength of green light, 546 nm, $\sin^2[2\pi(\Delta_c/2)/\lambda] = .024$. Because this value is small, and Ip is often 2.5% or less of the light intensity from the illuminator (see below), intense illumination is needed for DIC microscopy of cellular fine structural detail and objects like individual microtubules which require high magnification to be resolved by the detector. In addition, monochromatic and not white light gives the highest contrast with DIC because it is an interference contrast method.

We routinely use a 100-watt mercury lamp, which has an intense line at 546 nm wavelength for high-resolution imaging at video rates (Walker *et al.,* 1990). This lamp has a very small arc, about .3 mm when new. The arc is very inhomogeneous with bright spots near the electrode tips. To achieve uniform illumination of the condenser and objective aperture with this arc, we use a fiber optic light scrambler (Fig. 2) as developed by Ellis (1986) and available from Technical Instruments (Marine Biological Laboratory, Woods Hole, MA 02543). The standard collector in the lamp housing is adjusted to project the light to infinity. The mirror in the back of the light housing is adjusted so the image of the arc it produces is aligned with the primary image produced by the collector lens. A second collector lens focuses this light into the end of a 1-mm-diameter quartz fiber optic. The fiber optic is bent in several 10-cm circles to scramble the light and produce uniform illumination emanating from the other end. A third collector lens at the base of the microscope (Fig. 2) projects an image of end of the fiber optic in-focus and centered on the condenser diaphragm. The focal length of this collector lens is selected so that the image of the fiber optic end matches the size of the wide open condenser diaphragm and thus fully fills the objective aperture when the microscope is properly adjusted for Köhler illumination.

Other intense light sources we have used for VE-DIC include a 200-watt mercury lamp, which has a large arc which can be aligned at 45° along the DIC shear direction and fill the objective aperture in that direction without the need for an optical fiber (Walker *et al.,* 1988; Salmon *et al.,* 1989). However, flickering, promoted by the large electrode separation, substantially interferes with making high-quality VE-DIC video recordings, because the background light intensity fluctuates over time. We have also used a 200-watt Metal Halide lamp with a 5-mm-diameter liquid-filled optical fiber made by Nikon (Waterman-Storer *et al.,* 1995). This light source is about one-half the brightness of the 100-watt mercury lamp, but much more stable and easy to align. Unfortunately, it is not readily available from Nikon. For the finest structural detail, like pure tubulin 25-nm-diameter microtubules (Salmon *et al.,* 1989), 21-nm filaments of sickle cell hemoglobin (Samuel *et al.,* 1990), or 15-nm-diameter bacterial flagella (Block *et al.,* 1991), the high intensity of the 100-watt mercury lamp is needed. The use of lasers as light sources has been inhibited by the mottle and speckle induced in the image by the coherence of the laser light.

For imaging with longer exposures (e.g., 600 ms, Yeh *et al.,* 1995) or lower magnifications (Skibbens *et al.,* 1993), we use a standard 100-watt quartz halogen illuminator that has a mirror housing. This light source is less than 1/10th the brightness of the 100-watt mercury source at green wavelengths. After establishing Köhler illumination, we use the lamp housing controls to position the primary image of the lamp filament on one side of the objective aperture and the mirror image of the filament on the other side so together they just cover the objective aperture providing illumination of both the center and the edges of the aperture. Then we slide into the light path a ground glass filter to diffuse the light and give nearly uniform illumination of the aperture. The ground glass filters are specially designed by the microscope manufacturer to efficiently diffuse the light without producing a large reduction in light intensity.

The above descriptions assume the Köhler illumination method. There is another method of achieving full and even illumination of the objective aperture with the mercury light source, termed "critical" illumination. By this method, the image of the arc is focused on the specimen. Illuminating light at the condenser diaphragm plane is out-of-focus and uniform so the illumination criteria for high resolution (Eq. 2) is meet. However, the uniformity of intensity in the specimen plane depends on the homogeneity of the arc. If only a very small field of view is required, then the center of the arc may be homogeneous enough to work for VE-DIC as shown by Scholey and coworkers (Scholey, 1993). This method must be applied carefully, because the DIC optics are designed for condenser alignment with illumination using the Köhler method.

2. Filters

We use a combination of Zeiss heat cut and infrared interference reflection filters to remove infrared light from the illumination. This is important for preventing damage to living cells and native preparations.

Green light wavelengths are typically used for VE-DIC. Resolution is better for shorter wavelengths of light (Eq. 1), but live cells and native preparations are often damaged by illumination with blue or shorter wavelengths. In addition, light scattering by out-of-focus regions of thicker specimens is much more of a problem for shorter wavelengths of light. The eye is also most sensitive to green light. For these reasons, green-light illumination is a good compromise between the resolution advantages and disadvantages of shorter wavelengths. Narrow-band green interference filters are often used for DIC microscopy to achieve the highest contrast images. We use an interference filter with a central maximum at 546 nm and half-maximum bandwidth of 20 nm to capture the 546 mercury line (Omega 540DF40). For the quartz-halogen illumination, we use a 540-nm filter with a 40-nm bandwidth. These interference filters have peak transmission efficiencies of 85% or greater.

The heat absorption and green filters are placed in the illumination light path where the light is most parallel to the axis of the microscope. We have placed them either between the two collector lenses in the lamp (Walker et al., 1990) or just before polarizer (Fig. 2).

3. Polarizer and Analyzer

Polarizers and analyzers are made from materials which polarize light. In either case, polars are required with:

1. high transmission of plane polarized light at green wavelengths (85% or greater);
2. a high extinction factor (1000 or greater);
3. high optical quality and sufficient size to match the physical size of the condenser or objective apertures.

Ideal polars can transmit 100% of incident polarized light when the plane of polarization is aligned parallel to the polarization direction of the prism and transmit 50% of incident unpolarized light as plane polarized light.

The extinction factor, EF, measures how well a polarizer produces or and analyzer rejects plane polarized light:

$$EF = I_c/I_p, \qquad (4)$$

where I_c is the light transmitted by a pair of "crossed polars" and I_p is the light transmitted when their planes of polarization are parallel.

Optical quality refers to how much the polar interferes with the quality of image formation in the microscope. The polarizer and analyzer are usually placed in positions in the microscope just below the condenser and just above the objective where the light is more paraxial and imperfections in the polars are out-of-focus in the image. The most critical is the analyzer because its shape, thickness, and imperfections can induce geometrical aberrations which destroy image resolution.

The condenser aperture is typically about 2 cm in diameter for high resolution ($NA_{cond} = 1.4$) illumination. The objective aperture is typically much smaller, 7 mm or less.

Polars with the highest transmission efficiency are Glan-Thompson prisms (Karl Lambrecht, Chicago, IL), which polarize light by double-refraction in calcite prisms (Inoué, 1986). These prisms can have extinction factors of 10,000 to 100,000 (Inoué, 1986). To match a condenser aperture of 2 cm requires a large prism of about 2.5 cm diameter and 6 cm long and space along the microscope axis before the condenser to insert the prism. On many conventional microscope stands this is difficult, but possible on inverted stands (Schnapp, 1986) and on optical bench stands (Inoué, 1981; Walker et al., 1990). The thickness of the prism requires moving the field diaphragm further away from the condenser than its normal position without the prism. Glan-Thompson prisms have been successfully used for the analyzer in VE-DIC (Inoué, 1981), but their thickness requires special stigmating lenses and very careful alignment. As a consequence, we use on our optical bench VE-DIC microscope, a Glan-Thompson prism for the polarizer, but a high-quality polaroid polar for the analyzer (Walker et al., 1990).

Polars made with high-quality polaroid are typically 1–3 cm in diameter and 2–3 mm thick. These polars are made with two dichroic polaroid films about 50 μm or so thick immersed between two thin optical glass flats for high optical quality. A pair of polaroid films are used so that pinholes in one film overlie polarizing material in the other. Polaroid polars with the highest extinction factors (10,000 or greater) usually have poor transmission efficiency, 15% or less of unpolarized light. Polarization contributes 50% of the attenuation while absorption of light in the vibration direction contributes 35% of the attenuation. When a polarizer and analyzer made from this material are aligned with their vibration directions parallel, then the maximum intensity is only $0.15 \times 0.65 = 9.8\%$ of the incident light. This reduction in light intensity often makes the light intensity at the camera too low for decent signal-to-noise in the image.

High transmission polaroid polarizers have reduced extinction factors, 1000 or less. However, a typical extinction factor for one of our VE-DIC microscopes (Waterman-Storer et al., 1985) with a 60X/NA = 1.4 Plan Apochromat objective and matching condenser illumination is about 100 (because of the rotation of the plane of polarization at the highly curved surfaces of the 1.4-NA objective and condenser lenses; Inoué, 1986; Inoué and Oldenbourg, 1993). Thus, polars with extinction factors of about 1000 are more than sufficient and we have selected polaroid polarizers with 35% transmission efficiency extinction factors. These are termed "high transmission" polarizers and they should be requested from the microscope manufacturer.

4. Objective and Condenser

For high-resolution VE-DIC we have used either (1) Zeiss 100X/1.25 NA Plan Achromat or Nikon 100X/1.4 NA Plan Apochromat objectives and matching Zeiss 1.4 NA oil condenser (Salmon et al., 1989; Walker et al., 1990; Skibbens et al., 1993) or (2) Nikon 60X/1.4 NA Plan Apochromat Objective and matching 1.4-NA oil condenser (Waterman-Storer et al., 1995). In all cases the objectives and condensers were selected to be free of birefringence caused by strain or glues in the lenses. This birefringence produces background noise and mottle which can obscure fine structural detail no matter how much contrast enhancement is used. The Plan Achromat objectives do not quite have the resolution of the Apochromat objectives, but they have been important for our microtubule and motor protein assays at 37°C. Heating the optics induces birefringence in the lenses, particularly the objective lenses. Eventually, heating can induce permanent birefringence which makes the lens unsuitable for VE-DIC of fine structural detail like individual microtubules. The Plan Achromat lenses are 1/4 the price of the Plan Apochromat lenses and thus less expensive to replace. We have tried to restrict our use of the expensive Plan Apochromat lenses to studies at room temperature.

Most of our specimens we image by VE-DIC are either living cells or isolated organelles in aqueous media and usually focus is within 5 μm of the inner coverslip surface. This is fortunate because for oil immersion objectives, resolution deteriorates as focus is shifted away from the inner coverslip surface. At 5 μm depth resolution is noticeably poorer than near the coverslip and by 20 μm it is substantially deteriorated. This deteriorization in image quality occurs because of spherical aberration induced by refraction at the glass–water interface, where refraction is greater for the higher aperture rays (Keller, 1995). For specimens thicker than 10 μm, it is better to switch to water immersion objectives which have correction collars for coverslip thickness and the spherical aberration. Unfortunately, water immersion objectives with high numerical apertures (NA = 1.3) are expensive.

5. Beam–Splitting DIC Prisms

It is critical for VE-DIC that the effects of the condenser DIC prism be matched by the objective DIC prism. In the Nomarski design as described in Section II,A, the prisms are designed to be mounted at or near the condenser diaphragm plane and just above the objective. The prisms must be oriented correctly, with both their wedge directions at 45° to the polarizer, analyzer vibration directions. When the two prisms are aligned to extinguish the background light, no dark bar should be visible in the center of the prisms when they are viewed by a telescope by focusing on the objective back focal plane. If there is a dark bar, then the prisms are not properly matched. It is important to perform this test for pairs of objective and condenser DIC prisms to insure that their markings and orientations are correct.

6. Stage Rotation, Centering, and Drift

A rotatable stage is useful for VE-DIC microscopy because DIC contrast is directional. Often ±45° of rotation is sufficient. It is important that either the objectives or stage be centerable so that when the stage is rotated the specimen stays centered.

Another critical parameter is the stability of the stage. For long-term high-resolution VE-DIC recordings, stage drift should be minimal because z-axis depth-of-focus is often near 300 nm (Inoué, 1989). The stage on our optical bench microscope (Walker et al., 1990), for example, is fixed and the objective is moved to focus. This is a very stable configuration and drift is less than 0.3 μm/hr. On our upright microscope stands, the objective is fixed and the stage is moved to focus. Drift has been a major problem, and on one stand, can be several micrometers per hour. Nevertheless, for VE-DIC it is critical to check that focus is stable for long-term recordings.

7. Compensation Methods

Our microscope stands currently use translation of the objective DIC prism (see Section II,A) to vary retardation and image brightness and contrast. This works fine, but the amount of compensation must always be judged by eye. The preferred method is to use deSenarmont compensation with degrees of rotation of the polar, marked so that the operator can accurately set the amount of retardation.

8. Slides, Coverslips, and Other Specimen Chambers

For dry objectives, number 1.5 coverslips are used because these are near the 0.17-mm thickness specified in objective lens design to correct for geometrical aberrations (Inoué and Oldenbourg, 1993). For oil immersion objectives, we use number 1 coverslips which are a bit thinner than number 1.5 coverslips and give the objective more working distance between its front lens element and the coverslip outer surface.

For the best images, coverslips should be cleaned to remove dirt and oil. Coverslips are placed in a beaker full of detergent solution (Alconox, Alconox, Inc., New York, NY) which is in turn placed in an ultrasonic cleaner (Mettler Electronics model ME2.1, Anaheim, CA). After 15 min, the coverslips are rinsed for 20 min in running tap water. The coverslips are then placed in a specially designed box which holds the coverslips upright and separate so that their surfaces do not touch. This box is placed in the ultrasonic cleaner, agitated for 15 min, and rinsed with five washes of distilled water. The agitation and rinse is repeated twice and the coverslips are then placed in a beaker of 100% ethanol and returned to the ultrasonic washer for 15 min of agitation. Coverslips are then rinsed twice in ethanol, then stored one by one in a covered beaker of fresh ethanol. Prior to use, the coverslips are placed on a custom spinner (Inoué, 1986, p. 143) and spun dry. Slide surfaces are cleaned by wiping them with Kimwipe soaked in 70% ethanol and then air dried.

We often use Scotch double stick tape to make perfusion chambers or as spacers in standard slide coverslip preparations. The tape is about 70 μm thick. A razor blade or scissors is used to cut strips which are placed on opposites sides of the slide. The coverslip and preparation are added, and then the coverslip is sealed along the open edges using VALOP (1:1:1 mixture of vasoline, lanolin and parafin).

A key parameter to consider for specimen chambers for high resolution VE-DIC is that the optical path through the chamber and specimen must be equivalent to that of a standard slide coverslip preparation. The slide is 1-mm-thick glass of refractive index 1.52 so its optical path, OP, equals 1.52 × 1 mm. Chambers we have used in addition to the double-stick tape chambers include Berg and Block perfusion chamber (Berg and Block, 1984, Walker *et al.*, 1991), and a McGee-Russel perfusion and fixation chamber (Waters *et al.*, 1996).

For temperature control we have used either temperature-controlled rooms for temperatures between 20–28°C or placed a large plastic bag around the stage of the microscope excluding the cameras and illuminators and warmed the stage to 37°C with an air-curtain incubator (Walker *et al.*, 1988).

9. Magnification to the Camera Detector

The total magnification between the objective and detector in the camera must be sufficient so that resolution is not limited by the detector, video camera, digital image processor, or recording device. Detectors vary in their size and hence the size of their pixel elements. For the 1-inch diagonal Newvicon and

Chalnicon video detectors, we use a total magnification of 3.5 to 4× produced by the combination of the magnifier in the Optivar in the body tube and a projection adaptor to the cmount for the camera. For the smaller CCD video detectors, the necessary magnification is less than 4× depending on chip size: only 2× is needed for the 1/2-inch chip.

Often, however, it is not the camera that is resolution limiting, but the digital image processor or video recorder. Thus, it is essential to test each video system to determine the suitable magnification required using the diatom test plate or cheek cell test specimens described in Section V,A.

Remember, as magnification increases, the field of view decreases as $1/M$ and the intensity decreases as $1/M^2$.

10. Vibration Isolation

Any vibration of the image reduces resolution. Air-tables (Fig. 2, Newport Corp., Irvine CA, Model No. VW-3660 with 4-inch-high table top) or other means of vibration isolation (Salmon *et al.*, 1990) are essential for high resolution VE-DIC microscopy.

C. Video Components

1. Camera and Analog Processor

Cameras with 800 or more horizontal TV line resolution, low lag, low shading, low geometrical distortion, high image stability, high linearity, and superior signal-to-noise characteristics are desired for VE-DIC microscopy. Such cameras are usually distinct from the intensified video cameras or slow-scan cooled CCD cameras used for low-light-level imaging to render visible images generated by only a few photons (Inoué, 1986). We have found that research-grade cameras using either the Newvicon or Chalnicon tubes (DAGE-MTI model 1000 and Hamamatsu model C2400) give excellent performance and low noise for the very-high-contrast VE-DIC needed to see single microtubules (Salmon *et al.*, 1989; 1990).

The horizontal resolution of these cameras, 800 TV lines, is two- to threefold better than the vertical resolution of a video image with 480 active TV lines because of alising as described in Section III,A. To maximize resolution in the video image in the direction of maximum contrast in the DIC image, the camera should be rotated to align the horizontal line scan direction in the camera with the direction of shear in the DIC image.

Research video rate CCD cameras (DAGE-MTI model 72 or Hamamatsu model 75i) also give good images with VE-DIC, but their horizontal resolution has not been as good and their noise level has been somewhat higher than the research-grade Newvicon or Chalnicon tube cameras. Resolution in CCD cameras is determined by the number of detectors in the horizontal and vertical direction; typically 860 × 480 in the research-grade chips and horizontal resolution is 0.7 × 860 = 600 TV lines, somewhat less than the 800 TV lines for the tube cameras for the same field of view. New chips with more pixels and better noise characteristics will eventually replace the Newvicon and Chalnicon detectors.

One advantage of the video CCD cameras is that they have much lower lag than the tube cameras, but on-chip integration can be used to improve sensitivity when time resolution can be sacrificed (see Shaw *et al.*, 1995; Cole *et al.*, 1996). This requires some form of computer control either by electronics built into the camera itself (Cole *et al.*, 1996) or by associated image processing systems (Shaw *et al.*, 1995).

The research-grade camera controllers can have several other features useful for VE-DIC. An "idiot" light tells the operator that the video signal is in the right range: not too dim and not so bright that it saturates the detector input amplifier. The analog gain and pedestal controls are easily adjustable to achieve optimum analog contrast. Some controllers also have an automatic gain control to eliminate image flicker resulting from changes in line voltage, lamp flicker, and so on. In addition, controls are available to manually correct for shading at high contrast produced by uneven illumination or by slightly mismatched DIC prisms. We needed to use these shading correctors when we used a Nikon 100X/ NA = 1.4 Plan Apochromat objective with Zeiss 100X Plan Achromat DIC optics in our high-resolution studies of kinetochore motility in vertebrate epithelial cells (Skibbens *et al.*, 1993). We called this "Zikon" DIC optics. When this shading is corrected by these analog controllers, then much higher analog contrast enhancement can be achieved before the bright and dark regions of the image exceed the dynamic range of the video electronics (Fig. 8).

2. Digital Image Processor

Initially, we built our own image processing equipment to perform background subtraction and exponential averaging algorithms (Fig. 8) at video rates (30 frames/sec) (Walker *et al.*, 1988; Cassimeris *et al.*, 1989; Salmon *et al.*, 1989). Now, there are a number of commercial devices that provide similar functions. These include the Argus 10 and Argus 20 processors from Hamamatsu, DSP2000 from DAGE-MTI, and MetaMorph from Universal Imaging. These real-time processors need to have several frame buffers and at least two arithmetic processors; simple image capture boards or processors with one arithmetic unit won't do the job at video rates. The processors from Hamamatsu and DAGE-MTI do not require computer workstations, but they can output images either by video cable or digital images by SCSI cable to digital imaging workstations for digital photography (Shaw *et al.*, 1995) or storage to disk. The MetaMorph system resides in a digital imaging computer workstation and video images are captured and processed using a plug-in video card and digital imaging software in the computer.

The resolution of the digital image processing depends on the number of pixels per horizontal line in the frame store buffers in the processors. Many processors, like the Argus 10, generate digital images which are 512×480 pixel images, and the 640×480 processors used in the PC computers have about half the horizontal resolution of video cameras. The Argus 20 has higher horizontal resolution, producing 1000×480 pixel images, so that the digitization process does not reduce image resolution. Images can be transferred to a computer through a SCSI port.

In addition to video contrast enhancement, some of the image processors have numerous other useful features including position, length and area measurement, intensity measurement, and annotation of images.

3. Time–Date Generator

We use a Panasonic model WJ-810 time-date generator to write the date, hour, minute, and second on each image. The size and position in the video frame of the time-date display is adjustable. A stopwatch can also be displayed which shows on each frame hours, minutes, seconds, and milliseconds from the time the timer is started. The millisecond display allows identification of each frame when recorded video is played back frame-by-frame.

4. VCRs, OMDRs, Monitors, and Video Printers

VE-DIC requires monochrome recording and display devices; color is not required. Standard 1/2-inch VHS cassette recorders have too-poor resolution (about 230 TV line horizontal resolution) for high-resolution VE-DIC images.

S-VHS divides the video signal during recording into the luminance (monochrome intensity) and chromance (color information) signals. We use the luminance channel for recording and playback. S-VHS 1/2-inch cassette recorders have about 400–450 TV lines of horizontal resolution. These are the preferred video recorders in our lab at this time, and we use a Panasonic model AG1980P for real-time recording and model AG6750A for time-lapse recording.

Previously we used monochrome 3/4-inch U-Matic tape recorders (Sony VO 5800H). These recorders also have about 450–500 TV lines of horizontal resolution and somewhat lower noise in the tape recordings (Salmon *et al.*, 1989; 1990). The 3/4-inch video cassettes are much larger and more expensive than the 1/2-inch S-VHS cassettes.

For time-lapse recording, we use an Optical Memory Disk Recorder (OMDR). Ours is an older model, a monochrome device with 450 TV line horizontal resolution (Panasonic Model TQ2028). It can store about 16,000 images on one disk at video rates or at any slower speed. It can play back the images at any speed. The OMDR is usually controlled for either recording or playback by a host computer which also controls shutters on the microscope (Shaw *et al.*, 1995). Both Sony and Panasonic offer newer models of OMDRs which use disks which store about 75,000 video images. A PC computer and custom software controls image acquisition and a shutter placed before the filters and polarizer as described by Skibbens *et al.* (1993) and Yeh *et al.* (1995).

OMDRs are very useful for measuring object motility from successive video images either by using a cursor overlaid on the video image and controller by a computer mouse (Walker et al., 1988; Gliksman, 1993) or using semiautomatic tracking algorithms and computer processing (Gelles *et al.*, 1988; Skibbens *et al.*, 1993; Yeh *et al.*, 1995).

We routinely use Panasonic WV-5410 B&W video monitors. Video printers capture a video image and produce a print. An inexpensive printer we have used often in analysis of VE-DIC micrographs is the Sony model UP870MD, which produces a thermal print about 4 × 6 in size for 1 cent per print. Other printers are available from a number of companies which produce either larger thermal prints or photographic prints.

5. Digital Storage and Digital Photography

Methods for acquiring and manipulating VE-DIC images for publication are described by Shaw *et al.* (1995).

V. Test Specimens for Microscope and Video Performance

A. Diatom Test Slide

This slide tests the resolution of the microscope and video system. Diatoms have a silica shell shaped like a pillbox. There are pores in the shell arranged in a lattice specific for each species of diatom. Figure 9A shows a low magnification view of the eight diatoms in the center of the diatom test slide. Number 6, *Pleurosigma angulatum*, has a triangular pore lattice with spacings of 0.61 μm between rows (Fig. 9D). Number 5 is *Surrella gemma*, which has rows of pores where the rows are separated by 0.41 μm (Fig. 9C). Number 8 is *Amphipleura pellucida*, which has horizontal rows of pores. The rows are separated by about 0.25 μm (Fig. 9B) and the pores within rows are separated by about 0.19μm. For illumination with 546 nm green light, a properly aligned high-resolution VE-DIC system should readily resolve the rows of pores in *Amphipleura* (Fig. 9B) and the individual pores should be barely detectable since Eq. (1) predicts a resolution limit of $r = .546/(1.4 + 1.4) = .195$ μm for $NA_{obj} = NA_{cond} = 1.4$. The pores can be seen distinctly by using blue light at 480 nm where $r = .17$ from Eq. (1).

The diatoms are not low-contrast specimens. The pore lattice of each species if resolvable by the microscope optics is usually distinctly visible by eye. The main exception is visualizing the individual pores in the shell of *Amphipleura*. Because they are just at the limit of resolution for the highest NA optics, their contrast is very low and contrast enhancement by the video methods is necessary.

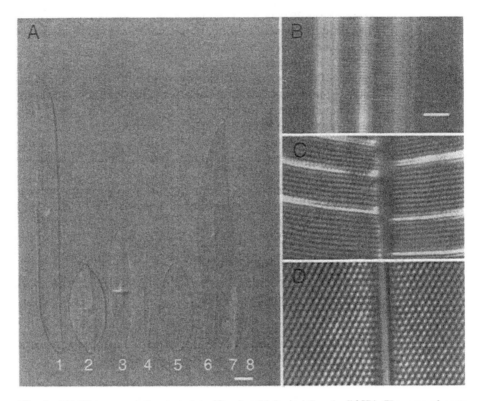

Fig. 9 (A) Diatom resolution test plate (Carolina Biological Supply, B25D). The rows of pores in the silica shell are spaced by about 0.25 μm in *Amphipleura pellucida* (A8, B), 0.41 μm in *Surrella gemma* (A5, C), and 0.62 μm in *Pleurosigma angulatum* (A6, C). Scale bar in A is 10 μm and in B, C, and D is 2.5 μm.

An important use of the diatom test slide is to test the resolution of the video system. The eye usually has much better resolution than the video cameras for the same field of view. At low magnifications, it is common for the pore lattice of diatoms to be distinctly visible by eye but invisible by the video camera because of the poorer resolution of the video system. The diatom test slide is a convenient way to determine how much projection magnification is needed for the video camera so that it does not limit resolution. Magnification beyond this point is deteriorates image quality because intensity drops off as $1/M^2$.

B. Squamous Cheek Cell

Another resolution test for the video camera and the alignment of the DIC optics are squamous epithelial cells scraped from your cheek (Fig. 10). The preparation is made by scraping the inside of your cheek with the tip of a plastic

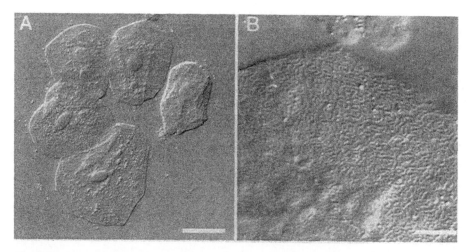

Fig. 10 Human cheek cell test specimen. (A) Low magnification of cheek cell preparation taken with 20X objective. Scale = 20µm. (B) High resolution view of the surface of the cell at top of (A) using 60X/NA = 1.4 Plan Apochromat objective and matching condenser illumination. Ridges at cell surface are often diffraction limited in width. Scale = 5 µm.

pipette or similar tool and spreading the cells and saliva on the center of a clean number 1.5 22 × 22-mm coverslip. The coverslip is quickly inverted on a clean slide and pressed down to spread the cell preparation into a thin layer. The edges of the coverslip are sealed with a thin layer of nail polish. As seen in the low magnification view in Fig. 10A, the cheek cells are large and flat with the nucleus in the center. The upper and lower surfaces have fine ridges, which swirl around much like fingerprints. Many of the ridges are diffraction limited in width and separated by about 0.5–1 µm. If the microscope is aligned properly, these ridges should be distinctly visible by eye and appear in high contrast by VE-DIC using high-resolution optics (Fig. 10B). The cheek cells are also a good test for the optical sectioning capabilities of the microscope since they are about 2–3 µm thick near the periphery.

C. Embedded Skeletal Muscle Thin Section

This specimen (Inoué, 1986) tests both the resolution and video contrast enhancement capabilities of the VE-DIC system and is available from the Microscope Facility (Marine Biological Laboratory, Woods Hole, Massachusetts 02543).

Frog sartorious muscle is fixed and embedded in epon and 100- or 300-nm thin sections cut with a microtome. These sections are placed on the inner surface of a clean number 1.5 coverslip in the center between two black India ink dots which are also on the inner surface of the coverslip. The coverslip is inverted onto a tiny drop of unpolymerized epon in the center of a clean glass slide, and the whole preparation is placed in an oven to polymerize the epon. In this process, the epon spreads into a thin layer and the muscle becomes embedded in a near uniform layer of epon. The thin muscle sections are usually not visible by eye in the DIC microscope (Fig. 11A) because the refractive index of epon

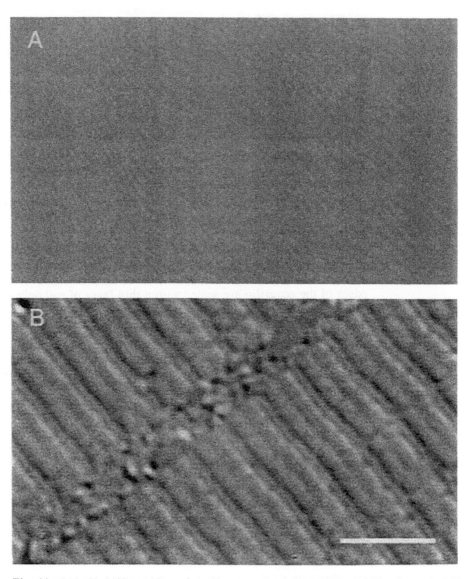

Fig. 11 Embedded 100-nm-thin section of frog sartorius skeletal muscle. (A) View by eye in DIC microscope. (B) VE-DIC image of same field as A. See text for details. Scale = 5 μm.

nearly matches the refractive index of the protein in the muscle. Sometimes, wrinkles in the sections are visible by eye and this helps in finding the sections. The best way to find the sections is to first focus on the inner edge of one of the black dots. Then, using VE-DIC, scan the coverslip in a direction between the dots until the sections are found. When the DIC compensation, orientation of the specimen, and video-enhanced contrast are adjusted properly, you should be able to see clearly the sarcomere structure of skeletal muscle (Fig. 11B). The sarcomeres repeat at about 2.5-μm intervals (dependent on muscle stretch) along the length of the 1-μm-diameter myofibrils. In the middle of each sarcomere are the bipolar thick filaments which appear thicker toward their ends because of the presence of the myosin heads. At the ends of the sarcomere are the z-disks. The actin thin filaments attach to the z-disks and extend into and interdigitate with the myosin thick filaments. There appears to be a gap between the z-disks and the ends of the myosin thick filaments because the actin thin filaments are lower in density than the myosin thick filaments.

References

Allen, R. D., and Allen, N. S. (1983). *J. Microsc.* **129**, 3–17.

Allen, R. D., Allen, N. S., and Travis, J. L. (1981a). *Cell Motil.* **1**, 291–302.

Allen, R. D., David, G. B., and Nomarski, G. (1969). *Z. Wiss. Mikf. Microtech.* **69**, 193–221.

Allen, R. D., Travis, J. L., Allen, N. S., and Yilmaz, H. (1981b). *Cell Motil.* **1**, 275–290.

Bennett, H. S. (1950). *In* "Handbook of Microscopical Technique" (C. E. McClung, ed.), pp. 591–677. Hoeber, New York.

Berg, H. C., and Block, S. M. (1984). *J. Gen. Microbiol.* **130**, 2915–2920.

Block, S. M., Blair, D. F., and Berg, H. C. (1991). *J. Bact.* **173**, 933–936.

Cassimeris, L. U., Pryer, N. K., and Salmon, E. D. (1989). *J. Cell Biol.* **107**, 2223–2231.

CLMIB Staff (1995). *Am. Lab.* **April,** 25–40.

Cole, R. W., Khodjakov, A., Wright, W. H., and Rieder, C. (1995). *J. Micros. Soc. Amer.* **1**, 203–215.

Ellis, G. W. (1985). *J. Cell Biol.* **101**, 83a.

Gelles, J., Schnapp, B. J., and Sheetz, M. P. (1988). *Nature* **331**, 450–453.

Gliksman, N. R., Parsons, S. F., and Salmon, E. D. (1992). *J. Cell Biol.* **5**, 1271–1276.

Inoué, S. (1981). *J. Cell Biol.* **89**, 346–356.

Inoué, S. (1986). "Video Microscopy." Plenum, New York.

Inoué, S. (1989). *Methods in Cell Biol.* **30**, 85–112.

Inoué, S., and Oldenbourg, R. (1993). *In* "Handbook of Optics" (Optical Society of America, ed.), 2nd ed. McGraw Hill, New York.

Inoué, S., and Salmon, E. D. (1995). *Mol. Biol. Cell* **6**, 1619–1640.

Inoué, S., and Spring, K. (1997). "Video Microscopy," 2nd ed. Plenum Press, New York.

Keller, H. E. (1995). *In* "Handbook of Biological Confocal Microscopy" (J. B. Pawley, ed.), pp. 111–126. Plenum, New York.

Kuo, S. C., and Sheetz, M. P. (1992). *Trends Cell Biol.* **2**, 116–118.

Kuo, S. C., and Sheetz, M. P. (1993). *Science* **260**, 232–234.

Pluta, M. (1989) "Advanced Light Microscopy; Vol. II: Specialized Methods." Elsevier, Amsterdam.

Salmon, E. D. (1995). *Trends Cell Biol.* **5**, 154–158.

Salmon, E. D., Walker, R. A., and Pryer, N. K. (1989). *BioTechniques* **7**, 624–633.

Samuel, R. E., Salmon, E. D., and Briehl, R. W. (1990). *Nature* **345**, 833–835.

Schnapp, B. J. (1986). *Meth. Enzymol.* **134**, 561–573.

Scholey, J. M. (ed.) (1993). *Methods Cell Biol.* **39**, 1–292.

Schotten, D. (1993). "Electronic Light Microscopy." Wiley-Liss, New York.

Shaw, S. L., Salmon, E. D., and Quatrano, R. S. (1995). *BioTechniques* **19,** 946–955.

Skibbens, R. V, Skeen, V. P., and Salmon, E. D. (1993). *J. Cell Biol.* **122,** 859–875.

Smith, F. H. (1955). *Research* (*London*) **8,** 385–395.

Spencer, M. (1982). "Fundamentals of Light Microscopy." Cambridge University Press, London.

Svoboda, K., and Block, S. M. (1994). *Cell* **77,** 773–784.

Svoboda, K., Schmidt, C. F., Schnapp, B. J., and Block, S. M. (1993). *Nature* **365,** 721–727.

Thomas, C., DeVries, P., Hardin, J., and White, J. (1996). *Science* **273,** 603–607.

Walker, R. A., Gliksman, N. R., and Salmon, E. D. (1990). *In* "Optical Microscopy for Biology" (B. Herman and K. Jacobson, eds.), pp. 395–407. Wiley-Liss, NY.

Walker, R. A., Inoué, S., and Salmon, E. D. (1989). *J. Cell Biol.* **108,** 931–937.

Walker, R. A., O'Brien, E. T., Pryer, N. K., Soboeiro, M., Voter, W. A., Erickson, H. P., and Salmon, E. D. (1988). *J. Cell Biol.* **107,** 1437–1448.

Waters, J. C., Mitchison, T. J., Rieder, C. L., and Salmon, E. D. (1996). *Mol. Biol. Cell* **7,** 1547–1558.

Waterman-Storer, C. M., Gregory, J., Parsons, S. F., and Salmon, E. D. (1995). *J. Cell Biol.* **5,** 1161–1169.

Weiss, D. G., Maile, W., and Wick, R. A. (1989). *In* "Light Microscopy in Biology: A Practical Approach" (A. J. Lacey, ed.), pp. 221–278.

White, J. (1996). *Science*

Yeh, E., Skibbens, R. V., Cheng, J. W., Salmon, E. D., and Bloom, K. (1995). *J. Cell Biol.* **130,** 687–700.

CHAPTER 10

A High-Resolution Multimode Digital Microscope System

**E. D. Salmon, Sidney L. Shaw, Jennifer Waters,
Clare M. Waterman-Storer, Paul S. Maddox, Elaine Yeh,
and Kerry Bloom**

Department of Biology, University of North Carolina
Chapel Hill, North Carolina 27599

I. Introduction

In this chapter we describe the development of a high-resolution, multimode digital imaging system based on a wide-field epifluorescent and transmitted light microscope and a cooled CCD camera. Taylor and colleagues (Taylor *et al.,* 1992; Farkas *et al.,* 1993) have reviewed the advantages of using multiple optical modes to obtain quantitative information about cellular processes and described instrumentation they have developed for multimode digital imaging. The instrument described here is somewhat specialized for our microtubule and mitosis studies, but it is also applicable to a variety of problems in cellular imaging including tracking proteins fused to the green fluorescent protein (GFP) in live cells (Cubitt *et al.,* 1995; Olson *et al.,* 1995; Heim and Tsien, 1996). For example, the instrument has been valuable for correlating the assembly dynamics of individual cytoplasmic microtubules (labeled by conjugating X-rhodamine to tubulin) with the dynamics of membranes of the endoplasmic reticulum (ER, labeled with $DiOC_6$) and the dynamics of the cell cortex [by differential interference contrast (DIC)] in migrating vertebrate epithelial cells (Waterman-Storer and Salmon, 1997). The instrument has also been important in the analysis of mitotic mutants in the powerful yeast genetic system *Saccharomyces cerevisiae.* Yeast cells are a major challenge for high-resolution imaging of nuclear or microtubule dynamics because the preanaphase nucleus is only about 2 μm wide in a cell about 6 μm wide. We have developed methods for visualizing nuclear and spindle dynamics during the cell cycle using high-resolution digitally enhanced DIC (DE-DIC) imaging (Yeh *et al.,* 1995; Yang *et al.,* 1997). Using genetic and molecular techniques, Bloom and coworkers (Shaw *et al.,* 1997a,b) have been able to label the cytoplasmic astral microtubules in dividing yeast cells by expression of cytoplasmic dynein fused to GFP. Overlays of GFP and DIC images of dividing cells has provided the opportunity to see for the first time the dynamics of cytoplasmic microtubules in live yeast cells and how these dynamics and microtubule interactions with the cell cortex change with mitotic cell cycle events in wild-type and in mutant strains (Shaw *et al.,* 1997a,b).

Our high-resolution multimode digital imaging system is shown in Fig. 1 and diagrammed in Fig. 2. The legend to Fig. 2 provides model numbers and sources of the key components of the instrument. There are three main parts to the system: a Nikon FXA microscope, a Hamamatsu C4880 cooled CCD camera, and a MetaMorph digital imaging system. First we will consider our design criteria for the instrument, then consider separately the major features of the

Fig. 1 Photograph of the multimode digital imaging system including the Nikon FXA upright microscope sitting on an air table (Newport Corp. Irvine CA, VW-3660 with 4-inch-high tabletop). Images are captured with a Hamamatsu C4880 cooled CCD camera. Image acquisition, storage, processing, analysis, and display are controlled by MetaMorph Digital Imaging software in a Pentium-based computer.

microscope components, the cooled CCD camera, and the MetaMorph digital imaging system. The reader is referred to other sources for general aspects of: microscope optics (Spencer, 1982; Inoué and Oldenbourg, 1993; Keller, 1995); DIC microscopy (Salmon, 1998); fluorescence microscopy (Taylor and Salmon, 1989); video cameras, slow-scan cameras, and video microscopy (Inoué, 1986; Aikens *et al.*, 1989; Shotten, 1995; CLMIB staff, 1995; Inoué and Spring, 1997); and 3D imaging methods (Hiroka *et al.*, 1991; Carrington *et al.*, 1995; Pawley, 1995).

II. Design Criteria

A. Fluorescence Considerations

When we began building our instrument 4 years ago, our primary objective was to obtain quantitative time-lapse records of spindle microtubule dynamics and chromosome movements for mitosis in live tissue culture cells and in *in vitro* assembled spindles in *Xenopus* egg extracts. Fluorescence optical components

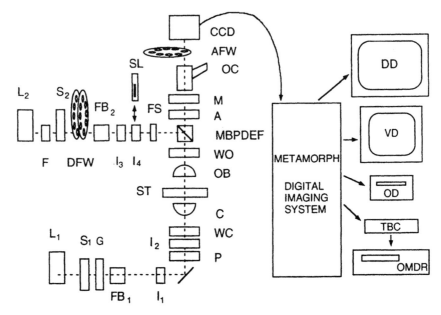

NIKON FXA LIGHT MICROSCOPE

Fig. 2 Component parts are: L1, 100-watt quartz halogen lamp; S1, Uniblitz shutter (No. 225L2A1Z523398, Vincent Associates, Rochester, NY); G, ground glass diffuser; FB1, manual filter changers including KG4 heat cut and green interference filters; I_1, field iris diaphragm; I_2, condenser iris diaphragm; P and A, high transmission Nikon polaroid polarizer and removable analyzer; WC, and WO, Wollaston prisms for condenser and objective; C, Nikon NA = 1.4 CON A, Achr-Apl condenser; ST, rotatable stage with focus position controlled by z-axis stepper motor (Mac2000, Ludl Electronic Products, LTD., Hawthorne, NY); OB, 20X/NA = .75 or 60X/NA = 1.4 Nikon objectives; MBPDEF, epifilter block with multiple bandpass dichromatic mirror and emission filter (Nos. 83101 and 83100, Chroma Technology Corp., Brattleboro, VT); L2, 100-watt HBO mercury lamp; F, KG4 heat cut filter; S2 shutter; DFW, dual 8-position filter wheel (Metaltek, Raleigh, NC), one wheel containing neutral density filters (No. FNQ011, Melles Griot, Irvine, CA) and the other a series of narrow bandpass excitation filters (Nos. 83360, 83490, and 83570, Chroma Technology Corp.); FB2, manual filter changer; I_3, epicondenser iris diaphragm; I_4, epifield diaphragm slider; SL, slit (25 μm width, No. 04PAS004, Melles Griot, Irvine, CA) cemented to Nikon pinhole slider (No. 84060) for photoactivation; FS, filter slider; M, optivar magnification changer, 1–2×; OC, oculars; AFW, 4-position filter wheel (Biopoint filter wheel No. 99B100, Ludl Electronic Products, LTD., Hawthorne, NY) with one position containing a Nikon high transmission polarizer for the analyzer and another position an optical glass flat of the same thickness as the analyzer); CCD, cooled CCD camera (No. C4880, Hamamatsu Photonics, Bridgewater, NJ); DD, 1024 × 768 pixel, 20-inch digital graphics display monitor (No. 2082, Viewsonic); VD, RGB video display monitor (No. PVM1271Q, Sony); MetaMorph digital imaging system (Universal Imaging Corp., West Chester, PA) with a 166-MHz Pentium computer having both EISA and PCI bus, 128 MByte RAM memory, Imaging Technology AFG digital and video image processing card, Hamamatsu C4880 CCD controller card, Matrox MGA Melenimum 4 Meg Ram graphics display card, SVGA graphics display to S-VHS converter card (Hyperconverter, PC video conversion Corp., San Jose, CA), 1.4-MByte floppy drive, 2 GByte hard drive, 1 Gbyte Iomega Jaz drive, Hewlett Packard SureStore CD Writer 4020I, ethernet card, and parallel port cards for controlling shutter S1 and driving laser printer; 8 serial port card for controlling MetalTek filter wheel, Ludl z-axis stepper, CCD camera, and OMDR; OD, Pinnacle Micro1.3 GByte optical drive; TBC, time-base corrector; OMDR, Panasonic 2028 Optical Memory Disk Recorder. Video is also recorded on a Panasonic AG-1980P S-VHS recorder and a Panasonic AG-6750A time-lapse S-VHS recorder. Modified from Salmon *et al.* (1994).

were chosen, in part, based on the fluorophores which were available for labeling microtubules, chromosomes, and other microtubule- or spindle-associated components. Microtubules could be fluorescently labeled along their lengths by incorporating X-rhodamine-labeled tubulin into the cytoplasmic tubulin pool (Fig. 3; Salmon *et al.*, 1994; Murray *et al.*, 1996). In addition, we needed to use fluorescence photoactivation methods to produce local marks on spindle microtubules in order to study the dynamics of their assembly (Mitchison and Salmon, 1993; Waters *et al.*, 1996). This is accomplished by the addition of a caged-fluorescein-labeled tubulin to the cytoplasmic pool and fluorescence photoactivation with a 360-nm microbeam as described by Mitchison and co-workers (Mitchison, 1989; Sawin and Mitchison, 1991; Sawin *et al.*, 1992). In extracts and some living cells, chromosomes can be vitally stained with the DNA intercalating dyes DAPI or Hoescht

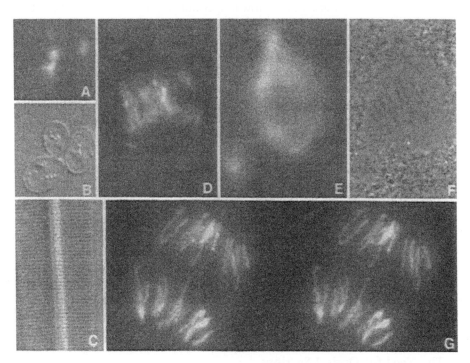

Fig. 3 Views of a living, dividing yeast, *Saccharomyces cerevisiae*, by (A) fluorescence of GFP protein bound to nuclear histones and (B) DIC.* (C) Image in DIC of the 0.24-μm spacing between rows of surface pores of the diatom *Amphipleura* illuminated with green light.* Images of a spindle undergoing *in vitro* anaphase in *Xenopus* cytoplasmic egg extracts: (D) DAPI-stained chromosomes, (E) X-rhodamine-tubulin-labeled spindle and aster microtubules, and (F) phase contrast (Nikon 20X/NA = 0.75 Fluar Phase 3 objective). See Murray *et al.* (1996) for details. Stereo-pair images (F) of DAPI-stained chromosomes generated from a stack of 0.5-μm optical sections through a *Xenopus* spindle fixed in the extracts in mid-anaphase.* (*, Nikon 60X/NA = 1.4 Plan Apo DIC objective and NA = 1.4 condenser illumination for DIC.) With permission from Salmon *et al.* (1994).

33342 (Sawin and Mitchison, 1991; Murray *et al.*, 1996). Thus, in fluorescence modes, we needed to be able to obtain images in a "red channel" for X-rhodamine microtubules, a "green channel" for photoactivated fluorescein marks, and a "blue channel" for chromosomes.

B. Live Cell Considerations

Photobleaching and photodamage are a major concern during time-lapse imaging of fluorescently tagged molecules in live cells or in *in vitro* extracts. Minimizing these problems requires that the specimen be illuminated by light shuttered between camera exposures, that the imaging optics have high transmission efficiencies, that the camera detectors have high quantum efficiency at the imaging wavelengths, and that light can be integrated on the detector to reduce illumination intensity and allow longer imaging sessions without photodamage. This later point is true not only for epifluorescence, but also for transmitted light (phase-contrast or DIC) imaging. In our studies, the detector also needed a high dynamic range (12 bits = 4096 gray levels) since we wanted to be able to quantitate the fluorescence of a single microtubule or the whole mitotic spindle, which could have 1000 or more microtubules. In addition, the red, green, and blue images needed to be in focus at the same position on the detector so that the fluorophores within the specimen could be accurately correlated.

C. Phase Contrast and DIC Imaging

Phase contrast or DIC transmitted light images allow the localization of fluorescently tagged molecules in the cells to be correlated with overall cell structure or the positions of the chromosomes, nucleus, cell cortex, and so on (Fig. 3). DIC has much better resolution than phase contrast (Inoué and Oldenbourg, 1995) and excellent optical sectioning abilities (Inoué, 1989), but unlike phase contrast, an analyzer must be placed in the light path after the objective. Conventional analyzers typically transmit 35% or less of incident unpolarized light, and thus for sensitive fluorescence methods, the analyzer must be removed from the light path for fluorescence imaging and inserted for DIC imaging.

D. The Need for Optical Sections

In some of our studies, 3D images from "z-axis" series of images or 4D microscopy where a z-axis image series is recorded at each time point in a time-lapse series have been required. For example, proteins localized to the kinetochore of one chromosome may be 5 μm out of focus with the kinetochore of another chromosome in the same cell. In order to see all of the kinetochores in the cell, several focal planes are acquired along the z-axis and projected into a single 2D representation of the 3D data.

III. Microscope Design

The Nikon FXA upright microscope (Figs. 1 and 2) can produce high-resolution images for both epifluorescent and transmitted light illumination without switching the objective or moving the specimen. The functional aspects of the microscope set-up can be considered in terms of the imaging optics, the epi-illumination optics, the transillumination optics, the focus control, and the vibration isolation table.

A. Imaging Optics

Light collected by the objective passes through the body tube, the filter cube holder of the epifluorescence illuminator, a magnification changer, then either straight to the camera or by a prism into the binocular tube for viewing by eye. The cooled CCD camera is mounted at the primary image plane of the microscope.

1. Objectives and Matching Condenser

In epifluorescence microscopy, the intensity of the image varies as the fourth power of objective numerical aperture (NA) and inversely with the square of the magnification (M) between the objective and the detector (Taylor and Salmon, 1989):

$$I_{sp} = I_{il}NA_{obj}^4/M^2, \tag{1}$$

where I_{il} is the intensity of the epi-illumination light entering the objective aperture. For high-resolution work, we typically use the highest NA possible (NA = 1.4) to maximize fluorescence intensity. Most of our specimens are less than 10 μm thick, so that lower NA objectives or water immersion objectives which have better performance for thicker specimens (Keller, 1995) have not been necessary. For high-resolution work, we use a Nikon 60X/NA = 1.4 Plan-Apochromatic objective, either the one designed for phase-contrast or for DIC. The specimen is illuminated in the transmitted light mode through a Nikon NA = 1.4 condenser containing a turret for selecting matching phase annulus or DIC Nomarski prism and a polarizer for DIC. For the highest-resolution work, immersion oil must be used with the condenser as well as the objective. For lower-resolution and magnification, we have also found the Nikon Fluar 20X/NA = 0.6 and 40X/NA = 1.3 to be very useful lenses. The Fluar lenses have better transmission efficiency at 360 nm in comparison to the PlanApochromat lenses so they are better for photoactivation experiments.

2. Magnification

The magnification changer is an important feature of the imaging path because the camera is fixed in position at the primary image plane above the objective and different specimens often require different magnifications. It provides 1.0, 1.25, 1.5, and 2.0× selectable intermediate magnifications to the camera detector. The body tube above the objective provides additional magnification of 1.25×. Therefore, for the 60X objective, the final possible magnifications to the detector are 75, 93.75, 112.5, and 150×.

How much magnification is necessary so that resolution is not limited by the pixel size of the cooled CCD detector? As discussed below, the Hamamatsu cooled CCD detector uses a chip with a pixel size of 12 by 12 μm. To resolve two adjacent specimen points as distinct points, their images at the detector must be *at least two* pixel elements apart, or 24 μm (Fig. 4). With the specimen points separated by three pixel elements, or 36 μm, the problem of aliasing (Inoue and Spring, 1997) is avoided. The diffraction limited lateral resolution, r, of two adjacent points in the specimen for the Nikon 60X/NA = 1.4 objective is about:

$$r = 0.61 \ \lambda/NA_{obj} = 0.24 \ \mu m \tag{2}$$

for λ = 550 nm green light. The magnifications needed so that 0.24 μm in the specimen is magnified to 24 or 36 μm on the detector is 24/.24 = 100× or 36/.24 = 150×. Thus, for resolution limited by the objective and not the detector we must use at least the 1.5× setting of the magnification changer and the 2× setting for the best resolution. Magnifications less than 100× produces brighter images as predicted by Eq. (1), but resolution in these images is limited by the resolution of the detector.

3. Determining Focus

We have found that an effective way to focus the specimen image on the camera is with dark-adapted eyes. Adjustments are available on the FXA microscope so

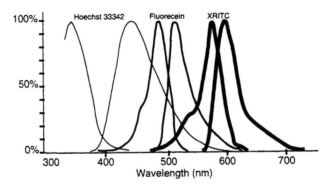

Fig. 4 Excitation and emission spectra for the blue, green, and red fluorophores we commonly use. The excitation and emission spectra for S65TGFP is similar to the fluorescein spectrum and we have had great success using the multiband pass filter (Fig. 5) and the fluorescein excitation filter.

that the oculars can be made parfocal with the camera. During an experiment, the specimen is viewed by eye and quickly brought into focus. Then the light is switched to the camera for recording. The Hamamatsu C4880 camera has a fast scan mode which can produce displayed video images as fast as 10 Hz. We have also used this camera feature for achieving focus, but for most situations, we have found that achieving focus with the dark-adapted eye is quicker than switching the camera's mode of operation back and forth between the fast scan mode of focusing and the slow scan mode for image recording.

4. Phase Contrast

When phase images are recorded along with fluorescent images, we use the phase objective and the matching phase annulus on the condenser turret. The DIC objective Nomarski prism and the DIC analyzers are removed from the imaging path. The phase ring in the back aperture of the objective produces some loss of fluorescent image brightness, but it has not proven significant.

5. DIC and the Analyzer Filter Wheel

For DIC, a Normarski prism is inserted just above the objective and the condenser turret is switched to the Nomarski prism that matches the objective. The Nikon polarizer beneath the condenser uses high transmission (35%) Polaroid materials mounted between optical glass flats.

For focusing by eye, a high-transmission DIC analyzer is slid temporarily into the light path just above the objective Normarski prism. For the best DIC image contrast, the bias retardation of the objective Nomarski prism is typically adjusted to about 1/15th to 1/10th the wavelength of light (see Salmon *et al.,* 1998) by the following procedure. First, the prism is adjusted to extinguish the background light. Then it is translated further brightening the background and making one side of the structure of interest maximally dark. At this position, this side of the structure will be darker than the background while the other side of the structure will be brighter than the background, producing the "shadowed" appearance of the DIC image (Fig. 3).

For DIC image recording, the focusing analyzer is removed from the light path. A filter wheel just beneath the camera (Fig. 2) is used to insert a high-transmission DIC analyzer into the light path. For fluorescence imaging, the filter wheel is rotated to an adjacent position which contains an optical glass flat of equivalent optical thickness to the analyzer. The Ludl filter wheel was chosen in part because it was sufficiently thin to fit between the top of the microscope and the camera without changing the position of focus of the primary image on the detector. It was also selected because it can change filter positions without inducing noticeable vibration of the objective. This filter wheel is relatively slow requiring about 1 sec to change filter positions; a time which has not proven a problem for our applications. Finally, the filter wheel is rotated by a stepping motor. There are 4000 steps between adjacent filter positions and the ability of the filter wheel to repeatedly bring the analyzer back to the exact same position (crossed with the polarizer) is excellent.

B. Epi-illumination Optics

There are a number of important features of the epi-illumination pathway for our multimode microscope including the mercury lamp housing, dual filter wheel, condenser, condenser diaphragm, field stop (diaphragm or slit), optical flat slider, and filter cubes.

1. Mercury Lamp and Housing

The lamp housing contains a mirror in the back so that both the direct light from the arc and the reflected light from the mirror pass through the lamp collector lens and become projected into the objective aperture by the epi-illumination optics. The mirror image helps obtain a more intense and even illumination of the objective aperture.

2. Heat Reflection Filter

This filter prevents harmful infrared light from heating the shutters, filters, or the specimen. Many conventional fluorescence filter sets do not block infrared light, which shows up as a hot spot on cameras though it is invisible to the eye.

3. Köhler Illumination

Nikon has a mirror within an objective mount which can be used for focusing and aligning the images of the arc lamp centered and in focus at the position of the objective back aperture. This is the procedure for Köhler illumination, the alignment that gives the most even illumination of the specimen. See the microscope manual for detailed alignment procedures.

4. Dual Filter Wheel and Epishutter

There are two eight-position filter wheels mounted near the lamp collector lens where the illumination light is out of focus. The wheels hold 1-inch diameter filters. The filter wheel assembly includes a 1-inch electronic shutter. The sides of the blades facing the lamp are silvered to reflect the light when the shutter is closed and prevent overheating of the shutter mechanism. The filter wheel position and the shutter are under computer control.

The first wheel has a series of neutral density filters including 0, 0.3, 0.5, 1.0, 1.3, 2.0, 2.3, and 3.0 O.D. The second filter wheel has one open position and the rest contain a series of bandpass filters. The open position is used with conventional filter cubes. Three of the bandpass filters are excitation filters used to excite the blue, green, and red fluorophores (Fig. 4) through a multiple bandpass dichromatic filter cube (Fig. 5). These have central wavelengths of 360, 480, and 570 nm. Other filter positions contain filters for 340, 408, and 546 nm.

5. Condenser Diaphragm

The epi-illumination condenser diaphragm is often useful for making manual "on-the-fly" adjustments of illumination intensity. Closing this diaphragm can reduce illumination intensity by about 1 O.D.

6. Field Stops

We use two field stops. One is a diaphragm which is used during fluorescence imaging to restrict the field of illumination to only the region of interest. As this diaphragm is closed down, the amount of fluorescent scattered light from the specimen decreases, and this can significantly enhance the contrast in fluorescence images. Thus, we take care to adjust this diaphragm for each specimen.

The other field stop which can be exchanged easily for the diaphragm is a strip of slits and spots which we use for fluorescence photoactivation. There are translation screws for centering the position of the field stops.

7. Correction for Longitudinal Chromatic Aberration

Between the field stop and the filter cube holder is a slider for 1.5-cm filters. For fluorescence imaging we use an open position. For photoactivation at 360 nm, we insert a 3-mm-thick optical glass flat to make the 360-nm excitation light image of the field spot or slit parfocal with the green or red light images we use for focusing the slit on the specimen as described in the photoactivation section below.

8. Filter Cube Turret

The Nikon FXA epi-illuminator has four positions for filter cubes. These filters must be rotated into position by hand. We usually leave one of these positions open for the highest-resolution DIC imaging when fluorescence images are not required. We use conventional filter cubes, or specialized cubes, in the other positions.

We initially tried to use a motor-controlled filter turret to select different filters. However, when different filter cubes were used, they had slightly different optical properties (tilted optical elements, nonparallel surfaces, etc.) so that the different-colored images were deflected to different positions on the CCD detector. Their alignment was often not repeatable each time a particular filter was rotated into place. Taylor and colleagues have had better repeatable position

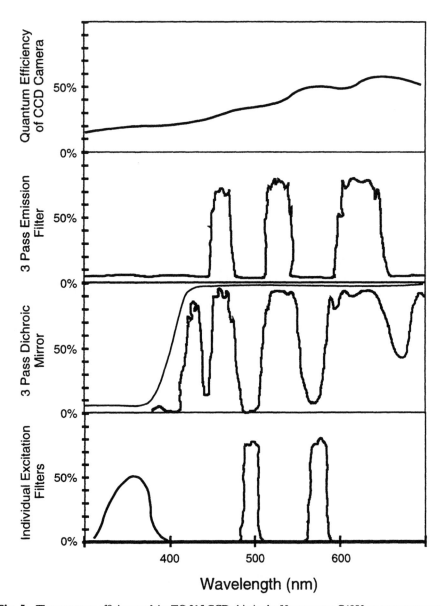

Fig. 5   The quantum efficiency of the TC-215 CCD chip in the Hamamatsu C4880 camera compared to the spectra of the Chroma excitation filters, multiple bandpass dichromatic mirror, and multiple bandpass emission filter. For the filters, the *y*-axis is percentage transmission. The thin line in the curves for the three-pass dichromatic mirror is produced by a coating on the backside for reflecting 360-nm excitation light.

accuracy using a linear filter slider (Taylor *et al.,* 1992; Farkus *et al.,* 1993); however, this does not eliminate the problem of image registration on the detector.

9. Multiple Bandpass Dichromatic Mirror and Emission Filter

This problem was solved by using a multiple bandpass dichromatic mirror and emission filter set manufactured by Chroma similar to one described by Sawin *et al.* (1992). In combination with the appropriate excitation filters, these have excitation and emission spectra similar to DAPI or Hoescht DNA intercalating dyes (360 ex/450 em), fluorescein and GFP (480 ex/510–530 em), and X-rhodamine (570 ex/620–650 em) (Fig. 5). This has been an important filter cube for our work because the image for the three different fluorophores project to the same focus on the detector. This is because they all use the same dichromatic mirror and emission filter.

Another important aspect of the Chroma filter set is the high transmission efficiency within each bandpass (Fig. 5). This is important for maximizing sensitivity and minimizing the effects of photobleaching and photodamage. The same principle applies to the selection of any filter cube for live cell experiments.

C. Transillumination Optics

1. Quarts-Halogen Lamp and Housing

The 100-watt quartz-halogen lamp is mounted in a housing that contains a rear mirror. The combination of the primary filament image with the mirror filament image produces more even illumination of the condenser aperture than occurs for the primary image of the filament alone.

2. Heat-Reflecting Filter

As described for the epi-illumination pathway, it is important to remove infrared light from the illumination beam. A 1.5-inch-diameter heat-reflecting filter is inserted in front of the lamp collector lens.

3. Thermal Isolation

We initially had considerable trouble with the stage drifting downward over time. Stage drift is a significant problem in time-lapse recording because the specimen continually goes out of focus. Stage drift was reduced substantially for the FXA stand by isolating the heat of the lamp from the base of the stand using a ceramic connector between the lamp housing and the base of the microscope stand.

4. Köhler Illumination and the Ground-Glass Diffuser

To achieve the highest resolution in DIC it is essential that the objective aperture be uniformly illuminated (Inoué, 1989; Inoué and Oldenbourg, 1993) and the procedures outlined below are designed to achieve that goal:

1. Initially, the slide is oiled to the condenser (without air bubbles!) and viewed with a low-power (10X or 20X) objective.

2. The condenser focus and horizontal position screws are adjusted so that the image of the field diaphragm in the transillumination light path is in focus and centered on the specimen when the specimen is in focus to the eye.

3. The objective is moved away and a small drop of oil (without air bubbles) is placed on the coverslip.

4. The 60X objective is rotated into place. The specimen is focused and the centration of focus of the field diaphragm readjusted. For best contrast, the field diaphragm should be closed till it just surrounds the specimen of interest.

5. The telescope (Bertrand lens) in the magnification changer is then used to focus on the objective back aperture. In that position the condenser diaphragm plane is also in focus. The condenser diaphragm is opened all the way and the ground-glass diffuser is removed from the light path and the position controls for the bulb and the mirror are adjusted so that the primary and mirror filament images sit side by side in focus at the condenser diaphragm plane. The condenser diaphragm is adjusted to just be visible at the periphery of the objective aperture.

6. The ground-glass diffuser is then placed in the light path to produce more uniform illumination of the 1.4 NA condenser aperture.

5. Bandpass Filters

The base of the FXA stand contains a filter changer. For live cell illumination we use a wide-band (40 nm) green 546-nm bandpass filter.

6. Transshutter

Transillumination of the specimen is controlled by a 1-inch Vincent electronic shutter which is under computer control. The shutter sits on top of the base of the microscope underneath the condenser carrier.

D. Focus Control

Z-axis focus is controlled by a Ludl stepping motor attachment to the fine focus knob. Coarse focus can be done by hand using the inner focus ring on the microscope. The stepping motor has a manual focus knob with three sensitivity positions: coarse, medium, and fine. The sensitivity is selected with a toggle switch on the knob box. We use the coarse setting for focusing at low magnification and the fine setting for focusing at the highest resolution.

During image recording, focus can be controlled by the computer system to yield a stack of images at equal z-axis steps through the specimen (0.1-μm steps are possible but we typically use either 0.5- or 1-μm intervals).

We also use the stepping motor to correct for the shift in focus as the stage drifts. Often stage drift occurs at a uniform rate in one direction. During time-lapse recordings, focus is stepped at each recording interval to compensate for the stage drift.

E. Vibration Isolation

To prevent vibration from destroying image fine structural detail, the FXA microscope is mounted on a Newport vibration isolation table (Fig. 1). An attractive feature of this table is that it provides arm rests and shelves for computer monitors and other equipment which are supported independently of the air table.

IV. Cooled CCD Camera

A. Camera Features

Our current camera is a Hamamatsu C4880 dual mode cooled charge couple device (CCD). The CCD detector is a rectilinear array of photodetectors (photodiodes) deposited on a silicon substrate. When the shutter is opened, incoming photons at each photodiode are converted to electrons which are accumulated in potential wells created by an electrode array on the surface of the chip. Thermal motion also knocks electrons into the potential well, and the C4880 is cooled by a Peltier device to about $-30°C$ to reduce "thermal noise." At the end of a light exposure, the shutter is closed. The C4880 camera uses a Texas Instrument TC-215 CCD chip which has a 1024×1024 pixel array where each pixel element is 12×12 μm in size. The well capacity is 60,000 electrons; beyond this number, the electrons will flow into adjacent wells. CCD detectors are graded for quality and defects. A grade-1 chip like the one in the C4880 camera has few defective pixel elements and few "hot pixels" (see below).

Both photon and thermally accumulated electrons in the rows and columns of the CCD array are "readout" by a combination of parallel and serial transfers of charge to a readout amplifier and analog-to-digital (AD) converter. The C4880 uses two readout amplifiers to increase readout speed. "Readout noise" depends on rate; the slower the better, and the C4880 uses a 500-KHz rate to readout pixel values in its precision "slow scan" mode. The C4880 also has a "fast scan" mode (7 frames/sec) which is much (about 5$\times$) noisier, but useful for imaging brighter light sources and for focusing.

The C4880 is not a conventional video-rate camera equipped with a Peltier cooler. Video-rate CCD chips readout at 30 frames/sec and the CCD chips are constructed differently from full-frame, slow-scan CCD chips like the TC-215 in the C4880 camera (CLMIB staff, 1995; Shaw *et al.,* 1995; Inoué and Spring, 1997).

We use the slow-scan mode of the C4880 for most of our applications and several significant features of its operation are described in more depth next. Details on the construction and properties of CCD chip cameras, including the various ways of obtaining images from slow-scan CCDs, are available elsewhere (Aikens *et al.,* 1989; Inoué and Spring, 1997).

B. Quantum Efficiency

Quantum efficiency measures the percentage of incident photon which yield measurable photoelectrons in a pixel well. Quantum efficiency depends on wavelength as shown in Figs. 5 and 6D. The TC-215 chip has a peak quantum efficiency of 55% at 650 nm, the emission wavelength for the X-rhodamine fluorescence for our labeled microtubules in living cells. At 510 nm, the emission wavelengths for fluorescein and GFP, the quantum efficiency is about 35%. In general, CCD chips are more quantum efficient toward red and infrared wavelengths and significantly less toward the blue and violet wavelengths (Fig. 6D). Our chip has been coated with a fluorophore which enhances the sensitivity in the blue-to-UV region of the spectrum (Fig. 6D).

The quantum efficiency of the TC-215 CCD is 1.5 to 2 times better than most slow-scan, full-frame read-out CCDs in the visible wavelengths and much better than conventional interline video-rate CCD cameras (Fig. 6D). In full-frame read-out CCDs arrays such as the T1215, the potential wells are illuminated through the metal electrode array, which reduces the quantum efficiency. Back-thinned, rear-illuminated CCD chips are much harder to manufacture, but recently they have become available for almost reasonable prices. These can have quantum efficiencies as high as 80% (Fig. 6D).

Video-rate CCD cameras which allow "on-chip integration" are very economical ($800–2000) and useful for bright fluorescence applications like immunofluorescence as reviewed by Shaw *et al.* (1995). Compared to the full-frame read-out chips, the quantum efficiency of conventional interline transfer chips is poor (15% peak) in part because half of the chip is masked (Fig. 6D). Sony has developed more sensitive chips by fabricating microlenses over each pixel element which collect light hitting the masked regions and direct it into a pixel element. This technique was developed for the "hyperhad" or HAD video-rate CCD chips by Sony. The quantum efficiency is still only 30% for most of these chips, and the readout noise is much higher than slow-scan, cooled CCD chips. Recently, a new Sony HAD chip has appeared which exhibits 55% quantum efficiency near 510 nm (HAD ICX-061 in Fig. 6D), the peak emission wavelength

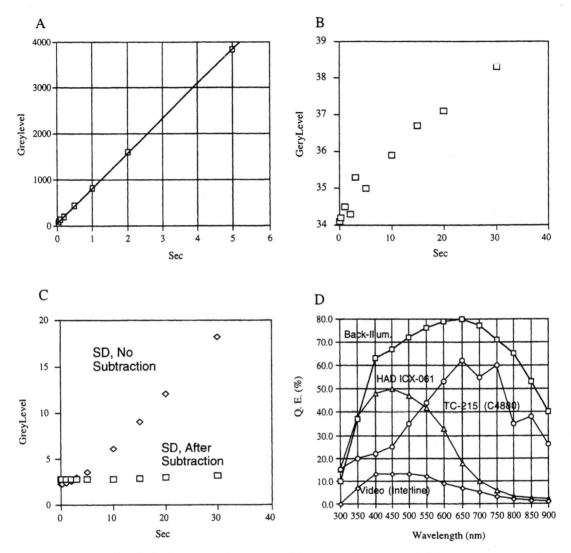

Fig. 6 Performance of the C4880 cooled camera with TC-215 CCD. (A) Measured linearity where 1 graylevel count equals 5 electrons read from the CCD. Light intensity was held constant and integration time (sec) varied. (B) Measured average values of "dark" graylevel verses integration time (sec) when the camera shutter is closed. The offset in converting electrons to graylevel is the intercept at zero time. (C) Measured values for camera noise obtained from the standard deviations (SD) of the dark signal verses integration time (open diamonds) and the SD after pairs of images taken for the same time were subtracted from each other to remove the "hot pixel" contributions (open squares). (D) Comparison of the quantum efficiency of the TC-215 CCD to other types of detectors. Data taken from manufactures specifications.

for GFP fluorescence imaging. For longer wavelength probes, these chips are much less sensitive than the conventional full-frame transfer chips. Because it is an interline chip, it supports video or near video readout rates and does not need a shutter. The pixel detectors are about 9 μm in size with a well capacity of 15,000 electrons. The readout and thermal noise is remarkably low and comparable to the high-quality slow-scan full-frame transfer devices. Hence, the gap between slow-scan "scientific-grade" cameras and more commercial grade video-rate cameras continues to narrow.

C. Readout

The C4880 has two readout modes: slow scan and fast scan. In either mode, a shutter is used to control light exposure to the CCD chip. We mainly use the slow-scan mode because it has much better signal-to-noise and higher dynamic range (4096 versus 1024 gray levels). Potentially, the fast-scan mode is useful for focusing, since it can run as fast as 10 frames/sec. But, in practice, we have found it easier to focus using the dark-adapted eye as described earlier. The following describes the operation of the camera in the slow-scan mode.

1. A/D Conversion

The number of accumulated electrons generates a voltage which is scaled and converted into a digital signal. The analog-to-digital converter is 12 bits or 4096 gray levels and the pixel value is transferred and stored in the computer as a 16-bit or 2-byte (1 byte = 8 bits) value. There are three gain factors (GF) selectable with this camera. One gray level equals 25, 5, and 1 accumulated electrons at low, high, and super high values of GF, respectively. We use the high gain setting normally because $5 \times 4096 = 20,480$, about a third of the full well capacity for the TI-215 detector.

2. Transfer Time

It takes about 4.5 sec after the chip is exposed to transfer a 1024×1024 pixel image into the computer display. About 4 sec is the time of the "slow scan" readout from the chip at a rate of 500,000 bytes/sec. The other 0.5 sec is the time required to move the image from the frame buffer in the Imaging Technology AFG Card in the computer to the computer's own RAM memory. This later delay can now be greatly eliminated using PCI bus adapter cards. Transfer time can be substantially shortened by reducing image size since it depends on the number of pixels in the image.

3. Size

In most of our applications, we do not need the full-chip image. The time for image acquisition depends on the size of the image transferred and this time can be substantially reduced by transferring a smaller region of the chip. For example, we often use a 300 × 300-pixel image size in our mitosis studies. The total number of pixels for these images is about 1/10 the number for the full chip, and the rate of transfer is proportionately faster.

The other big advantage of smaller-sized images is that they take less memory in RAM and/or disk recording. For example, a 300 × 300 pixel image requires 1/10 the storage capacity as the full-chip image. We often record 500 images or more in time lapse. A 300 × 300-pixel image requires 90,000 pixels * 2 bytes = 0.18 Mbytes of memory or disk space. A full-chip image requires 1,048,576 pixels * 2 bytes = 2.1 Mbytes of storage, 23 times more than the 300 × 300-pixel image.

4. Binning

During readout, the C4880 camera controller can combine adjacent pixel values into one value. This "binning" increases the sensitivity of the camera and reduces the image size in proportion to the number of binned pixels. Binning of 2 × 2, 4 × 4, or 8 × 8 produces increases in sensitivity and reduces image size by factors of 4-, 16-, or 64-fold, respectively. The enhanced sensitivity is shown in Fig. 7 for a field of immunofluorescently stained microtubules where the excitation light intensity is reduced by insertion of neutral density filters as binning is increased. Note that binning also reduces resolution in proportion to the binned pixels in each direction. We often use 2 × 2 or 4 × 4 binning in our fluorescence applications where resolution is limited not by optical limitations, but by too few photons. The 4- or 16-fold gain in sensitivity by binning can make substantial improvements in image quality.

D. Linearity, Noise Sources, and the Advantages of Cooling

CCD detectors are ideal for quantitative 2D intensity measurements and position measurements because the signal (S) for each pixel is proportional to light intensity over a wide range and there is no geometrical distortion. Fig. 6A shows the output linearity of the C4880 camera measured for a region of a fluorescent specimen where the light intensity was held constant, but the exposure time was varied from 20 ms to 5 sec.

Essentially, the CCD detector is a geometrical array of pixel detectors, where the pixel gray level value (0–4096 gray levels for 12-bit digitization) acquired by the computer, $P_{i,j}$, for a single readout of the chip is:

$$P_{i,j} = (I_{i,j}Q_{i,j}t + N_{T_{i,j}}t + N_{R_{i,j}})GF \tag{3}$$

where $I_{i,j}$ is the light intensity, $Q_{i,j}$ is the quantum efficiency, $N_{T_{i,j}}$ is the thermally generated electrons/sec, t is the exposure time, $N_{R_{i,j}}$ is the electron equivalent error in the readout process, and GF is the gain factor in the conversion of accumulated electrons to gray level (see above).

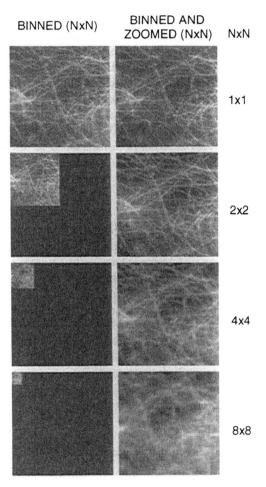

BINNED (NxN)   BINNED AND ZOOMED (NxN)   NxN

1x1

2x2

4x4

8x8

Fig. 7 Binning increases sensitivity, reduces image size, and reduces resolution. Images of a field of microtubules in a cell fixed and stained with fluorescent antibodies to tubulin. Images were taken with $N \times N$ binning at constant integration time and excitation light intensity. Image brightness increased as N^2. The final images were scaled to the same maximum brightness in the print.

Thermal noise N_T is usually specified as the number of electrons per second at a given temperature. This number decreases by about 50% for each 8 degrees of cooling (Fig. 8). So, for example, if a detector element accumulates 400 thermal electrons/sec at 20°C (about room temperature), there will be only 6 at −28°C. Our C4880 camera has a specified average value of $N_T = .1$ electrons/pixel/sec at −40°C (Fig. 6B). Some pixel elements on the chip are much worse than most and these are termed "hot" pixels. For a chip cooled to −30°C these hot pixels only become apparent for exposures longer than several seconds, as seen in Fig. 8. For uncooled chips, these hot pixels are not apparent at video rates (1/30-sec

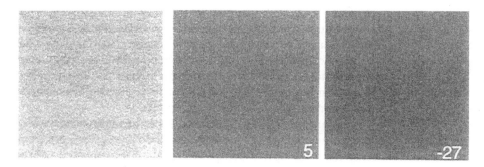

Fig. 8 Dark images from the C4880 camera as a function of temperature of the CCD chip. Exposures were for 5 sec and temperature in degrees centigrade is indicated on each frame.

exposures), but they can become very apparent even for exposures of .1 to 1 sec (Shaw *et al.*, 1995). The new HAD Sony chips, however, have remarkably low thermal noise and few hot pixels.

The other major source of noise is readout noise, N_R. Unlike the photon and thermal noise, it does not depend on integration time, but occurs once when the chip is readout to the computer. N_R depends on how fast and the way readout is achieved (Aikens *et al.*, 1989; Inoué and Spring, 1997).

We measured the thermal and readout noises by measuring the standard deviations (SD) of "dark" images obtained by capping the camera and recording for durations between 20 ms to 40 sec when the camera was cooled to $-30°C$ (Fig. 6B). Both thermal and readout noise are governed by random number statistics where SD is a measure of their noise contributions. For exposures less than 5 sec, readout noise dominates (Fig. 6C). Readout noise is constant, independent of exposure time, at $N_R = 2.6$ gray levels, or 5 (electrons per gray level) $\times 2.6 = 11.3$ electrons. Thermal noise increases with exposure time and begins to be more significant than readout noise for exposures longer than 5 sec, mainly because of the "hot pixels."

E. Dark and Background Reference Image Subtraction

The average values of the contributions of thermal and readout noise can be subtracted from a specimen image by subtracting a "dark" reference image. The dark reference image is obtained by blocking light to the camera for the same exposure time t used for the specimen image and their difference is given from Eq. (3) by:

$$\begin{aligned}
P_{i,j} &= \text{specimen image} - \text{dark reference image} \\
&= (I_{i,j}Q_{i,j}t + (N_T \pm \Delta N_T)_{i,j} + (N_R \pm \Delta N_R)_{i,j})\text{GF} \\
&\quad - ((N_T \pm \Delta N_T)_{i,j} + (N_R \pm \Delta N_R)_{i,j})\text{GF} \\
P_{i,j} &= (I_{i,j}Q_{i,j}t \pm \Delta N_{T_{i,j}} \pm \Delta N_{R_{i,j}})\text{GF},
\end{aligned} \qquad (4)$$

where $\Delta N_{T_{i,j}}$ and $\Delta N_{R_{i,j}}$ represent the random noise in the thermal and readout contributions. For bright specimens, these noise values are small and insignificant. For low-intensity specimens near the "noise floor" of the camera, these values are significant and generate poor image quality and large uncertainty in image measurements. Dark image subtraction reduces substantially the dark image noise from hot pixels (Fig. 6C) and this subtraction is essential for good image quality for weakly fluorescent specimens.

There are other sources of noise from the microscope including out-of-focus fluorescence, reflections from the coverslip and optical components, and fluorescence from the optical components. The average values of these contributions to the specimen image can be subtracted out by obtaining a "reference" image from a region of the slide lacking the specimen or by focusing up or down so the specimen is out of focus.

F. Shading

Quantum efficiency is remarkably constant between different pixel elements for the highest-quality CCD chips and does not have to be corrected except in the most critical applications. Variation in sensitivity between different pixels is termed "shading." This shading can be corrected for any signal by multiplying each pixel value with a scaling factor ($SF_{i,j}$). $SF_{i,j}$ is obtained by acquiring a bright image (B) with constant light intensity (I) across the detector and acquiring a dark image (D) with the shutter closed on the camera. In the computer, the average values for all the pixels in the bright image (B_{avg}) and the dark image (D_{avg}) are calculated, and the scale factor for each pixel ($SF_{i,j}$) is obtained from the ratio:

$$SF_{i,j} = (B_{avg} - D_{avg})/(B_{i,j} - D_{i,j}) \qquad (5)$$

G. Signal-to-Noise

The ratio of signal (S) to noise (N) is a measure of image quality and the reciprocal (N/S) represents the uncertainty in making an image measurement. After background subtraction and shading correction, the ability to see image detail and the uncertainty of intensity measurement depends on random sources of noise in the integrated signal ($I_{i,j} Q_{i,j} t$) as well as from thermal noise (ΔN_T) and readout noise (ΔN_R). Noise in the specimen signal (called photon noise or shot noise) depends on random number statistics just as the other sources of noise in the camera. For S photon-produced electrons, the standard deviation (ΔS) is a measure of the noise of S, just as the standard deviations are a measure of thermal and readout noise. For example, if the other sources of noise are neglected and the average number of signal electrons recorded for one pixel is 100, then the ΔS is the square root of 100 or 10. $S/N = 10$ and the uncertainty in measurement = 10%. There is a 63% chance of measuring $S \pm$ square root of S electrons in a subsequent measurement made the same way.

In general, the more photoelectrons accumulated by photon absorption by silicon in the CCD chip the better the image quality. This is shown in Fig. 9. An image of a fluorescent specimen was acquired for 20, 200, and 2000 ms exposure time at constant illumination intensity then scaled to the same peak brightness for printing. At 20 ms exposure, the image is very "grainy" and the fluorescent actin filament bundles in the cell are indistinct and poorly resolved because of too few photons. The image is much better at 200 ms and beautiful at 2 sec. Quantitatively, signal-to-noise in the camera is given by:

$$S/N = S/(\Delta S^2 + \Delta N_T^2 + \Delta N_R^2)^{.5} \qquad (6)$$

Note from Fig. 6B that at short exposure times (less than 2 sec), thermal noise is insignificant in the C4880 camera and image quality is determined by photon statistics and readout noise.

V. Digital Imaging System

A Microscope Automation and Camera Control

1. Computer Control of the Microscope

Essential to using the multimode microscope system is a flexible software package which allows control of the microscope components. Commercial software packages operating on Windows, Macintosh, and UNIX platforms are now available. We have selected the MetaMorph software package from Universal Imaging Corporation running under Windows95 for the majority of our image acquisition and processing. Figure 2 details the hardware components that run the software to control the illumination, focus, and camera.

The software contains instructions (drivers) for controlling auxiliary devices attached to the microscope. As an example, a driver within MetaMorph controls

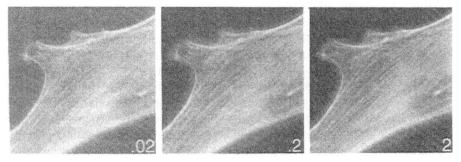

Fig. 9 Too-few photons makes a noisy image. Images of actin filaments in a cell that was fixed and stained with rhodamine-labeled phaloidin taken at the same illumination light intensity but different integration times as indicated in seconds on each frame.

the dual eight-position filter wheels containing O.D. and fluorescence excitation filters. An excitation wavelength and light intensity can be selected from a pull-down menu for a dim rhodamine-labeled specimen and a second set of filters chosen for brighter DAPI staining. The two sets of selections are saved as preset filter combinations and are automatically selected during time-lapse experiments when rhodamine and DAPI images are taken sequentially. In addition to filter wheels, the MetaMorph software contains drivers for controlling the focusing motor, epi- and transillumination shutters, and output to devices such as an optical memory disk recorder (OMDR).

The MetaMorph camera driver allows us to take advantage of several properties specific to the Hamamatsu C4880 CCD camera. The time of exposure, area of CCD chip readout, binning of pixels, and camera gain and offset can all be saved as a preset file. Having these variables saved affords us the ability to obtain images of a specimen brightly illuminated for DIC or dimly illuminated for fluorescence without manually reconfiguring the camera.

Critical to our use of the software is a scripting language (or journaling) for customized control of the parameters, sequence, and timing of image acquisition. Journals are a sequence of instructions to the MetaMorph imaging software generated by recording a sequence of selections from the program menus. For example, to take phase contrast, DAPI, and rhodamine images of the same spindle in time lapse, journals are recorded which control image acquisition for each optical mode. For phase contrast, the transmitted shutter is opened, the CCD camera instructed to open its shutter for 600 ms, close the camera shutter, and then transfer the 12-bit image to computer memory. Similarly, journals are written to control image acquisition for DAPI and rhodamine epifluorescence which specify the exposure time of the camera, the neutral density filter, the excitation filter, and the opening and closing of the epifluorescence light shutter. A time-lapse journal then calls the phase contrast, DAPI, and rhodamine journals in sequence at a specified time interval.

2. Image Processing and Analysis

In addition to automated control of the microscope, the software contains routines for manipulating the size, scale, and contrast of images and allows handling of multi-image stacks. Image stacks are created in circumstances such as multimode time-lapse experiments where all DAPI images in a series remain in the DAPI stack and all rhodamine or DIC images in their respective stack. The ability to perform contrast enhancements or arithmetic (such as background image subtraction) on all images in a stack rather than performing the function on each image individually saves tremendous amounts of labor. Our camera and digital-to-analog converter generate images with greater than 4000 gray levels (12 bits). The MetaMorph software has tools for linear and nonlinear modification of image contrast as well as scaling from 12 bits to 10- or 8-bit images. The contrast enhancement allows presentation of very bright specimen information in the same image as very dim information, as is often the case when microtubules in a spindle are shown next to astral microtubules (Fig. 3E).

The software has excellent routines for creating 24-bit RGB image stacks and for overlaying fluorescent colors onto grayscale DIC or phase-contrast images. For example, an RGB stack can be easily obtained from three stacks taken in a time-lapse series where one stack is the rhodamine image, the other a fluorescein or GFP image, and the third a DAPI or DIC or phase-contrast image. Separate gain and contrast adjustment controls are available for each channel of the RGB image. The RGB image stack can be treated like any other image stack and displayed by a movie player.

Many of our experiments require quantitative analysis of images for relative fluorescence intensity or for motion. Analysis is helped tremendously by the use of calibration files. The MetaMorph software package contains routines for saving distance measurements, made from images of a stage micrometer, and intensity measurements from florescence standards. These measurements can be called upon to provide calibrated values instantly. The time and date of image acquisition are contained in the header of all MetaMorph image files allowing accurate rates to be calculated from distance measurements of time-lapsed images. The ability to couple image processing with analysis greatly increases the quantity and variety of measurements available. The journaling language described above for automating image acquisition can also be applied to image processing routines to automatically link repetitive processes.

B. Other Useful Software and Devices

We have found several other software packages useful for image presentation and production of movies. These include Adobe Photoshop for making complicated montages of images generated in MetaMorph and saved as TIFF-formatted files and Adobe Premiere for making digital movies in .mov or .avi file formats. For making slides and labeling prints, images are copied into Microsoft Power-Point, where lettering and annotation of images is convenient. Black-and-white draft prints of images are obtained with a Hewlett Packard Laser Jet 4M printer and photographic-quality prints are generated with a Tektronics Phaser SDXII Dye Sublimation printer. Other ways of image processing and presentation are described by Shaw *et al.* (1995) in a paper concerning a low-cost method of digital photography using NIH-Image software and a Macintosh computer.

VI. Example Applications

A. DE-DIC

Our digitally enhanced (DE-) DIC images are recorded typically with 600-ms on-chip integration. The standard test specimen for high resolution in the light microscope is the frustrule pore lattice of the diatom *Amphipleura*. These pores in the silica shell are aligned with about 0.19 μm separation along rows and 0.25 μm separation between rows. As seen in Fig. 3C, the rows of pores are clearly resolved by our DE-DIC system and the individual pores are almost resolvable with green light illumination. A major advantage of this system for living cells is the low illumination intensity and the computer control of shutter, which illuminates the specimen only during camera exposure. As a result, cells are viable for long periods of time-lapse recording.

Using this system we have obtained high-resolution real-time images of yeast nuclear motion in the cell division cycle. In one series of experiments (Salmon *et al.*, 1994), GFP histones were expressed to visualize the nucleus (Fig. 3A) and cellular structural detail recorded by DIC (Fig. 3B). The native GFP fluorescence excited at 490 nm was photoactivated using a several-second pulse of 360-nm light, then images were recorded using the fluorescein channel and excitation filter. In other studies (Yeh *et al.*, 1995; Yang *et al.*, 1997), we discovered that the cooled CCD camera produces high-contrast and high-resolution DE-DIC images, such that the positions of the spindle pole bodies and the anaphase elongation of the mitotic spindle within the 2-μm diameter preanaphase nucleus can be clearly seen (Fig. 10). To achieve the images shown in Fig. 10, it is necessary to reduce the refractive index mismatch between the cell wall and the media. This is achieved by embedding the cells in 25% gelatin-containing LPD yeast growth medium (Yeh *et al.*, 1995). Slides are prepared with a 10-μm-thin film of the 25% gelatin, a 5-μl drop of cell suspension is added a no. 1.5 22 × 22-mm coverslip is pressed down on the preparation, and the preparations are then sealed with VALAP (1:1:1 vasoline, lanolin, and paraffin).

B. Multimode Imaging of Anaphase *in Vitro*

The Multimode microscope system has also been particularly useful for recording anaphase spindle dynamics and chromosome segregation for the first time *in vitro* (Salmon *et al.*, 1994; Murray *et al.*, 1996) by reconstituting mitosis in a test tube from sperm nuclei and cytoplasmic extracts prepared from *Xenopus* eggs (Shamu and Murray, 1992; Holloway and Murray, 1993) (Figs. 3D, 3E, and 3F). Here phase and/or DAPI fluorescence image stacks record chromosome movements while X-rhodamine-labeled tubulin fluorescence shows spindle microtubule assembly dynamics. The spindles in cytoplasmic extracts are held in a chamber between a slide and a coverslip separated by 12-μm latex beads and sealed with halocarbon oil and VALAP. For time-lapse, sequential 600-ms exposures of rhodamine and DAPI and/or phase images are recorded to their respective image stacks at either 30- or 60-sec intervals. The *Xenopus* egg cytoplasm is very optically dense, making the rhodamine and DAPI fluorescent images the most useful for recording spindle and chromosome dynamics. Three-dimensional images derived from a single time-point stack of optical sections have also proved very useful for determining the behavior of all of the chromosomes and their

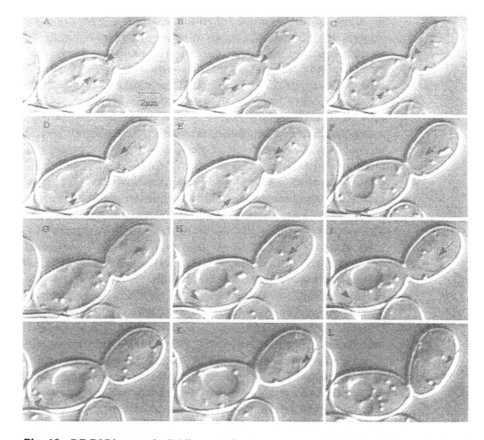

Fig. 10 DE-DIC images of a dividing yeast *Saccharomyces cerevisiae.* The spindle extends through the center of the nucleus and is formed from an overlapping bundle of microtubules extending from opposite spindle pole bodies (arrows). (A) Preanaphase. (B–K) Anaphase. (L) Cytokiness. The preanaphase nucleuse in A is about 2.5 μm in diameter. Modified from Yeh *et al.* (1995).

kinetochore regions within the spindle (Salmon *et al.*, 1994; Murray *et al.*, 1996) (Fig. 3G). Figure 11 (see color figures) shows frames from an RGB digital movie of *in vitro* anaphase where the DAPI blue chromosome images are overlaid on the rhodamine tubulin red image of the spindle microtubules (Murray *et al.*, 1996).

C. Fluorescence Photoactivation

Another application of our multimode system has been in the analysis of the dynamics of kinetochore microtubules within mitotic tissue cells using fluorescence photoactivation methods (Mitchison and Salmon, 1993; Waters *et al.*, 1996). Figure 12 (see color figures) shows a metaphase newt lung epithelial cell which has been microinjected with X-rhodamine-labeled tubulin to label all the spindle microtubules and C2CF caged-fluorescein-labeled tubulin for photoactivation of tubulins within the spindle fiber microtubules as developed by Mitchison (1989). For photoactivation, the field diaphragm in the epi-illuminator is replaced with a 25-μm slit cemented to a Nikon pinhole slider (Fig. 2). The spindle is marked with 350-nm illumination for 2 sec using the Chroma DAPI exciter filter within the excitation filter wheel and no neutral density filters. Image stacks for rhodamine, fluorescein, and phase are acquired at 30-sec intervals. The exposures are 600 ms using two OD-neutral filters in front of the 100-watt Hg lamp and using a 2 × 2-pixel binning of pixels in the Hamamatsu Cooled CCD detector. The 2 × 2-pixel binning reduces the 300 × 300 central image capture area to 150 × 150 pixels and increases image sensitivity fourfold. This has proven to be important in allowing lower excitation light intensity, which reduces photobleaching and has not significantly reduced useful resolution in these low-light-level images. The images in Fig. 12 show that photoactivated marks on the kinetochore fibers flux poleward, demonstrating that at metaphase the kinetochore microtubules which connect the chromosomes to the poles have a net addition of tubulin subunits at their kinetochore ends and a net disassembly of subunits at their polar ends. In anaphase, this poleward flux of kinetochore microtubules persists as chromosomes move poleward (Mitchison and Salmon, 1993; Zhai and Borisy, 1995; Waters et al., 1996). Photoactivation has also been used to show that poleward flux of spindle microtubules occurs in other cell types (Sawin and Mitchison, 1991). This has important implications for how kinetochore microtubule dynamics are coupled to force generation-for-chromosome movement and proper alignment of the chromosomes for accurate segregation during mitosis (Mitchison et al., 1993; Waters et al., 1996).

D. Microtubule and ER Dynamics in Migrating Tissue Cells

The sensitivity of the imaging system has made it possible to track with great clarity the assembly dynamics of individual microtubules and endoplasmic reticulum (ER) in migrating epithelial cells (Waterman-Storer and Salmon, 1996). Microtubules are fluorescently labeled red by microinjecting X-rhodamine-labeled tubulin into cells. Endoplasmic reticulum is fluorescently labeled green using the membrane-intercalating dye DiOC$_6$ (Waterman-Storer et al., 1993) and the leading edge lamellopodia and the lamella of the cell are recorded by DIC. Sets of image stacks are recorded every 7 sec. Figures 13A–13C (see color figures) show a pair of images, one from the rhodamine channel showing the microtubules and the other by DIC showing the leading edge of the lamella. Figures 13D–13F show a pair of images, one from the rhodamine channel and the other from the DiOC$_6$ channel. One significant advance in the prevention of photobleaching in a sealed slide–coverslip preparation is the oxygen-scavenging enzyme system Oxyrase (Waterman-Storer et al., 1993), used in these experiments.

E. GFP–Dynein-Labeled Microtubules in Yeast

In a recent study (Shaw *et al.*, 1997a), the dynamics and organization of the cytoplasmic microtubules in live budding yeast have been imaged (Fig. 14). GFP–dynein expressed in yeast binds all along the lengths of the cytoplasmic but not the nuclear microtubules. The fluorescence is barely detectable with the dark-adapted eye with no attenuation of the Hg illumination to the specimen. Images with distinct astral microtubules are recorded using exposures of 3 sec, 1 OD illumination attenuation, and the field diaphragm closed down around the specimen to reduce the out-of-focus fluorescence from the gelatin layer which holds the yeast against the coverslip surface. We found that a single optical section was not sufficient to see the organization of astral microtubules in these small cells which are about 5 μm in diameter. As a result, we developed an image acquisition procedure where for each time-lapse point (1-min intervals), a series of five fluorescent optical sections is taken through the yeast at 1-μm intervals along the z-axis (Shaw *et al.*, 1997b). In the position of the middle section, a DIC image is taken by inserting the analyzer before the camera with the emission filter controller described earlier. After the completion of the time lapse over the yeast cell cycle of about 90 min, the fluorescent image stack is processed so that at each time interval, the fluorescent microtubule images in each five-plane stack are projected onto one plane. This projected image can

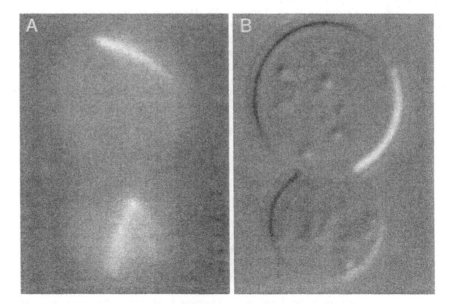

Fig. 14 (A) Fluorescent image of GFP-dynein bound to astral microtubules and (B) the corresponding DIC image of an anaphase yeast *Saccharomyces cerevisiae.* The elongated nucleus is mostly out-of-focus in the DIC image. See text for details.

then be displayed sided by side or overlaid with the corresponding DIC image (Fig. 14). Movies of these images show the organization and dynamics of astral microtubules relative to the cell nucleus and other structural features of the cell visible by DIC.

F. Immunofluorescent Specimens

We have found that the cooled CCD image acquisition system produces images of immunofluorescent-stained specimens with great clarity. Often, our specimens do not have high levels of out-of-focus fluorescence and the advantages of confocal imaging are not required. Figure 15 (see color figures) shows an example where mammalian PtK1 tissue cells are stained with antibodies for a protein implicated in the control of the anaphase onset during mitosis, MAD2 (Chen *et al.*, 1996). The hypothesis is that when bound to kinetochores, MAD2 signals the cell to "stop" the cell cycle. When kinetochores become properly attached to microtubules emanating from opposite poles, MAD2 antibody staining of kinetochores disappears (all the "red stop lights" are off) an the cell enters anaphase.

References

Aikens, R. S., Agard, D., and Sedat, J. W. (1989). *"Methods in Cell Biology,"* Vol. **29**, pp. 292–313. Academic Press, San Diego.

Carrington, W. A., Lynch, R. M., Moore, E. D. W., Isenberg, G., Forarty, K. E., and Fay, F. S. (1995). *Science* **268**, 1483–1487.

Chen, R. H., Waters, J. C., Salmon, E. D., and Murray, A. W. (1996). *Science* **274**, 242–246.

CLMIB staff (1995). *Am. Lab.* April, 25–40.

Cubitt, A. B., Heim, R., Adams, S. R., Boyd, A. E., Gross, L. A., and Tsien, R. Y. (1995). *Trends Biochem. Sci.* **20**, 448–455.

Farkas, D., Baxter, G., DeBiasio, R., Gough, A., Nederlof, M., Pane, D., Pane, J., Patek, D., Ryan, K., and Taylor, D. (1993). *Annu. Rev. Phys.* **55**, 785.

Heim, R., and Tsien, R. (1996). *Curr. Biol.* **6**, 178–182.

Hiroka, Y., Sedlow, J. R., Paddy, M. R., Agard, D., and Sedat, J. W. (1991). *Semin. Cell Biol.* **2**, 153–165.

Holloway, S. L., Glotzer, M., King, R. W., and Murray, A. W. (1993). *Cell* **73**, 1393.

Inoué, S. (1986). "Video Microscopy," Plenum, New York.

Inoué, S. (1989). *"Methods in Cell Biology,"* Vol. 30, pp. 112. Academic Press, San Diego.

Inoué, S., and Oldenbourg, R. (1993). *In* "Handbook of Optics," (Optical Society of America, eds.), 2nd edition. McGraw-Hill, New York.

Inoué, S., and Salmon, E. D. (1995). *Mol. Biol. Cell* **6**, 1619–1640.

Inoué, S., and Spring, K. (1997). "Video Microscopy," 2nd edition. Plenum, New York.

Keller, H. E. (1995) *In* "Handbook of Biological Confocal Microscopy" (J. B. Pawley, ed.), pp. 111–126. Plenum, New York.

Mitchison, T. J. (1989). *J. Cell Biol.* **109**, 637.

Mitchison, T. J., and Salmon, E. D. (1993). *J. Cell Biol.* **119**, 569.

Murray, A. W., Desai, A., and Salmon, E. D. (1996). *Proc. Nat. Acad. Sci. USA* **93**, 12327–12332.

Olson, K., McIntosh, J. R., and Olmsted, J. B. (1995). *J. Cell Biol.* **130**, 639.

Pawley, J. (ed.) (1995). "Handbook of Biological Confocal Microscopy," 2nd edition. Plenium, New York.

Salmon, E. D., Inoué, T., Desai, A., and Murray, A. W. (1994). *Biol Bull.* **187**, 231–232.

Sawin, K. E., Theriot, J. A., and Mitchison, T. J. (1992). *In* "Fluorescent Probes for Biological Activity in Living Cells." Academic Press.

Sawin, K. E., and Mitchison, T. J. (1991). *J. Cell Biol.* **112**, 941.

Schotten, D. (1993). "Electronic Light Microscopy," Wiley-Liss New York.

Shamu, C. E., and Murray, A. W. (1992). *J. Cell Biol.* **117**, 921.

Shaw, S. L., Salmon, E. D., and Quatrano, R. S. (1995). *BioTechniques* **19**, 946–955.

Shaw, S. L., Yeh, E., Maddox, P., Salman, E. D., and Bloom, K. (1997a). *J. Cell Biol.* **139.**

Shaw, S. L., Yeh, E., Bloom, K., and Salman, E. D. (1997b). *Curr. Biol.* **7**, 701–704.

Shotten, D. M. (1995). *Histochem. Cell Biol.* **104**, 97–137.

Spencer, M. (1982). "Fundamentals of Light Microscopy," Cambridge Univ. Press, London.

Taylor, D. L., and Salmon, E. D. (1989). *"Methods in Cell Biology,"* Vol. **29**, pp. 207–237. Academic Press, San Diego.

Taylor, D. L., Nederhof, M., Lanni, F., and Waggoner, A. S. (1992). *Am. Sci.* **80**, 322–335.

Waterman-Storer, C. M., and Salmon, E. D. (1997). *J. Cell Biol.* **139**, 417–434.

Waterman-Storer, C. M., Sanger, J. W., and Sanger, J. M. (1993). *Cell Motil. Cytoskel.* **26**, 19.

Waters, J. C., Mitchison, T. J., Rieder, C. L., and Salmon, E. D. (1996). *Molec Biol Cell.* **7**, 1547–1558.

Yang, S. S., Yeh, E., Salmon, E. D., and Bloom, K. (1997). *J. Cell Biol.* **136**, 1–10.

Yeh, E., Skibbens, R. V., Cheng, J. W., Salmon, E. D., and Bloom, K. (1995). *J. Cell Biol.* **130**, 687–700.

Zhai, Y., Kronebusch, P., and Borisy, G. (1995). *J. Cell Biol.* **131**, 721.

CHAPTER 11

Ratio Imaging Instrumentation

Kenneth Dunn

Department of Medicine
Indiana University Medical Center
Indianapolis, Indiana 46202-5116

Frederick R. Maxfield

Department of Biochemistry
Cornell University Medical College
New York, New York 10021

I. Introduction

With the continuing development of new fluorescent probes, sensitive microscope imaging equipment, and affordable image processing systems, the use of fluorescence microscopy in cell biology is flourishing. In addition to providing multiparameter characterizations of tissue organization and cellular structure, fluorescence microscopy is increasingly used as a quantitative tool in studies of cell physiology, genetics, and biochemistry.

The major technical difficulty in using fluorescence microscopy quantitatively is that the amount of fluorescence detected in a given microscopic volume is a function of several parameters. These include the amount of fluorophore in the volume, the molecular and physical environment of the fluorophore, the background fluorescence of the sample, the intensity and spectrum of illumination, the light collection properties of the objective and optical train, the optical corrections of the optical train, and the sensitivity of the detector to the fluorophore emission. The underlying principle of ratio microscopy is that the quantification of fluorescence as a ratio has the effect of isolating the variable of interest by providing an internal control for many of the extraneous parameters.

For example, when an investigator uses BCECF fluorescence to measure cytosolic pH, while the fluorescence of BCECF excited by 490-nm illumination is sensitive to pH and to all of the other parameters listed above, the fluorescence excited at 450-nm illumination is relatively insensitive to pH while being similar to the fluorescence excited at 490 nm in its sensitivity to the extraneous parameters. By calculating the ratio of one to the other, one obtains a quotient that is sensitive to pH, but largely insensitive to other aspects of the physical environment, probe concentration, and optical path length. In a well-corrected digital microscopy system, these are the only parameters that are variable within and between fields, and so the other parameters comprise constants for the system, affecting absolute measurements of fluorescence power but not ratio measurements.

Ratio microscopy has been most widely used for ion measurements using special environment-sensitive dyes (see Chapter 12 of this volume), but it is increasingly being applied to a variety of new applications (Dunn et al., 1994a). For example, relative amounts of two (or more) probes within the same volume can be calculated to address questions of endocytic sorting (Mayor et al., 1993), cytosol permeability (Luby-Phelps and Taylor, 1988), localized actin polymerization (Giuliano and Taylor, 1994), genome mapping (Reid et al., 1992), and genetic screening (Kallioniemi et al., 1992). Fluorescence ratios can also be calculated between different regions of cells and have been used to characterize intracellular protein distributions (Adams et al., 1996), drug distributions (de Lange et al., 1992), calcium permeability of the nuclear membrane (O'Malley, 1994), and membrane potentials (Loew, 1993).

Molecular associations can be assessed on the basis of resonance energy transfer, in which the ratio of donor to acceptor fluorescence provides a sensitive measure of proximity between the two fluorophores. This property has been used in numerous studies of living cells, addressing such questions such as lipid-protein sorting (Uster and Pagano, 1986), cAMP levels (Adams et al., 1991), lipoprotein catabolism (Myers et al., 1993), transmembrane electrical potentials (Gonzalez and Tsien, 1995) as well as various aspects of protein structure. Molecular concentration and intermolecular associations can also be characterized from fluorophore–fluorophore self-interactions which shift the intensity or fluorescence spectrum of certain fluorophores at high concentrations. Ratios of shifted to unshifted fluorescence have been used to analyze lipid endocytosis (Chen *et al.*, 1997) and to reveal local differences in the structure of the plasma membrane (Terzaghi *et al.*, 1994).

Finally, ratio microscopy can be conducted on the basis of fluorescence lifetimes rather than color. Much as the color spectrum of certain fluorophores is environmentally sensitive, so is the fluorescence lifetime. Fluorophores that are indistinguishable on the basis of fluorescence spectra frequently have distinctive fluorescence lifetimes. Furthermore, fluorescence lifetimes are frequently environmentally sensitive. For example, ratios of fluorescence emissions occurring during two intervals following pulsed illumination have been used to measure cytosol pH (Sanders *et al.*, 1995) and endosome fusion (Oida *et al.*, 1993). This review will primarily address spectroscopic ratio microscopy. Fluorescence lifetime imaging is described in Chapter 15 of this volume.

Fluorescence ratio microscopy can be divided into several categories. These include excitation ratio microscopy, in which fluorescence ratio images are collected by sequentially exciting the sample with two different wavelengths of light and sequentially collecting two different images; emission ratio microscopy, in which the sample is illuminated with a single wavelength of light and images formed from light of two different emission wavelengths are collected; and a combination of the two, in which the sample is illuminated with two wavelengths of illumination and images are formed from light of two different emission wavelengths.

II. Choosing an Instrument for Fluorescence Measurements

A. Microscope, Fluorometer, or Flow Cytometer?

The first question to consider when designing a fluorescence study is whether the experiment is best addressed by imaging microscopy as opposed to flow cytometry or fluorometry in cuvettes or microplates. Even with advances in microscopy instrumentation and software, these latter techniques are often much easier to implement than a quantitative fluorescence microscopy study.

The obvious value of quantitative imaging microscopy is that it yields spatial information not available through the other approaches. Resolution of subcellular structures can only be achieved by imaging microscopy. Similarly, time-course measurements of changes in single cells can only be obtained by microscopy. A less obvious reason to use imaging is that for studies of dynamic events in single cells it provides enhanced temporal resolution and sensitivity when compared with fluorometric techniques. Unless a cell culture is synchronized with respect to the measured parameter, cell–cell variability has the effect of masking transient changes when using methods that do not measure single cells. For example, rises in intracellular free calcium, $[Ca^{2+}]_i$, in response to a stimulus frequently occur asynchronously within a cell culture, and averaging over many cells can reduce the apparent size of the cellular response and prolong its apparent duration.

In many cases, subcellular resolution or time-course information about single cells may not be required, and the nonimaging fluorescence techniques may be more appropriate. Flow cytometry can provide quantitative measurements on thousands of individual cells, allowing the distribution of fluorescence intensities in a population to be measured conveniently. Furthermore, the time course of a response can be measured by measuring the change in a parameter as a function of time after a stimulus. Flow cytometry cannot provide information about the time course of a response in a single cell (e.g., repetitive increases in $[Ca^{2+}]_i$), but it can easily measure the time course of response in a population of cells. In many studies the ability of flow cytometry to measure properties of thousands of cells may be more valuable than detailed information on a smaller number of cells obtained by microscopy.

Fluorometry, either in a cuvette or in a microplate reader, is usually the simplest type of fluorescence measurement to carry out. In a cuvette, cells can be either in suspension or attached to a coverslip held at an angle to the incident beam (Salzman and Maxfield, 1989). Average fluorescence intensity from thousands of cells can be measured rapidly by these methods. Some potential problems occur in fluorometry that are less serious for microscopy or flow cytometry studies. One such problem is that some cells may become extremely bright during fluorescent labeling (e.g., dead cells in some immunofluorescence experiments). These are easily excluded in analyses of flow cytometry data or in selecting fields for analysis by microscopy, but they are included in fluorometer experiments. Such cells may dominate the total signal in some unfavorable cases. Also, extracellular fluorophore will contribute to the signal in fluorometer experiments. This is a serious problem with low-molecular-weight indicator dyes which can exit cells by diffusion or by active transport (Maxfield *et al.*, 1989).

B. Whole-Cell or Subcellular Microscopy?

An important early consideration is to determine whether subcellular resolution is needed or if measurements on whole cells will provide the desired information. For many studies, such as changes in cytoplasmic pH or $[Ca^{2+}]_i$, whole-cell measurements are typically sufficient. Until recently, microfluorometry using a photometer mounted on the microscope was often used to quantify fluorescence from a region of the microscope field. This method is still useful if very fast response times are required or if repeated measurements with very short time intervals are to be made. However, sensitivity equivalent to that of a microscope fluorometer can be obtained with cooled CCD detectors, and image processing software can be used to independently measure changes in fluorescence in the regions corresponding to individual cells. Using low-power objectives, many cells can be imaged simultaneously, and integrated fluorescence power of each cell can be recorded. For several reasons, measurements of fluorescence power from whole cells are typically much easier than measurements of subcellular objects. First, the fluorescence power from the whole cell is greater and this improves the signal-to-noise values of the measurements. Also, the identification of whole cells in the image is usually easily carried out by image analysis software.

C. Confocal or Wide–Field Ratio Microscopy?

To date, most ratio microscopy has been conducted using wide-field microscope systems. Confocal microscopy offers the potential for enhanced resolution, which is frequently necessary for data interpretation. Furthermore, the superior rejection of out-of-focus fluorescence provides the confocal microscope with the capacity to image specimens whose background fluorescence is so high that it prevents adequate imaging by conventional, wide-field microscopy. Such samples include cells immersed in fluorescent labels (Dunn *et al.,* 1994b), thick samples such as polarized epithelia, and ramifying systems such as nerve cells within tissues.

The scanning nature of the confocal microscope also gives it an advantage for emission ratio microscopy. In a wide-field microscope, the simultaneous collection of ratio images requires two separate cameras. Even slight differences in the alignment or geometry between the two cameras will result in a lack of pixel-for-pixel registration between ratio images. This misregistration can cause problems for ratio calculations. Misregistration is avoided in confocal microscopes since the scanning process maps photons from a particular position to the same location in multiple images.

The enhanced resolution of a confocal microscope comes at a price, however. By imaging through a pinhole the confocal microscope trades field-of-view for depth discrimination. Consequently, the image of a microscope field is constructed by scanning the illumination and the detector over the field. The scanning nature of the confocal microscope means that images are assembled relatively slowly; for most microscope designs, this means that image collection requires several seconds. This compares poorly with wide-field ratio microscopy in which it is frequently possible to acquire pairs of images within fractions of a second. This slow image collection hampers imaging dynamic events and makes confocal systems susceptible to problems of sample movement and stage drift (either of which will yield spurious ratio measurements).

In addition, in order to assemble an image in a reasonably short period of time, the sample must be scanned very rapidly and intense illumination is needed to generate sufficient fluorescence in the diffraction-limited spot during the exceedingly short period of time that the spot is being imaged (in the range of microseconds). For most designs, illumination is provided by lasers. These lasers have a limited number of emission wavelengths, which in turn limits the choice of fluorophores. The problem is compounded for excitation ratio imaging, which requires excitation at two specific wavelengths.

These problems have been addressed in alternative designs of confocal micro-scopes. For example, several designs have been developed to provide very high scan rates that are capable of providing video-rate confocal images. However, the fluorescence of most samples is insufficient to provide a useable signal during the exceedingly brief pixel dwell-time of these systems. Consequently, reasonable signal-to-noise is frequently obtained only after averaging several frames, and image collection still requires seconds rather than milliseconds. Another solution is to simultaneously scan multiple points in the sample as in tandem scanning disk confocal systems. These designs provide for video rate output and have the added advantage that they utilize "white" light sources, such as mercury lamps, providing for flexible choice in excitation wavelengths. However, this illumination is utilized inefficiently, resulting in relatively low levels of sample illumination and, therefore, fluorescence. New instruments utilizing a larger fraction of the source illumination have recently been developed. These types of microscopes offer the further advantage of allowing for direct visual observation. Acquisition times can also be shortened by employing laser line-scanning rather than point scanning detection, which sacrifices some resolution in exchange for video-rate acquisition rates.

As will be discussed below, confocal microscopy places some special require-ments on the optical train of the microscope which also must be considered when deciding between confocal and wide-field ratio microscopy.

III. Different Methodological Approaches to Ratio Imaging

A. Excitation Ratio Microscopy

This approach has been the most widely used method of ratio microscopy. A variety of fluorophores have been developed whose fluorescence excitation properties change in response to the presence of specific ions, allowing character-izations of the ionic environment of the cytosol, organelles, or extracellular space. The best known of these are fura-2 and BCECF, which are used to characterize intracellular $[Ca^{2+}]_i$ and pH, respectively. The specifics of the excitation ratio technique used to measure intracellular $[Ca^{2+}]_i$ are detailed in Chapter 12 of this volume.

Fluorescence excitation ratio measurements have been largely limited to im-ages collected by wide-field microscopy rather than confocal microscopy. The white-light sources used in wide-field microscopy allow flexible selection of opti-mal excitation wavelengths, and the rapid, sensitive image collection provided by these systems allows accurate characterization of dynamic events.

A general problem with excitation ratio imaging of living cells is that the delay between collection of the two images in a ratio pair can result in a lack of correlation between the two images. Spurious ratios can result from collecting images from different time periods during a dynamic cellular process or from a sample that is moving. Small amounts of stage drift or movements in the sample can produce profound effects on ratio images especially at the edges of objects. The consequences of movement are more severe in confocal microscopy due to the enhanced vertical discrimination which causes intensities to shift if an object moves into or out of focus.

Despite these difficulties, wide-field excitation ratio imaging has been used to measure properties such as local changes in $[Ca^{2+}]_i$ in rapidly moving leukocytes (Marks and Maxfield, 1990) or the pH of single endosomes and lysosomes (Yamashiro and Maxfield, 1987). The major obstacle to measuring localized $[Ca^{2+}]_i$ changes in moving cells is the change in cell thickness imaged to any given pixel as the cell moves. Especially at the edge of cells, the cytoplasmic thickness can change significantly within the time (typically about 1/4 sec) required to obtain an image pair. There are various strategies to deal with this. One is to produce a series of ratio images in which the order of acquisition is reversed for each sequential pair. With fura-2, excitation wavelengths of 340 and 380 nm are used. The first ratio image can be produced from the first 340-nm image divided by the first 380-nm image. The second ratio would then be produced by the second 340-nm image divided by the *first* 380-nm image. If increasing thickness at the front of a moving cell was affecting the ratio, the 340/380 ratio at the front of the cell would be low in the first ratio image and high in the second ratio image. This type of alternation is a warning that movement is a major contributor to ratio differences at the cell edge, and appropriate changes in the experimental design need to be made. Another way to reduce the impact of edge effects is to take into account the reduced intensity at the edges in producing output images (Marks and Maxfield, 1990; Tsien and Harootunian, 1990). This can be done by blanking all pixels that fall below an arbitrary threshold above background. Alternatively, the intensity of each pixel in the ratio image can be weighted by the intensities in the underlying single-wavelength images. If this type of correction is not made, ratio images tend to overweight the pixels with low intensity (and poor signal-to-noise).

The problems of excitation ratio measurements in the confocal microscope can, to some extent, be circumvented. For example, multiple lasers may be combined to produce the illumination wavelengths appropriate for certain excitation ratio dyes. Kurtz and Emmons (1993) combined two separate lasers to provide the 488- and 442-nm excitation wavelengths used for confocal excitation ratio measurements of intracellular pH of cells loaded with BCECF. Another approach is to use excitation ratio dyes that have been designed to be used with particular laser lines such as NERF, a pH-sensitive dye whose fluorescence is such that it permits excitation ratio measurements using the 488- and 514-nm lines of the argon laser. Due to the low pKa of NERF, NERF has been used to measure the pH of especially acidic compartments such as lysosomes (Dunn *et al.*, 1991) and *Xenopus* yolk platelets (Fagotto and Maxfield, 1994). In each of these studies, the problem of sample movement was minimized, to varying degrees, by rapidly acquiring image pairs. Rapid switching of laser lines can also be accomplished by means of acoustooptical tunable filters (AOTFs) which may be used to rapidly alternate sample illumination between two or more wavelengths of light. These devices are capable of switching excitation lines many times during the dwell time of illumination of a given volume so that by synchronizing collection in two detectors with the illumination with one or the other wavelength, the two images can be collected effectively simultaneously (Carlsson *et al.*, 1994).

B. Emission Ratio Microscopy

In emission ratio microscopy, the pair of images can be collected either sequentially with a single camera or simultaneously with two cameras. For living cells simultaneous image collection has the advantage of providing perfect temporal correlation between the two ratio images. For fixed cells, sequential or simultaneous image collection is a matter of convenience.

In order to collect different emission images, different optical filters and/or dichromatic reflectors are placed in the emission path for each image of a ratio pair. These optical filters can generate linear and nonlinear geometric differences between the images. When using an imaging detector in wide-field microscopy, this can lead to distortions and relative displacement of pixels. A major source of pixel misregistration has been slight differences in light path through different dichroic beam splitters. This problem can be avoided by using polychromatic mirrors that pass and reflect at multiple wavelengths and thus permit ratio images to be collected with a single reflector. The use of high-quality plane-parallel emission filters also reduces distortion and displacement of images. With improved machining and alignment of optical elements, the displacement of an object when switching between two reflecting dichroics can be almost undetectable in some recently manufactured microscopes.

When two imaging detectors are used to simultaneously collect ratio images, further misregistration can be generated by geometric differences between two detectors. This is especially true for tube-type cameras which have inherently poor geometric linearity, but will to some extent be true with any system using two-dimensional detectors due to unavoidable optical differences in the two light paths. The two detectors are also likely to show spatial differences in sensitivity, so while the ratio technique controls for spatial heterogeneity in the detector when using a single camera, this benefit is reduced with two cameras. This heterogeneity may occur at the level of single pixels or over a larger scale. To a large extent, differences between detectors can be characterized and corrected via image processing (as will be discussed later) (see also Chapter 6 by Wolf).

With most current systems, laser scanning confocal microscopy is better suited to fluorescence emission ratio microscopy than to fluorescence excitation ratio microscopy. First, since emission ratio images can be collected simultaneously with two photometers, the slow collection time of the confocal smears dynamic events, but it doesn't contribute to a lack of correlation between the images in a ratio pair. Second, since for some applications only a single excitation wavelength is needed, the emission ratio technique is less affected by the limited number of laser wavelengths available. Unlike the excitation ratio method, the limited number of excitation wavelengths limits the efficiency of excitation, but it does not limit the dynamic range of the ratio, which is set by the optical division of light between the two detectors.

Scanning has some advantages over wide-field microscopy for emission ratio imaging. Linear scanning systems, such as those used in scanning confocal systems avoid geometric distortion by digitally mapping photons collected from a particular volume in the sample to the equivalent location of two images. Lateral inhomogeneity is less of a problem with scanning systems than with two-dimensional detectors since detector variability is not spatially correlated.

In many applications of emission ratio microscopy, the sample is illuminated with a single wavelength of light, and images from two different ranges of emission wavelengths are collected. This is the approach typically used in studies of fluorophore proximity as assessed by resonance energy transfer, in which the ratio of donor-to-acceptor fluorescence is measured, and by excimer formation, in which the ratio of excimer-to-monomer fluorescence is measured. This technique is also used to image certain fluorescent indicators whose emission spectra are sensitive to the ionic environment of the fluorophore such as Indo-1 and SNARF (and its relatives), which are used to measure cytosolic $[Ca^{2+}]$ and pH, respectively. In some cases, however, it is necessary to change both the excitation and emission wavelengths. By utilizing polychromatic reflectors (or AOTFs to select laser lines), it is possible to illuminate a sample with two different excitation wavelengths and simultaneously collect two images in different emission wavelengths. This approach can be used to extend the dynamic range of ratio measurements of certain indicators whose excitation and emission spectra are both environmentally sensitive. For example, since both the excitation and emission spectra of SNAFL change with pH, it may be used in either an excitation or emission ratio mode. However, the most sensitive response is obtained by combining the two approaches, by exciting the dye with 488- and 568-nm light while collecting images centered at 530 and 620 nm, respectively. More generally, simultaneous multiwavelength excitation and emission permits efficient simultaneous imaging of two different fluorophores.

C. Ratio Imaging with Multiple Fluorophores

In many cases it is necessary to obtain the ratio of fluorescence from two fluorescent dyes. For example, this can provide quantification of the relative abundance of molecules within an organelle (Mayor *et al.*, 1993). Ion levels can be measured using an ion-sensitive dye and an ion-insensitive dye both coupled to a macromolecule (e.g., fluorescein–rhodamine–dextran for pH; Dunn *et al.*, 1994a,b). Ratio quantification of separate fluorophores incurs a unique set of difficulties not encountered with single fluorophores. A general problem with using more than one fluorophore is the susceptibility of the ratio to photobleaching. Since two fluorophores are unlikely to bleach at the same rate, a measured ratio will reflect not only the parameter of interest, but also the illumination history of the sample. Meaningful data can be collected by scrupulously maintaining a consistent protocol for image collection (to make the effects of bleaching uniform in all samples) and by minimizing illumination intensity and duration. If necessary, it may be possible to correct for rates of photobleaching in a sample. However, it has been shown that rates of bleaching can vary within a cell, and this may affect the validity of the corrections (Benson *et al.,* 1985).

When the two fluorophores are not covalently linked, differential access of the probes to the imaged volume is also a potential problem. For example, in immunofluorescence, differential access of two antibodies to their epitopes in a given structure can complicate interpretations of the ratio with respect to the distributions of the two epitopes. Accessibility problems are difficult to control, and this has limited the effective use of quantitative immunofluorescence for measuring the relative abundance of molecules in cellular structures. A solution to this is to use directly labeled molecules or intrinsically fluorescent molecules such as constructs containing green fluorescent protein (GFP). For example, using fluorescently labeled transferrin and LDL, endocytic trafficking parameters could be measured and an overall kinetic model for endocytosis could be developed (Dunn and Maxfield, 1992; Ghosh *et al.,* 1994; Mayor *et al.,* 1993). Size-dependent differences in permeability of macromolecules has been exploited to measure the regional properties of the cytoplasm (Luby-Phelps and Taylor, 1988).

The final consideration for ratio studies of multiple probes is that of chromatic aberration, which can have a significant impact on ratio measurements, especially for fluorophores whose excitation and/or emission spectra fall outside the design parameters of the microscope optics. This issue is considered in depth below.

IV. Optical Considerations for Ratio Microscopy

The choice of an appropriate objective is essential for the success of a ratio-imaging project. As with any aspect of fluorescence microscopy, efficient light collection is very important in ratio microscopy, so high-numerical-aperture objectives are usually preferred. However, in some cases it may be preferable to image through a wide depth-of-focus (e.g., in making whole-cell ratio measurements), and a smaller numerical aperture may be beneficial. A limiting factor in many studies is the amount of light captured at one pixel in the imaging detector. This can be increased by increasing the numerical aperture of the objective. Since fluorescence power projected onto a given area of the detector decreases as the square of magnification, increased brightness can also be achieved by using the lowest-power objective appropriate to the level of resolution needed. The number and quality of optical elements in an objective can also affect its light-collection efficiency. For example, glass transmits ultraviolet light poorly, so efficient excitation of ultraviolet dyes requires the use of optics designed to be transparent to ultraviolet light.

As mentioned previously, the most widely practiced form of ratio microscopy is wide-field excitation ratio microscopy. Alternative forms of ratio microscopy include emission ratio microscopy and confocal ratio microscopy. It turns out that relative to these alternative approaches, wide-field excitation ratio microscopy is fairly forgiving of optical aberrations. Prior to conducting ratio microscopy experiments, particularly using one of these alternative approaches, it is important to appreciate the demands that they place on the microscope optical train.

A. Spherical Aberration

Spherical aberration in an objective results in the axial and peripheral rays being focused differently, so that light is focused to a broad region rather than to a single point. The blurring that results generates images with poor contrast and resolution. Although modern objectives are typically well corrected for spherical aberration, this correction applies to specific conditions. First, the correction is specific to particular wavelengths of light. Thus some objectives may not be corrected for ultraviolet light or for far-red light. Second, the correction for spherical aberration is accurate only when the light path is carefully controlled. For high-NA oil-immersion objectives this means that the light path will have a specific and more-or-less constant refractive index, which means either that the refractive index of the immersion and mounting media must match or that the object is imaged at the surface of the coverslip. This condition is frequently unobtainable in studies of live cells in a perfusion chamber or of thick tissues. Recently, objective manufacturers have responded to this problem with new high-NA water-immersion objectives which, in some cases, make it possible to collect images 220 microns into an aqueous medium, with negligible loss in contrast (Inoue, 1995). For water-immersion objectives designed for use with a glass coverslip, the thickness of the coverslip must be carefully controlled or the objective adjusted by means of an adjustment collar. Water-immersion objectives designed to be used without a coverslip avoid even this requirement, but they are largely limited to use with upright microscopes.

Spherical aberration results in decreased sensitivity in confocal microscopy which, in rejecting out-of-focus light, also rejects much of the signal from an object imaged with spherical aberration (Majlof and Forsgren, 1993). Sensitivity is also lost to spherical aberration as the illumination spot is spread beyond the ideal diffraction limited spot. Unfortunately, confocal microscopy is frequently applied to imaging thick samples in which spherical aberration is particularly manifest. With oil-immersion objectives spherical aberration becomes increasingly pronounced as focus penetrates into an aqueous medium such that image contrast is reduced to inadequate levels within 15 to 20 microns of the coverslip

(Hell and Stelzer, 1995; Keller, 1995). Careful matching of refractive index is clearly critical in order to exploit the potential for three-dimensional imaging by confocal microscopy. For studies of living cells or other aqueous samples, this requirement is nicely satisfied by the new generation of high-NA water-immersion objectives.

B. Chromatic Aberration

Although high-quality microscopic systems are largely corrected for chromatic aberration (apochromatic), residual chromatic aberration can still have profound effects in ratio microscopy where the quantitative imaging of two or more colors provides the most severe test. As described below, this is particularly true for ratio confocal microscopy and for any emission ratio microscopy.

Chromatic aberrations result in color-dependent distortions in the image both laterally and axially. Lateral chromatic aberrations result in different wavelengths of light being magnified to different degrees. As a consequence, a field of objects labeled with two fluorophores will produce a pair of images in which one color is offset relative to the other, with the offset increasing with distance away from the optical axis. In transillumination microscopy, such aberration is frequently observed in objects that have colored edges. Axial chromatic aberration, in which different wavelengths of light are focused to different depths, has a less-obvious effect in that the efficiency of collection of one or another wavelength of light is dependent upon the vertical position of the object relative to the focal plane. For some objectives, lateral chromatic aberration is left uncorrected, being corrected at the microscope eyepiece. Thus it is critical to ensure that imaging systems use matched apochromatic optics. It should be obvious that either form of chromatic aberration can have a serious impact on quantitative ratio microscopy. The exact consequences depend upon the nature of the ratio microscopy, as outlined below.

For conventional, wide-field excitation ratio microscopy, in which two wavelengths of light are used to excite a fluorophore which emits at a single wavelength, chromatic aberration will have minimal effect. The Koehler-illuminated light path of such a system ideally provides for parallel uniform illumination of the specimen plane, in which case chromatic aberration will have minimal consequence.

For wide-field emission ratio microscopy, in which one is collecting fluorescence emissions at two wavelengths, chromatic aberration will have a measurable effect on ratio measurements. In this case chromatic aberration will result in the two sets of emissions assigned to a given point in the image being collected from different volumes in the sample. Lateral chromatic aberration will result in horizontal displacements, which may be easily recognized and perhaps corrected through nonlinear "warping" of images using digital image processing (see discussion in Chapter 6 by Wolf). Axial chromatic aberrations will result in different efficiencies of collection for the two wavelengths, depending upon the vertical position of the sample volume relative to the objective. Ratios will thus reflect not only the relative amounts of fluorescence at each wavelength, but also the vertical position of the imaged object. In fact, axial chromatic aberration has been used in reflection microscopy to measure surface contours at high resolution (Boyde, 1987). For fluorescence microscopy this form of aberration is hard to recognize and difficult to correct in an image once collected. It should be possible to correct for this in a 3D data set using deconvolution algorithms if 3D point spread functions have been obtained at multiple wavelengths. Aside from adding variance to fluorescence ratios, axial chromatic aberration will also decrease detection sensitivity, as a given structure will necessarily be defocused in one or the other color.

As might be expected, accurate confocal ratio microscopy is especially vulnerable to chromatic aberration. If chromatic aberration results in the two colors of fluorescence emissions attributed to a given point in an image arising from two different volumes in the sample, the problem is aggravated by the enhanced horizontal and vertical discrimination of the confocal microscope. Figure 1 illustrates the problem. Cells were incubated simultaneously with fluorescein-transferrin (green) and Cy-5-transferrin (red). Both probes were endocytosed by the cells, and the intracellular distributions were imaged by dual-emission confocal microscopy with excitations at 488 and 647 nm. Single optical sections and projections through the cells are shown. One objective exhibited serious axial chromatic aberration in this specimen, and the second objective showed acceptable levels of axial chromatic aberration. Projection images obtained with both objectives show a uniform brown color indicating that the two transferrin derivatives were trafficked similarly following internalization. However, in single optical sections obtained with one of the objectives there was an apparent lack of colocalization (i.e., separation of the red and green objects) because individual endosomes appeared in different focal positions for the red and green images. Obviously, this would lead to erroneous ratio values if the red/green intensity values were measured.

A less-obvious way that chromatic aberration affects ratio measurements in confocal microscopy is through the effect that it has on detection efficiency (Bliton and Lechleiter, 1995). Because of the Stokes shift (the difference in the wavelengths of fluorophore excitation and emission), the wavelengths of excitation and emission may not be focused to the same volume. Unlike a two-dimensional detector system, lateral chromatic aberration in a point scanning system such as a confocal microscope can result in the illumination of one point in the object plane and the imaging of another, nonilluminated point. Since in a confocal microscope the focused spot moves away from the detector pinhole with increasing scan angle, fluorescence collection efficiency decreases with distance off-axis (Sandison et al., 1995). Since each wavelength will have a different degree of aberration, a ratio of two wavelengths will likewise vary spuriously with distance off-axis. For quantitative ratio imaging it is best to acquire images near the optical axis of the microscope even though "zoom and scroll" features in most scanning systems will allow the acquisition of data far from the optical axis. Axial chromatic aberration profoundly decreases detection efficiency because it causes excitation of a volume that is not parfocal with the detector aperture, and thus much of the excited fluorescence is rejected. Ratio measurements will be affected by how detection efficiency of each wavelength is affected by axial chromatic aberration. This will have a large, apparently random component for each individual object that is determined by the position of the object relative to the nominal focal plane.

The particular sensitivity of the confocal microscope to chromatic aberration is shown by the fact that levels of residual chromatic aberration in the ultraviolet range that were tolerated for wide-field microscopy proved to be excessive for confocal microscopy. Solutions recently introduced to this problem include placement of an additional lens into the excitation light path and the development of new objectives that are chromatically corrected into the ultraviolet range.

Another solution to chromatic aberration in the confocal microscope is the two-photon design, in which a high local density of long-wavelength photons is capable of exciting fluorescence typically associated with visible or ultraviolet excitation. Since the photon density outside a small region is insufficient to excite fluorescence, confocality is determined simply by the focal point of excitation. Since no detector aperture is needed, differences in the focal plane of the excitation (which now are in the infrared) and emission wavelengths have no effect on sensitivity. This should be a valuable method for emission ratio applications (Denk *et al.*, 1995). At present it is not generally feasible to excite at more than one wavelength in this mode.

V. Illumination and Emission Control

For wide-field microscopes the intense illumination needed for epifluorescence is typically provided by either mercury or xenon arc lamps. The spectrum of illumination of a xenon arc lamp is relatively uniform from the ultraviolet into the visible spectrum, while that of a mercury arc lamp includes strong peaks at several wavelengths in the ultraviolet and visible wavelengths. While it may seem that the uniform spectroscopic output of the xenon would make it a better choice, in fact several of the "mercury lines" correspond well to the excitation spectra of popular fluorophores and frequently yield better sensitivity.

For excitation ratio studies, temporal variation in lamp output generates variability in fluorescence which can adversely affect ratio measurements. Since "flicker" increases with the age of the lamp, output variation can be minimized by frequent lamp replacement. Flicker resulting from movement of the arc ("arc wander") can be minimized by "scrambling" lamp output through a coiled fiber-optic (Kam *et al.*, 1993). These systems also provide for a more spatially homogeneous field of illumination than can be obtained with lenses alone.

For confocal microscopy, the range of illumination wavelengths available has been broadened by the development of multiline lasers and the combination of multiple lasers. The high-powered pulsed lasers used in two-photon systems are typically tunable over a wide range of wavelengths, but provide only a single-emission wavelength at a time.

Each of these systems must have some means of accurately controlling the duration and the spectrum for both sample illumination and image collection. Typically, these functions are controlled by special peripheral devices controlled with personal computers. High-speed filter wheels and shutters may be placed in either the excitation or emission light path to provide rapid, consistent image collection parameters. As mentioned previously, AOTFs can provide rapid selection of laser lines. These devices have the additional advantage that they permit separate and continuously variable levels of attenuation for each excitation wavelength.

The correct choice of optical filters can have a large effect on the quality of the data obtained in ratio imaging experiments. Among the considerations are the wavelength, transmission, and bandwidth of the filters. In any ratio study, one seeks a sensitive ratio with a wide dynamic response. In emission ratio microscopy, bleedthrough of signal between channels limits the upper and lower bounds of the ratio. Narrower bandwidth emission filters can improve selectivity, but this may reduce the signal to the point where noise becomes limiting. On the other hand, a large overlap in transmission through two emission filters will reduce the dynamic range of a ratio measurement. In general, discrimination of a particular fluorophore from another is enhanced by selective illumination at particular wavelengths and collection of emission at particular wavelengths. In the case of dual excitation and emission ratio microscopy in which image pairs are collected simultaneously, some discrimination is lost because the sample is simultaneously illuminated at multiple wavelengths. As discussed previously, a solution to this problem may be provided by the use of an AOTF to synchronize specific illumination and collection while providing for effectively simultaneous image collection (Carlsson *et al.*, 1994). Typically, trade-offs such as dynamic range of the ratio versus signal-to-noise must be evaluated for each set of experiments.

When choosing emission filters, one should consider the possibility that the cellular environment will cause shifts in fluorophore emissions. Bright *et al.* (1989) and Carlsson and Mossberg (1992) give several examples of fluorophores whose emission spectra change once inside cells at different concentrations or in different solvents. Such environmental effects are invoked to explain the frequently observed lack of correlation between intracellular and *in vitro* calibrations in ion fluorescence ratio studies. Whenever possible one should use emission filters that encompass the range of emission spectra of such probes.

VI. Detector Systems

Discussion of detector systems may be found in Chapters 9 and 10. Although issues of sensitivity, signal-to-noise, spectroscopic sensitivity, and linearity are all important to ratio microscopy in this chapter we will consider only those aspects of detectors which are uniquely pertinent to ratio microscopy.

For many studies, the speed of image collection is paramount, either because of the speed of the dynamic process being characterized or because a pair of images is expected to freeze a particular point in time. SIT cameras and intensified tube cameras, which were long the standard for low-light level microscopy are particularly lacking in this respect due to the lag in their response time. Intensified CCDs show significantly less lag but with similar levels of sensitivity. Each of these systems is capable of collecting images at video rate, or 30 frames per sec. Cooled CCDs, while showing no lag, are limited by readout noise which can become a significant fraction of the total intensity at low signals and short acquisition times. Consequently, images for quantitative analysis are rarely collected any faster than two per second even though modern chip designs permit faster readout for focusing and choosing a field. Faster image collection can also be accomplished by frame-transfer, in which an image is collected onto a portion of the CCD array and the charges are then rapidly transferred to another portion of the array. Additional images can then be collected and transferred on the chip. Using this technique it is possible to rapidly collect multiple images, each of a restricted size, and then to read them out at a later time.

The second consideration for ratio imaging is that of dynamic range. For quantitative studies, the linear dynamic range of the detector must be sufficient to encompass not only the intrascene range of a single image, but the entire range of images that are to be compared. For ion ratio studies, the range of response is frequently such that sensitivity in one experimental condition is compromised in order to avoid signal saturation in another condition. This limited dynamic range places an unnecessary limitation on data collection and may impose a bias on the results. SIT cameras typically have very limited dynamic range and intensified CCDs and tube cameras show somewhat more. However, while images from these devices are typically digitized into 8 bits (256 gray levels), the images rarely contain more than 6 bits of data above the noise. At present the best solution is provided by cooled CCDs which are capable of linear detection at 12 to 14 bits with 8–11 bits of data above the instrument noise.

The limitations of 8-bit digitization for quantitative ratio imaging are illustrated in Fig. 2. Cells were incubated with dextran coupled to both fluorescein and rhodamine, and the dextran conjugates were delivered to lysosomes which are acidic organelles. These conjugates are useful for pH measurements of endosomes and lysosomes (Dunn *et al.*, 1994a, b) because the fluorescein fluorescence decreases at low pH whereas rhodamine is insensitive to pH changes. Figure 2A shows an image of live cells in which the fluorescein (green) emissions are quenched by the low pH of lysosomes so that when combined with the rhodamine (red) emissions the lysosomes appear yellow. When the pH gradients were collapsed by addition of an ionophore (Figure 7B), the fluorescein fluorescence of the lysosomal dextran increased to the extent that in many of the lysosomes it exceeded the 8-bit range, and saturated pixels are shown in red. Ratio data would be lost for all of these lysosomes. In effect, the 8-bit range of this system limits the range of lysosome pH values that can be measured by this technique.

Until recently direct storage of digital images presented a problem when acquiring data on moving cells because of the large amounts of data that images contain. It was possible to store data in analog form on videotapes or videodisks, but images were typically limited in terms of signal-to-noise and resolution by the quality of the video recorders. With the increase in speed and decrease in cost of digital storage devices, direct digital storage is almost always used in new systems. If an analog-to-digital converter is used to digitize a video signal, it is important to use a scientific-quality digitizer to maximize the signal-to-noise. Although the images require disk space, it is almost always advisable to store original images rather than processed ratio images. This will allow for later changes in the analysis such as improved algorithms used to correct for background.

VII. Digital Image Processing

Complete descriptions of image processing software and hardware are given in Chapters 6 and 7, respectively. In this chapter we will consider only those aspects of image processing that are uniquely pertinent to ratio microscopy.

At a minimum an image analysis system must be capable of distinguishing an object from the image background (segmentation), correcting the amount of fluorescence in the object for the amount of background fluorescence in the image (background correction), and then measuring the integrated intensity within the bounds of the object in each of two ratio images.

Image segmentation may be conducted in a variety of ways. For images with few objects, it is sufficient for the user to designate regions of interest; for example, by drawing a region around the perimeter of a cell. For images with more objects, it may be preferable to use an automated segmentation procedure. Automated segmentation may be based upon a threshold criterion or upon more elaborate morphological criteria. Certain systems are capable of "learning" these criteria on the basis of models designated by the user.

The intention of background correction is to isolate the fluorescence arising in the region of interest from whatever stray light may exist in that region and to correct for spatial variations in detected fluorescence. Typically, stray light is eliminated from images by collecting an image from a empty field and subtracting this image from the image to be measured. Shading correction circuitry is included in certain detectors to compensate for large-scale spatial variability in detector sensitivity. In addition, both cooled CCDs and intensified CCDs show some degree of pixel variation. Spatial variation in detected fluorescence (which reflects spatial variability in both illumination and detection) can be compensated by using an image of a thin, uniform film of fluorophore as a calibration standard (Jericevic *et al.,* 1989).

Because of image-registration problems (from a variety of sources described above), measurement of the integrated fluorescence within the bounds of an object in two images may not be the same thing as measuring the fluorescence in the corresponding pixel loci of the two images. To some extent image registration can be accomplished using image processing techniques capable of linear and nonlinear geometric transformations.

Image processing may be conducted either "off-line," at some time after image collection, or "on-line," in which case some minimal level of processing is conducted during image collection. For some studies, it is critical to be able to monitor the progress of the study, as displayed, for example, in the real-time display of ratio images. On the other hand, it can be argued that much of the subjective aspect of microscopy is avoided when the microscopist is unaware of the experimental outcome while collecting "representative" image fields.

VIII. Summary

Using ratio imaging to obtain quantitative information from microscope images is a powerful tool that has been used successfully in numerous studies. While ratio imaging reduces the effects of many parameters that can interfere with accurate measurements, it is not a panacea. In designing a ratio-imaging experiment all of the potential problems discussed in this chapter must be considered. Undoubtedly, other problems that were not discussed can also interfere with accurate and meaningful measurements. Many of the problems discussed here were observed in the authors' laboratories. In our experience there are no standard routines or methods that can foresee every problem before it has been encountered. Good experimental design can minimize problems, but the investigator must continue to be alert.

Progress in instrumentation continues to overcome some of the difficulties encountered in ratio imaging. CCD cameras with 12- to 14-bit pixel depth are being used more frequently, and several confocal microscope manufacturers are now also using 12-bit digitization. The dramatic increase in the use of confocal microscopes over the past decade is now causing microscope manufacturers to more critically evaluate the effect of axial chromatic aberration in objectives, and recent designs to minimize this problem are being implemented. Other developments such as the use of AOTFs to attenuate laser lines extend the applicability of ratio imaging.

Ratio imaging is clearly applicable to a wide range of cell biological problems beyond its widespread use for measuring ion concentrations. Imaginative but careful use of this technique should continue to provide novel insights into the properties of cells.

References

Adams, S. R., Harootunian, A. T., Buechler, Y. J., Taylor, S. S., and Tsien, R. Y. (1991). Fluorescence ratio imaging of cyclic AMP in single cells. *Nature* **349**, 694–697.

Adams, C. L., Nelson, W. J., and Smith, S. J. (1996). Quantitative analysis of cadherin–catenin–actin reorganization during development of cell–cell adhesion. *J. Cell Biol.* **135**, 1899–1912.

Benson, D. M., Bryan, J., Plant, A. L., Gotto, A. M., and Smith, L. C. (1985). Digital imaging fluorescence microscopy: Spatial heterogeneity of photobleaching rate constants in individual cells. *J. Cell Biol.* **100**, 1309–1323.

Bliton, A. C., and Lechleitner, J. D. (1995). Optical considerations at ultraviolet wavelengths in confocal microscopy. *In* "Handbook of Biological Confocal Microscopy" (J. B. Pawley, ed.), 2nd edition, pp. 431–444. Plenum, New York.

Boyde, A. (1987). Colour-coded stereo image from the tandem scanning reflected light microscope (TSRLM). *J. Microsc.* **146**, 137–142.

Bright, G. R., Fisher, G. W., Rogowska, J., and Taylor, D. L. (1989). Fluorescence ratio imaging microscopy. *In* "Methods in Cell Biology: Fluorescence Microscopy of Living Cells in Culture" (Y.-L. Wang and D. L. Taylor, Eds.), Vol. 29, Part B, pp. 157–192. Academic Press, New York.

Carlsson, K., and Mossberg, K. (1992). Reduction of cross-talk between fluorescent labels in scanning laser microscopy. *J. Microsc.* **167**, 23–37.

Carlsson, K., Aslund, N., Mossberg, K., and Philip, J. (1994). Simultaneous confocal recording of multiple fluorescent labels withimproved channel separation. *J. Microsc.* **176**, 287–299.

Chen, C.-S., Martin, O. C., and Pagano, R. E. (1997). Changes in spectral properties of a plasma membrane lipid analog during the first seconds of endocytosis in living cells. *Biophys. J.* **72**, 37–50.

de Lange, J. H., Schipper, N. W., Schuurhuis, G. J., ten Kate, T. K., van Heijningen, T. H., Pinedo, H. M., Lankelma, J., and Baak J. P. (1992). Quantification by laser scan microscopy of intracellular doxorubicin distribution. *Cytometry* **13**, 571–576.

Denk, W., Piston, D. W., and Webb, W. W. (1995). Two-photon molecular excitation in laser-scanning microscopy. *In* "Handbook of Biological Confocal Microscopy" (J. B. Pawley, Ed.), 2nd edition, pp. 445–458. Plenum, New York.

Dunn, K. W., and Maxfield, F. R. (1992). Delivery of ligands from sorting endosomes to late endosomes occurs by maturation of sorting endosomes. *J. Cell Biol.* **117**, 301–310.

Dunn, K., Maxfield, F., Whitaker, J., Haugland, R., and Haugland, R. (1991). Fluorescence excitation ratio measurements of lysosomal pH using laser scanning confocal microscopy. *Biophys. J.* **59**, 345a.

Dunn, K., Mayor, S., Meyer, J., and Maxfield, F. R. (1994a). Applications of ratio fluorescence microscopy in the study of cell physiology. *FASEB J.* **8**, 573–582.

Dunn, K., Park, J., Semrad, C., Gelman, D., Shevell, T., and McGraw, T. E. (1994b). Regulation of endocytic trafficking and acidification are independent of the cystic fibrosis transmembrane regulator. *J. Biol. Chem.* **269**, 5336–5345.

Fagotto, F., and Maxfield, F. R. (1994). Changes in yolk platelet pH during *Xenopus* laevis development correlate with yolk utilization: A quantitative confocal study. *J. Cell Sci.* **107**, 3325–3337.

Ghosh, R. N., Gelman, D. L., and Maxfield, F. R. (1994). Quantification of low density lipoprotein and transferrin endocytic sorting in HEp2 cells using confocal microscopy. *J. Cell Sci.* **107**, 2177–2189.

Giuliano, K. A., and Taylor, D. L. (1994). Fluorescent actin analogs with a high affinity for profilin *in vitro* exhibit an enhanced gradient of assembly in living cells. *J. Cell Biol.* **124**, 971–983.

Gonzalez, J. E., and Tsien, R. Y. (1995). Voltage sensing by fluorescence resonance energy transfer in single cells. *Biophys. J.* **69**, 1272–1280.

Hell, S. W., and Stelzer, E. H. K. (1995). Lens aberrations in confocal fluorescence microscopy. *In* "Handbook of Biological Confocal Microscopy" (J. B. Pawley, Ed.), 2nd edition, pp. 347–354. Plenum, New York.

Inoue, S. (1995). Foundations of confocal scanned imaging in light microscopy. *In* "Handbook of Biological Confocal Microscopy" (J. B. Pawley, Ed.), 2nd edition, pp. 1–17. Plenum, New York.

Jericevic, Z., Wiese, B., Bryan, J., and Smith, L. C. (1989). Validation of an imaging system: Steps to evaluate and validate a microscope imaging system for quantitative studies. *In* "Methods in

Cell Biology: Fluorescence Microscopy of Living Cells in Culture" (Y.-L. Wang and D. L. Taylor, Eds.), Vol. 29, Part B, pp. 47–83. Academic Press, New York.

Kallioniemi, A., Kallioniemi, O.-P., Sudar, D., Rutowitz, D., Gray, J. W., Waldman, F., and Pinkel, D. (1992). Comparative genomic hybridization for molecular cytogenetic analysis of solid tumors. *Science* **258,** 818–821.

Kam, Z., Jones, M. O., Chen, H., Agard, D. A., and Sedat, J. W. (1993). Design and construction of an optimal illumination system for quantitative wide-field multi-dimensional microscopy. *Bioimaging* **1,** 71–81.

Keller, H. E. (1995). Objective lenses for confocal microscopy. *In* "Handbook of Biological Confocal Microscopy" (J. B. Pawley, Ed.), 2nd edition, pp. 111–126. Plenum, New York.

Kurtz, I., and Emmons, C. (1993). Measurement of intracellular pH with a laser scanning confocal microscope. *In* "Methods in Cell Biology: Cell Biological Applications of Confocal Microscopy" (B. Matsumoto, Ed.), Vol. 28 pp. 183–193. Academic Press, New York.

Loew, L. M. (1993). Confocal microscopy of potentiometric fluorescent dyes. *In* "Methods in Cell Biology: Cell Biological Applications of Confocal Microscopy" (B. Matsumoto, Ed.), Vol. 28, pp. 195–209. Academic Press, New York.

Luby-Phelps, K., and Taylor, D. L. (1988). Subcellular compartmentalization by local differentiation of cytoplasmic structure. *Cell Motil. Cytoskeleton* **10,** 28–37.

Majlof, L., and Forsgren, P.-O. (1993). Confocal microscopy: Important considerations for accurate imaging. *In* "Methods in Cell Biology: Cell Biological Applications of Confocal Microscopy" (B. Matsumoto, Ed.), Vol. 28, pp. 79–95. Academic Press, New York.

Marks, P. W., and Maxfield, F. R. (1990). Ratio imaging of cytosolic free calcium in neutrophils undergoing chemotaxis and phagocytosis. *Cell Calcium* **11,** 181–190.

Maxfield, F. R., Bush, A., Marks, P., and Shelanski, M. L. (1989). Measurement of cytoplasmic free calcium in single motile cells. *Methodol. Surveys Biochem. Analysis* **19,** 257–272.

Mayor, S., Presley, J., and Maxfield, F. R. (1993). Sorting of membrane components from endosomes and subsequent recycling to the cell surface occurs by a bulk flow process. *J. Cell Biol.* **121,** 1257–1269.

Myers, J. N., Tabas, I., Jones, N. L., and Maxfield, F. R. (1993). β-VLDL is sequestered in surface-connected tubules in mouse peritoneal macrophages. *J. Cell Biol.* **123,** 1389–1403.

Oida, T., Sako, Y., and Kusumi, A. (1993). Fluorescence lifetime imaging microscopy (flimscopy): Methodology development and application to studies of endosome fusion in single cells. *Biophys. J.* **64,** 676–685.

O'Malley, D. M. (1994). Calcium permeability of the neuronal nuclear envelope: Evaluation using confocal volumes and intracellular perfusion. *J. Neurosc.* **14,** 5741–5758.

Reid, T., Baldini, A., Rand, T. C., and Ward, D. C. (1992). Simultaneous visualization of seven different DNA probes by *in situ* hybridization using combinatorial fluorescence and digital imaging microscopy. *Proc. Natl. Acad. Sci. USA* **89,** 1388–1392.

Salzman, N. H., and Maxfield, F. R. (1989). Fusion accessibility of endocytic compartments along the recycling and lysosomal endocytic pathways in intact cells. *J. Cell Biol.* **109,** 2097–2104.

Sanders, R., Draaijer, A., Gerritsen, H. C., Houpt, P. M., and Levine, Y. K. (1995). Quantitative pH imaging in cells using confocal fluorescence lifetime imaging microscopy. *Anal. Biochem.* **227,** 302–308.

Sandison, D. R., Williams, R. M., Wells, K. S., Strickler, J., and Webb, W. W. (1995). Quantitative fluorescence confocal laser scanning microscopy. *In* "Handbook of Biological Confocal Microscopy" (J. B. Pawley, Ed.), 2nd edition, pp. 39–53. Plenum, New York.

Terzaghi, A., Tettamanti, G., Palestini, P., Acquotti, D., Bottiroli, G., and Masserini, M. (1994). Fluorescence excimer formation imaging: A new technique to investigate association to cells and membrane behavior of glycolipids. *Eur. J. Cell Biol.* **65,** 172–177.

Tsien, R. Y., and Harootunian, A. T. (1990). Practical design criteria for a dynamic ratio imaging system. *Cell Calcium* **11,** 93–109.

Uster, S., and Pagano, R. E. (1986). Resonance energy transfer microscopy: Observations of membrane-bound fluorescent probes in model membranes and in living cells. *J. Cell Biol.* **102,** 1221–1234.

Yamashiro, D. J., and Maxfield, F. R. (1987). Acidification of morphologically distinct endocytic compartments in mutant and wild type Chinese hamster ovary cells. *J. Cell Biol.* **105,** 2723–2733.

CHAPTER 12

Ratio Imaging: Practical Considerations for Measuring Intracellular Calcium and pH in Living Tissue

Randi B. Silver

Department of Physiology and Biophysics
Cornell University Medical College
New York, New York 10021

I. Introduction

In order for a cell to maintain a steady-state there must be mechanisms in place to ensure the appropriate balance between the intracellular and extracellular milieu. The study of intracellular ion concentrations in the steady-state and the role of various ions in regulating cellular processes was until recently limited to systems that could be impaled with a microelectrode. This chapter focuses on the measurement of two intracellular ions, intracellular H^+ (pH_i) and Ca^{2+}(Ca_i), at the single-cell level with a relatively noninvasive technique, quantitative ratio

METHODS IN CELL BIOLOGY, VOL. 56

imaging. The development and availability of Ca_i- and pH_i-specific fluorescent probes coupled to major advances in the technology and design of low-light-level charge-coupled devices geared toward biological applications and improved microscope optics has made it possible to visualize a two-dimensional fluorescence signal that is related to the free concentration of Ca and pH inside a cell. This chapter describes the basis for using the dual excitation ratio imaging technique and will attempt to provide the reader with a framework for understanding and developing this technique for their own use.

II. Why Ratio Imaging?

The uncertainties associated with quantitating single wavelength fluorescence imaging are apparent from the standard fluorescence equation as presented by Bright et al. (1989):

$$F = f(\theta) \, g(\lambda) \, \phi F \, I_O \, \varepsilon\beta c,$$

where fluorescence emission (F) is related to a geometric factor $[f(\theta)]$, the quantum efficiency of the detector $[g(\lambda)]$, fluorescence quantum yield of the probe $[\phi F]$, excitation intensity $[I_O]$, extinction coefficient of the probe (ε), optical pathlength (β), and fluorophore concentration (c). From this equation it becomes apparent that the fluorescence emission of a probe may be influenced by a variety of factors including: compartmentalization of the probe in intracellular organelles; uneven or heterogenous loading of the dye in the cells (Heiple and Taylor, 1981); and instrumentation noise, sample geometry, and intrinsic properties of the probe itself (Bright et al., 1989). These factors can all lead to difficulty in interpreting the changes in a single wavelength fluorescence signal.

One way to overcome these limitations is to use an intracellular probe specific for a parameter of interest, which changes its fluorescence intensity in its bound and unbound form. This can be either in its excitation or its emission spectrum. The total fluorescence would represent two independent fluorescent forms of the dye originating from the same volume. If the emitted fluorescence from both forms of the dye is measured and ratioed from this volume then the fluorescence measurement will be independent of both path length and intracellular concentration of the dye (Heiple and Taylor, 1981). This principal has been realized with the development of specific fluorescent indicators or dyes for the measurement of intracellular Ca^{2+} and pH and are designed to shift their wavelength distribution of fluorescence excitation or emission upon binding the free ion. This chapter will limit its discussion to two of the most widely used ratiometric dyes, 2′,7′-bis (carboxyethyl)-5(6)-carboxyfluorescein (BCECF) and fura-2, for the measurement of pH_i and Ca_i, respectively.

III. Properties of the Intracellular Indicators BCECF and Fura-2

A. Structure and Spectral Characteristics

Figure 1 shows the chemical structures of BCECF and fura-2. BCECF is a derivative of fluorescein with three additional carboxylate groups; two are attached by short alkyl chains, which raises the pK from 6.4 (for fluorescein) to a more physiological pK of 7.0 (Rink *et al.*, 1982). Fura-2, developed from the Ca chelator BAPTA, is the double aromatic analog of EGTA with added fluorophores (Cobbold and Rink, 1987). Compared to its predecessor quin2, fura-2 is a larger fluorophore with a much greater extinction coefficient and a higher quantum efficiency translating to 30-fold higher brightness per molecule (Cobbold and Rink, 1987). Both dyes are hydrophilic pentaanions. As discussed in more detail in Section III,B, the structures were modified to make them membrane permeable. This was accomplished by masking the charges with acetoxymethyl esters to render the molecule neutral. These neutral structures are shown in Fig. 1 and denoted with the letters AM.

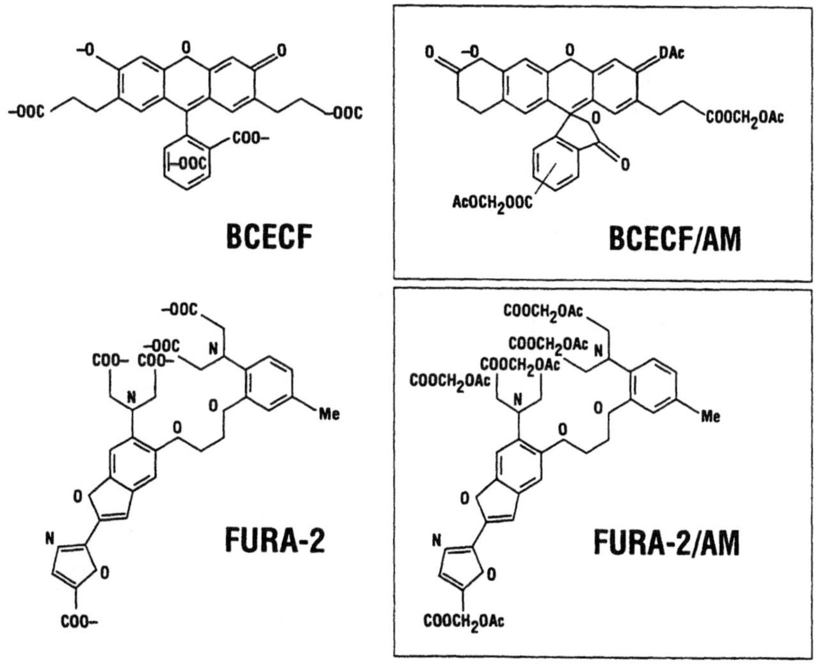

Fig. 1 Chemical structures of BCECF and fura-2. The free acid and membrane-permeant forms of each dye are represented in the figure. (Reprinted by kind permission from Molecular Probes, Inc.)

A fundamental requirement which must be met if an indicator is to be used for quantitative ratio imaging is differential sensitivity to its ligand at one specific wavelength relative to another wavelength. The measured fluorescence signal at the to given excitation or emission wavelengths would then result from the ratio or the relative proportion of both fluorescent forms of the dye. This differential sensitivity may be seen by looking at excitation spectra for BCECF and fura-2.

The pH sensitivity of BCECF is shown in Fig. 2, which is a series of excitation spectra (emission, 520–530 nm) generated on salt solutions each containing 2 μM BCECF titrated to various pHs. Between 490 and 500 nm the BCECF is strongly pH sensitive where the fluorescence intensity increases as the pH becomes more alkaline. In contrast the pH sensitivity of the dye at about 440 nm appears to be pH independent. Therefore, alternately exciting the dye molecules at 490 and 440 nm in a given volume yields the relationship of the changes in the BCECF fluorescence signal at each wavelength with respect to the other. In the example in Fig. 2, the ratio of the emission intensity at 490 nm excitation, the pH-dependent wavelength, to the emission intensity at 440 nm excitation, the pH-independent wavelength, increases as pH becomes more alkaline and decreases as pH becomes more acidic.

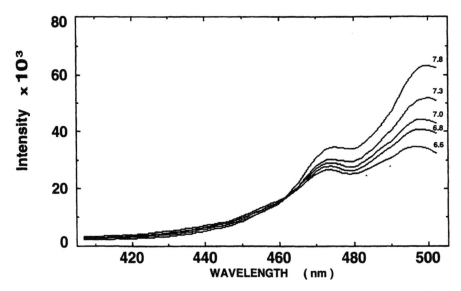

Fig. 2 Excitation spectra of 2 μM BCECF at various pHs. The salt solution contained 130 mM KCl, 1 mM MgCl$_2$, 2 mM glucose, 2 mM CaCl$_2$, and 25 mM HEPES. The spectra (emission: 520) were generated with a fluorimeter (Photon Technology Incorporated—Deltascan) scanning at 1-nm intervals and interfaced to an inverted microscope (Zeiss IM35) equipped with a Hamamatsu photomultiplier tube. Individual dishes placed on the stage of the microscope and aligned in the epifluorescent light path contained 250 μl of HEPES-buffered salt solution with 2 μM BCECF at the pHs indicated. The spectra have not been corrected for background noise. The peaks at 470 nm are due to xenon lamp output.

   The spectral characteristics of fura-2 are shown in Fig. 3. The shape of fura-2's excitation spectrum (emission 510 nm) and the large dynamic range of the dye at both 340 and 380 nm, as first shown and determined by Grynkiewicz and colleagues (1985), make it an ideal dye for ratio imaging. As shown in the figure, the fluorescence intensity at 340 nm is directly proportional to the free-Ca concentration and inversely proportional to the ionized Ca at 380 nm excitation. The ratio of the emission intensity at 340 nm excitation divided by the emission signal at 380 nm excitation can be used to determine the ionized-Ca concentration.

   It should be apparent from the spectra for BCECF and fura-2 that by choosing the appropriate excitation wavelengths it is possible to have enough emission intensity signal to measure the ratio of the fluorescent species of the dye bound and unbound.

   To summarize, ratiometric imaging with the dual excitation dyes BCECF and fura-2 normalizes the fluorescence signal for factors which could potentially influence the measured intensities yet be unrelated to actual changes in the parameter of interest. This principle is further illustrated in Fig. 4 which looks at the effect of varying the focus on the BCECF signal. For this maneuver 2 μM BCECF free acid in 250 μl salt solution was placed in a dish on the stage of an inverted microscope interfaced to an imaging system. The top graph shows the traces for the emitted fluorescence excitation intensities at 490 nm (open circles) and 440 nm (closed circles) and the lower graph is the corresponding 490/440 ratio (closed triangles). The focus was changed at the arrow. As shown in the figure both the 490- and 440-nm excitation intensities decreased proportionately

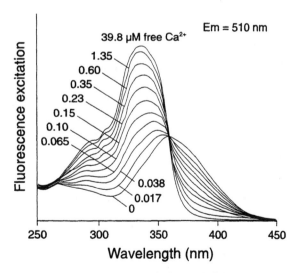

Fig. 3 Excitation spectra of fura-2 in a range of solutions containing 0 to 39.8 μM free Ca. Emission is at 510 nm. (Reprinted by kind permission from Molecular Probes, Inc.)

Randi B. Silver

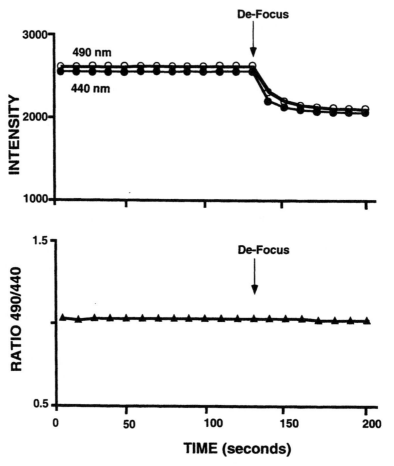

Fig. 4 The effect of changing focus on the fluorescence signal from BCECF. A dish containing 2 μM BCECF free acid in 250 μl salt solution was placed on the stage of an inverted microscope. The top graph shows the individual traces for the emitted fluorescence excitation intensities at 490 nm (open circles) and 440 nm (closed circles) and the bottom graph is the ratio of 490/440. The focus was changed at the arrow. As shown in the top graph both the 490- and 440-nm signals decreased with the change in focus; however, the 490/440 ratio remained the same.

in response to the change in focus while the 490/440 ratio remained constant. If one were only looking at the change of a single excitation wavelength the results would be difficult to interpret.

B. Incorporation into Intact Living Cells

Because fura-2 and BCECF are charged polyanions (see Fig. 1) they are membrane impermeant. In this pentaanion form the only way to introduce

BCECF and fura-2 into a cell is by microinjection, which can be disruptive to the plasma membrane. The synthesis of neutral and membrane-permeant analogs of fura-2 and BCECF was a major accomplishment and provided a noninvasive way for these dyes to enter a cell (Cobbold and Rink, 1987). The neutral membrane permeable analogs were made by replacing the negatively charged carboxylate groups with acetoxymethyl esters (Fig. 1), allowing the dye to enter a cell's cytosol through nonionic diffusion (Tsien, 1989). These membrane-permeant analogs are referred to as the AM form of the dye. Once the lipophilic AM form of the dye diffuses into the cell, intrinsic cytoplasmic esterases cleave off the hydrolyzable AM esters regenerating the membrane impermeant anionic form of the fluorophore. In this form the dye is trapped inside the cytosol and free to bind its ligand (e.g., Ca^{2+} or H^+). This strategy for loading the dye into a cell is summarized in Fig. 5 using BCECF as an example. The cells would be exposed at room temperature to BCECF AM. In the cytosol the AM esters are hydrolyzed yielding the free acid pentaanion form of the dye. For each molecule of BCECF hydrolyzed, 5 molecules of formaldehyde, 5 molecules of acetate, and 10 protons are formed.

C. Signal-to-Noise

The procedures used for loading dye into cells vary from cell type to cell type. Generally speaking, one begins by exposing cells from a concentration of 1–5 μM AM form of the dye in medium or Ringer's solution at room temperature for 30 to 60 min and then rinsing the cells free of dye. Simple buffered salt solutions should be used during an experiment to prevent extrinsic autofluorescence from phenol, amino acids, and vitamins found in some culture media. The fluorescence signal measured at each excitation wavelength should be at least two to three times greater than background noise from the equipment and optics as well as intrinsic autofluorescence of the tissue. Tissue autofluorescence should be measured before exposing the cells to the fluorophore with the imaging system exactly configured (neutral density filters and exposure conditions) for performing measurements. Optimal conditions for yielding a good fluorescence signal will come from balancing loading time, dye concentration, and the appropriate combination of neutral density filters and excitation exposure time. It is important that the experimental cells are not overloaded with dye for this can lead to toxic build-up of the end-products of hydrolysis (see Fig. 5) and buffering of the steady-state intracellular Ca^{2+} and H^+. One wants to ensure that the Ca_i and pH_i measurements represent unperturbed, unbuffered values. When analyzing the experimentally determined fluorescence images for each excitation wavelength of the ratio pair, the background and autofluorescence should be subtracted out on a pixel-by-pixel basis before determining the final ratio value.

D. Potential Problems

It has been reported in some tissues that fura-2 AM hydrolysis may be incomplete leaving partially cleaved forms of fura-2 which fluoresce at 380 nm (Tsien,

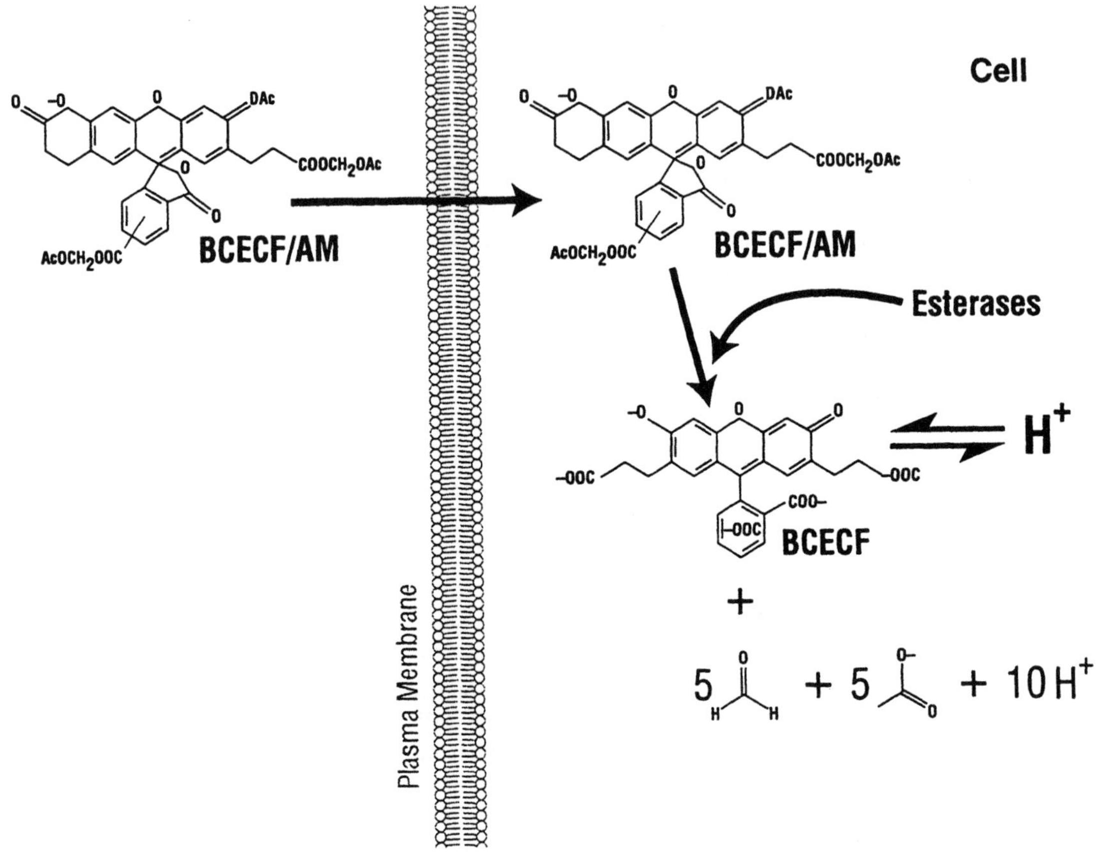

Fig. 5 Cartoon of BCECF AM entering a cell and undergoing hydrolysis. The neutral molecule of BCECF is membrane permeable and readily diffuses into a cell. Once in the cytosol, endogenous esterases cleave the acetoxymethyl esters from the molecule leaving the polyanion trapped inside the cell where it will combine with free H^+ A similar strategy applies for fura-2. The chemical reaction depicted in the figure shows that for every BCECF AM molecule hydrolyzed to its pentaanionic free acid, 5 molecules of formaldehyde, 5 molecules of acetate and 10 protons are formed.

1989). Overloading cells may also lead to compartmentalization of the fluorescence signal in intracellular organelles, thereby not reflecting a contribution of all the intracellular compartments which include the cytosol and the organelles (Tsien, 1989). It has also been shown that hydrolyzed dye can in certain cell types be extruded from the cell by anion transporters. This movement of dye out of the cell can be inhibited by blocking the anion transporter with probenecid (De Virgilio et al. 1988, 1990; Millard et al., 1989; Cardone et al., 1996).

IV. Calibration of the Fluorescence Signal

One way to calibrate the ratio signal is the *ex situ* method. Calibration curves are generated in which intensity ratios are measured in solutions containing the free acid form of the indicator and known pH or pCa values. The experimentally determined ratios are then extrapolated to fit these curves and a relative free ion concentration is generated. An underlying assumption of this technique is that the indicator acts in a similar fashion in salt solutions and cytoplasm, which may not be valid. This protocol often requires the use of viscosity correction factors to account for the difference in behavior of the dye inside the cell versus an external solution (Tsien, 1989).

Another calibration method which more accurately reflects the behavior of the dye inside a cell is the *in situ* method of calibration. It is performed at the end of an experiment on the same field of cells. Specific ionophores or carriers are incorporated into the plasma membrane of the cells allowing one to "clamp" the Ca_i or pH_i to known values. Equilibrium between the external solution and internal environment of the cell is assumed. The corresponding fluorescence signals from the two excitation wavelengths at these known values are then used to calibrate the experimentally determined ratios.

One commonly used technique for the *in situ* calibration of BCECF is the high potassium–nigericin clamp according to the method of Thomas *et al.* (1979). Briefly, the intracellular pH is clamped to known values at the end of an experimental protocol in the presence of nigericin, a K/H antiporter. Potassium solutions are titrated to various pHs in order to force $[K]_i/[K]_o$ equal to $[H]_i/[H]_o$. The slope and y intercept of this relationship can then be used to convert the experimentally determined ratios to intracellular pH values as shown in Fig. 6 (Silver *et al.* 1992).

Intracellular calibration of fura-2 can also be carried out *in situ* according to the technique of Grynkiewicz *et al.* (1985) with the Ca ionophore ionomycin at either saturating levels of Ca (2 mM) to give a maximum ratio (R_{max}) of 340/380 or minimum levels of Ca (2 mM EGTA and no added Ca) to give a minimum ratio (R_{min}) of 340/380. Intracellular Ca concentrations measured at intermediate ratios are calculated according to the relationship:

$$[Ca_i] = K_d \frac{R - R_{min}}{R_{max} - R} \frac{S_{f2}}{S_{b2}},$$

where K_d is the effective dissociation constant (224 nM) at 37°C, S_{f2} is the free dye signal at 380 nm, and S_{b2} is the signal of Ca-bound dye at 380 nm.

Figure 7 is an example of an *in situ* calibration of fura-2 performed at the end of an imaging experiment measuring Ca_i on cultured HeLa cells. An individual cell's response to the addition of ionomycin (10 μM) to the perfusing solution is shown in the graph. The open circles represent the 380-nm excitation response,

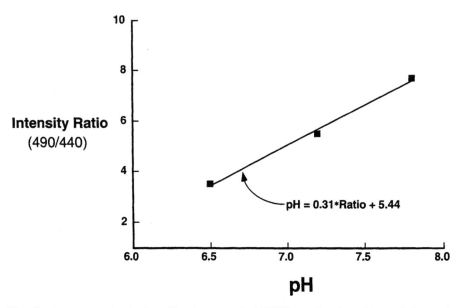

Fig. 6 A representative *in situ* calibration curve for BCECF as a function of intracellular pH in one cell from rabbit kidney cortical collecting duct. Intracellular and extracellular pH were equilibrated with the H/K exchanger nigericin (10 μM) added to potassium solutions titrated to the indicated pH. Experimentally determined ratios were transformed to the appropriate pH using the variables defining the linear function of the calibration curve. (Reproduced from *J. Membr. Biol.* **125**, 13–24 (1992) with kind permission from Springer-Verlag, New York.)

the closed circles represent the 340-nm excitation response, and the closed triangles the 340/380 ratio response. From this *in situ* calibration, the reciprocal relationship between 340 nm excitation and 380 nm excitation at saturating levels of Ca^{2+} (R_{max}) and the reverse of this relationship with a change to a perfusate containing no Ca^{2+} (R_{min}) is shown. The calculated Ca_i value for this cell using the experimentally determined values based on this *in situ* calibration and fitted to the relationship above is 91 nM.

Three pseudocolor ratio images obtained on the HeLa cells are also shown in Fig. 7 corresponding to three different time points. The pseudocolor ratio scale is the same for all three images on this panel, with red reflecting the highest (R_{max}) and dark blue the lowest (R_{min}) ratio value obtained. The pseudocolor image at the top (initial) is from one time point obtained before addition of the ionomycin. The middle image is taken from a point obtained during the period generating the R_{max} value and corresponds to the highest Ca_i bound to fura-2 value obtained. The bottom image is the opposite of this and reflects the least amount of Ca_i bound to fura-2 and was taken during the R_{min} level reached at the end of the trace.

Best results will be obtained if this type of calibration analysis is performed on every individual cell in the field, with each cell generating its own R_{max} and

R_{min} value fitted to the equation presented above. Currently, software allowing this type of calibration is available on some systems or modifications can be made.

V. Components of an Imaging Workstation

Figure 8 is a schematic of an imaging workstation. The fluorescence light path begins with a high-energy lamp (Fig. 8A) attached to the back of the microscope behind the filter wheel. High-pressure xenon and mercury lamps are commonly

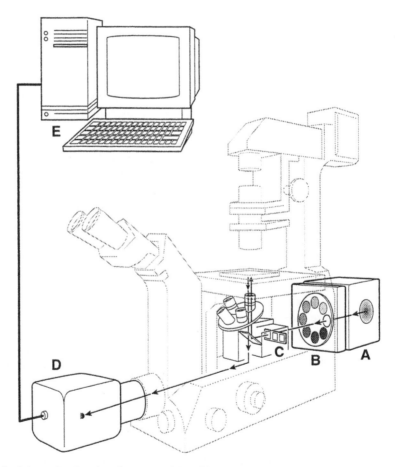

Fig. 8 Schematic of an imaging work station. The components include a high energy lamp (A) connected to a shutter-controlled filter wheel (B) attached to the back of the microscope. The dichroic mirror and emission filter are housed in the base of the nosepiece (C). The low-light-level imaging device (D) is interfaced to a port on the microscope. The image processor and operating system are housed in the computer (E) which controls the entire work station.

used for excitation ratio imaging. Xenon lamps provide a continuous light output from 270 to 700 nm (Lakowicz, 1983) except around 450–485 nm, where one sees some sharp lines similar to those seen in the BCECF spectra in Fig. 1. Mercury lamps produce a higher intensity than xenon lamps but this output is not continuous; rather it is concentrated in lines so these lamps are only suitable if the lines coincide with the peak excitation of the fluorophore. A high-energy light source providing the optimal output of intensity in the excitation range needed should be chosen. A 75-watt xenon lamp works well for both fura-2 and BCECF measurements. It is also important that the high-energy lamp be well focused and provide a flat and evenly illuminated field.

Output from the lamp passes through the appropriate excitation filters housed in a shutter-controlled filter wheel (Fig. 8B). This device changes excitation wavelengths and controls the duration of the illuminating light. To minimize photobleaching and photodynamic damage to the cells, fluorescence excitation can be shuttered off, except during the brief periods required to record an image. In many filter wheels a neutral density filter can be inserted in the same slot as the excitation filter, adding additional protection against photobleaching. Neutral density filters can also be chosen for equalizing the two excitation wavelengths of light.

The two alternating wavelengths of light enter along the optical axis of the microscope and encounter a filter cube housing an appropriate dichroic mirror and emission filter (Fig. 8C). The dual-purpose dichroic mirror reflects the exciting light up through the objective to the sample sitting in the chamber on the stage of the microscope. The emitted fluorescence travels back down through the objective and is transmitted through the dichroic mirror and emission filter to the detector (Fig. 8D). An inverted epifluorescence microscope outfitted with high numerical aperture (na) oil or glycerin immersion objectives (>1.25) is best for low-light-level imaging. The light collecting efficiency of an objective varies approximately with the square of the numerical aperture.

The basic requirements for a detector or camera used in low-light-level imaging are wide dynamic range and linear response within the range of detection. Intensified video cameras and frame transfer charge coupled devices (CCDs) can both be used for ratio imaging. An advantage of the frame transfer CCD is its wide dynamic range, useful and sometimes the only option for performing quantitative ratio imaging on all of the cells viewed in the field. For example, at a given camera gain with a fixed gray level scale of 0 to 255, a cell population may not have homogenous loading. This can be due to a variety of reasons like differences in intrinsic esterase activity or widely disparate subcellular compartment volumes. In experimental situations of this nature, a frame transfer camera with an expanded gray scale would be very helpful. For example, a frame transfer camera containing a frame transfer chip (EEV-37) with 12-bit readout has a gray scale 16 times greater than a standard digital camera and goes from 0 to 4095 gray levels.

Another advantage of using the frame transfer type device for ratio imaging is the speed at which it records an image pair. This is possible due to the chip in the camera which has half of the surface covered by an opaque mask. The first excitation fluorescence excitation wavelength signal may be recorded on the unmasked side and electronically transferred to the masked side. (Bright *et al.*, 1989) The second excitation wavelength is then recorded while the first is being read off the masked side of the chip which allows the recording of two images in rapid succession with on-chip transfer of a 14-bit image taking less than 2 ms (Bright *et al.*, 1989).

The best way to decide which detector is most suitable is to test various detection devices with your biological specimen. If possible try to perform an *in vivo* calibration to see how the detection device performs at the maximum and minimum range of signals. A discussion of the properties of different types of low-light-level cameras may be found elsewhere in this volume.

The entire workstation is controlled by a host computer (Fig. 8E) which includes the operating system and an image processor. The operating system controls the shutter and filter wheel and coordinates the acquisition and analysis of the ratio images. The image processor is required for data acquisition and decreases processing time over that of the computer, permitting read out of the experimentally acquired images in close to real time (Bright *et al*, 1989).

VI. Experimental Chamber and Perfusion System—A Simple Approach

As important as choosing the right components of the imaging workstation is finding an experimental chamber and perfusion system that is easy to use and assemble. Chapter 13 of this volume is devoted to chamber design. This section will be a brief description of a relatively inexpensive chamber and perfusion system routinely used in my laboratory.

Figure 9 is a diagram of a perfusion apparatus consisting of a Lucite rack from which glass syringes are suspended (Fig. 9A). Solutions are gravity fed into a six-port Hamilton valve with six individual inputs and one output (Fig. 9B). The solution exiting from the output port then goes directly into a miniature water-jacketed glass coil (Fig. 9C) (Radnotti Glass Technology, Monrovia, CA) for regulating solution temperature. The warmed solution (37°C) then enters the experimental chamber (Fig. 9D) which fits on the stage of the inverted epifluorescence microscope.

The experimental chamber which was designed to fit on the stage of an inverted microscope is diagrammed in more detail in the inset of Fig. 9. It consists of a piece of Lucite fitted on the bottom and top with interchangeable coverslips (number 1 thickness and 22-mm square) which are temporally attached to the chamber with Dow Corning high-vacuum grease. The specimen area of the chamber contains an entry port for accommodating the entering perfusate and an exit port to convey the perfusate to a reservoir sitting at a lower level. The chamber volume which is exchanged is roughly 600 μl at a rate of about 2 ml/min.

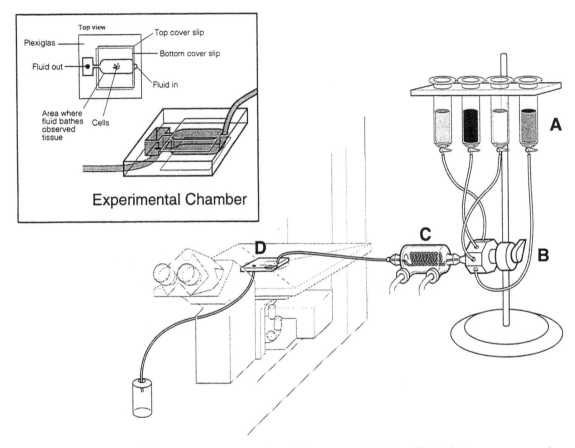

Fig. 9 Diagram of experimental perfusion set-up and chamber. The perfusion apparatus consists of a ring stand with a rack holding glass syringes (A). Each syringe is attached to an input on a Hamilton valve (B) via tubing. The output from the Hamilton valve goes into a miniature water-jacketed glass coil (C) (Radnotti Glass Technology, Monrovia, CA) where it is warmed to 37°C. Output from the coil enters the experimental chamber (D) mounted on the stage of an inverted microscope.

For an inverted microscope, the experimental cells must rest on the surface of the bottom coverslip. This is essential due to the small working distance of the high-na objectives which are required for performing low-light-level imaging experiments. For example, the Nikon Fluor 40X oil immersion objective with variable na (0.8–1.3) has a working distance of 0.22 mm when used with a number 1 size coverslip of 0.17-mm thickness. With this combination of objective and coverslip there is enough working distance to view cells stuck to the coverslip surface facing the chamber. A coverslip is also placed on the top of the chamber to maintain temperature, laminar flow, and optimal conditions for viewing the cells.

VII. Conclusion

In conclusion, the technique of quantitative ratio imaging for the measurement of pH_i and Ca_i has revolutionized the field of cell physiology. Using the proper equipment and choosing the right dyes for your experimental needs should provide reliable and reproducible results. More importantly, the amount of data produced from each experiment when analyzing pH_i and Ca_i in cells on an individual basis yields valuable information on the heterogeneity of cellular responses and will open up new areas for understanding the roles of pH and Ca in signal transduction.

Acknowledgments

This work was generously supported by (NIDDK) Grant DK-45828 and the Underhill and Wild Wings Foundations.

References

Bright, G. R., Fisher, G. W., Rogowska, J., and Taylor, D. L. (1989). Fluorescence ratio imaging Microscopy. *In* "Methods in Cell Biology" (D. L. Taylor and Y. Wang, eds.), Vol. 30, pp. 157–192. Academic Press, San Diego.

Cardone, M. H., Smith, B. L., Mennitt, P. A., Mochly-Rosen, D. Silver, R. B., and Mostov, K. E. (1996). Signal transduction by the polymeric immunoglobulin receptor regulates membrane traffic. *J. Cell. Biol.* **133**, 997–1005.

Cobbold, P. H., and Rink, T. J. (1987). Fluorescence and bioluminescence measurement of cytoplasmic free calcium. *Biochem. J.* **248**, 313–328.

Di Virgilio, F., Steinberg, T. H., and Silverstein, S. C. (1990). Inhibition of Fura-2 sequestration and secretion with organic anion transporter blockers. *Cell Calcium* **11**, 57–62.

Di Virgilio, F., Steinberg, T. H., Swanson, J. A., and Silverstein, S. C. (1988). Fura-2 secretion and sequestration in macrophages: A blocker of organic anion transport reveals that these processes occur via a membrane transport system for organic anions. *J. Immunol.* **140**, 915–920.

Grynkiewicz, G. M., Poenie, M., and Tsien, R. Y. (1985). A new generation of Ca^{2+} indicators with greatly improved fluorescence properties. *J. Biol. Chem.* **260**, 3440–3450.

Heiple, J. M., and Taylor, D. L. (1981). An optical technique for measurement of intracellular pH in single living cells. *Kroc Found. Ser.* **15**, 21–54.

Lakowicz, J. R. (1983). "Principles of Fluorescence Spectroscopy." Plenum, New York.

Millard, P. J., Ryan, T. A., Webb, W. W., and Fewtrell, C. (1989). Immunoglobulin *E. coli* receptor cross-linking induces oscillations in intracellular free ionized calcium in individual tumor mast cells. *J. Biol. Chem.* **264**, 19730–19739.

Rink, T. J., Tsien, R. Y., and Pozzan, T. (1982). Cytoplasmic pH and free Mg^{2+} in lymphocytes. *J. Cell Biol.* **95**, 189–196.

Silver, R. B., Frindt, G., and Palmer, L. G. (1992). Regulation of principal cell pH by Na/H exchange in rabbit cortical collecting tubule. *J. Membrane Biol.* **125**, 13–24.

Thomas, J. A., Buchsbaum, R. N., Zimniak, A., and Racker, E. (1979). Intracellular pH measurements in Ehrlich ascites tumor cells utilizing spectroscopic probes generated *in situ. Biochemistry* **18**, 2210–2218.

Tsien, R. Y. (1989). Fluorescent probes of cell signalling. *Annu. Rev. Neurosci.* **12**, 227–253.

CHAPTER 13

Perfusion Chambers for High-Resolution Video Light Microscopic Studies of Vertebrate Cell Monolayers: Some Considerations and a Design

Conly L. Rieder[*,†,‡] and Richard W. Cole[*]

[*]Division of Molecular Medicine,
Wadsworth Center for Laboratories and Research,
New York State Department of Health,
Albany, New York 12201-0509

[†]Department of Biomedical Sciences,
State University of New York,
Albany, New York 12222

[‡]Marine Biology Laboratory,
Woods Hole, Massachusetts 02543

I. Introduction

For the purposes of this chapter we will define a perfusion chamber as a vessel, mounted on the stage of a light microscope, that enables a living specimen to be followed continuously by high-resolution light microscopy (LM) while also allowing the fluid bathing the specimen to be rapidly exchanged during the observation period. Perfusion chambers are employed in a variety of biological and biomedical studies of cells and tissues. Since they allow the media bathing the specimen to be removed and replenished, they are widely used to maintain specimen viability during long-term investigations (e.g., Pouchelet and Moncel, 1974; Spring and Hope, 1978; Jensen *et al.*, 1994), to study the effects of environmental changes (e.g., Roberts and Trevan, 1961), and to periodically sample for the production of cellular products (e.g., neurotransmitters, Pearce *et al.*, 1981). Perfusion chambers are also employed to deliver tracers for uptake studies (e.g., Sprague *et al.*, 1987) and to assay the effects of drugs (e.g., Bajer *et al.*, 1982; Forscher and Smith, 1988) and other biologically active components (e.g., Inoué and Tilney, 1982; Sakariassen *et al.*, 1983; Varnum *et al.*, 1985; Forscher *et al.*, 1987; Kaplan *et al.*, 1996) on cell/tissue structure and function. Finally, perfusion chambers are required for correlative microscopic studies, where the specimen is followed in the living state and then suddenly fixed during the observation period in order to reconstruct the structural or cytochemical events underlying the process of interest (e.g., Bajer and Jensen, 1969; Rieder and Borisy, 1981; Rieder and Alexander, 1990; Cole *et al.*, 1991; Fig. 1).

The goals of this chapter are twofold. The first is to identify some of the variables that must be considered when constructing or choosing a perfusion chamber for high-resolution video-enhanced LM studies. The second is to describe a rugged and simple laminar-flow chamber used in our laboratory for short-term (≤2-day) studies of vertebrate cell monolayers. It is not our intent to provide a critical and exhaustive analysis of the vast literature on perfusion chambers and their uses in biomedical research. Rather, the narrow content of this chapter is in response to numerous queries over the years on how we conduct our correlative microscopic studies.

In general the perfusion chambers discussed in this chapter are amenable for studying specimens that can be immobilized by adherence to glass coverslips, by steric considerations (e.g., Inoué and Tilney, 1982), and/or by first embedding in a thin (agar) overlay (e.g., Mole-Bajer and Bajer, 1968). They are especially useful for short-term correlative microscopic studies of vertebrate cell monolayers at resolutions afforded by video-enhanced LM. These chambers are not designed for studying organs or slices of tissues (e.g., kidney, lung, or brain) that are many cells thick—these require chambers that meet more demanding environmental conditions (e.g., White, 1966; Schwartzkroin, 1975; Haas *et al.*, 1979; Strange and Spring, 1986). Similarly, without modification the chambers described in this chapter are not designed for studying the transport properties of epithelia sheets, which require that each side of the sheet be bathed by a different solution (e.g., Spring and Hope, 1978). Finally, because they are not "open" systems, where the specimen is readily accessible to physical probes and/ or micromanipulation, they are not amenable without modification to electrophysiological (e.g., Datyner *et al.*, 1985; Chabala *et al.*, 1985; Ince *et al.*, 1985),

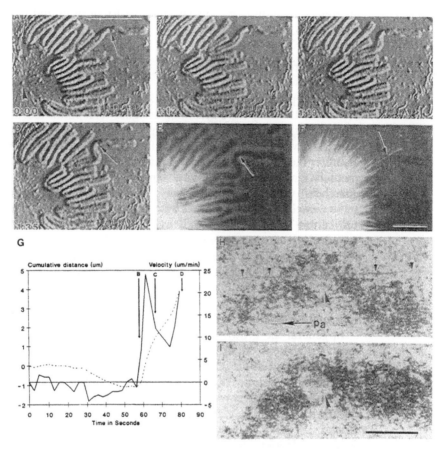

Fig. 1 (A–D) Selected frames from a time-lapse recording of a chromosome (A, arrow) in a newt lung cell being fixed by perfusion immediately after attaching to the spindle and initiating its poleward movement. The polar area (A, arrowhead) is clearly visible throughout the series. Time (min:sec) is on bottom left corner of each frame. Attachment occurs in B, and the primary constriction of the chromosome continues to move poleward (C) until the fixative reaches the cell, 23 sec later (D). This cell was then processed for the indirect immunofluorescent localization of microtubules as shown in E and F. These micrographs were printed to reveal the attaching chromosome (E, arrow) and the microtubule associated with its primary constriction. Note that this microtubule can be followed from within the aster to its termination well past the chromosome. A plot of this chromosome's motion until fixation, in velocity (solid line) and cumulative poleward distance moved (dashed line), is shown in G. The lettered arrows within this plot correspond to the frames pictured in B–D. This cell was subsequently embedded and sectioned for electron microscopy. Serial electron micrographs (H and I) confirm that the thin fluorescent line noted by the arrow in F corresponds to a single microtubule (H, small arrowheads) that is associated with and extends well past the surface of the kinetochore (H and I, large arrowhead). PA (in H), direction of the polar area. Bars (A–D) 20 μm; (D and F) 10 μm; (H and I) 1.0 μm (from Rieder and Alexander. Reproduced from the *Journal of Cell Biology*, 1990, **110**:81–95, by copyright permission of the Rockefeller University Press).

micromanipulation (e.g., Nicklas *et al.*, 1979), or air/liquid interface (e.g., Adler *et al.*, 1987) studies.

II. Historical Perspective

The first perfusion chambers appeared shortly after Harrison's advent of cell culture in 1907. By 1912 chamber designs were published that used a wick to convey perfusate through vessels constructed from glass (Romeis, 1912) or cork with glass windows (Burrows, 1912; for a concise review on the early perfusion chamber literature see Toy and Bardawil, 1958; Roberts and Trevan, 1961). As methods for cell and tissue culture continued to be developed, numerous "growth chambers" were constructed, many of which allowed the cells and tissues inside to be viewed by LM. Over the ensuing years the primitive chambers of the early cell culture pioneers (reviewed in Willmer, 1954) were replaced by optically superior growth chambers designed for the high-resolution live-cell studies afforded by phase- (Zernike, 1942) and differential interference contrast (Nomarski, 1955) LM. When combined with the application of time-lapse cinematography (reviewed in Bajer, 1969), these growth chambers enabled scientists to initiate systematic studies at high spatial and temporal resolution on the structure, function, and behavior of living cellular systems.

In general the growth chambers designed during the 1950s for maintaining and viewing cells under good optical conditions (e.g., Rose, 1954; Rose *et al.*, 1958; Sykes and Moore, 1959, 1960) were based on sandwiching two coverslips, separated, for instance, by a rubber O-ring between two metal plates. A number of perfusion chambers were subsequently constructed from modifying this basic "sandwich" design (e.g., Roberts and Trevan, 1961; Poyton and Branton, 1970; Friend, 1971; Dvorak and Stotler, 1971; Finch and Stier, 1988; Ince *et al.*, 1990; Pentz and Horler, 1992). Unfortunately, although useful for many studies (e.g. Spring and Hope, 1978; Soll and Herman, 1983; Varnum *et al.*, 1985), most of these chambers were not designed for laminar flow. In laminar flow the formation of chemical gradients is minimized because the solution existing within the chamber is rapidly, gently, and completely pushed out (i.e., replaced) by another solution with minimal mixing of the two solutions. By contrast, during nonlaminar and turbulent flow (e.g., Kaplan *et al.*, 1996) the two solutions become rapidly mixed during replacement, which creates an initial chemical gradient. However, regardless of how solutions are exchanged, because necessity remains the mother of invention, most perfusion chambers are usually custom designed to meet the specific requirements of a particular study or class of studies. As a result, new designs continue to appear regularly in the literature (e.g., Ince *et al.*, 1990; Pentz and Horler, 1992; Salih *et al.*, 1992; Sevcik *et al.*, 1993; Kaplan *et al.*, 1996).

III. Basic Perfusion Chamber Designs

Since most forms of LM require that the specimen be imaged between two thin parallel planes of glass for optimum resolution, perfusion chambers for high-resolution LM studies are designed to support two separated but horizontal and parallel glass planes, one of which bears the specimen (Figs. 2–4). For the most part these designs fall into one of three basic categories. The simplest (e.g., Fig. 2) are based on a glass microscope slide, a coverslip, and a spacer consisting, for example, of coverslip fragments (e.g., Cande, 1982; Forscher *et al.,* 1987), U-shaped plastic strips (e.g., Inoué and Tilney, 1982), or teflon tape (e.g., Roth *et*

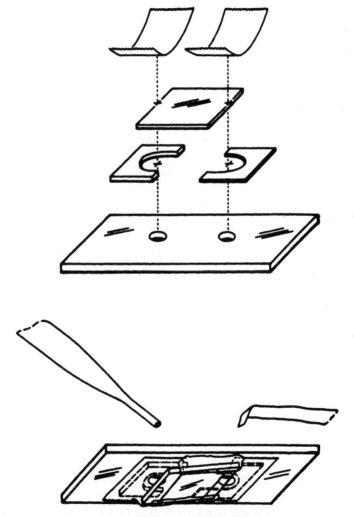

Fig. 2 Schematic diagram of a "microscope slide"-type perfusion chamber used by Inoué and Tilney (1982) to study the acrosomal reaction of *Thyone* sperm. Note that this chamber is designed to be used on an inverted microscope. (Top) The process of assembly. (Bottom) The inverted, assembled chamber as seen from below (from Inoué and Tilney; Reproduced from the *Journal of Cell Biology,* 1982, **93:**812–819, by copyright permission of the Rockefeller University Press).

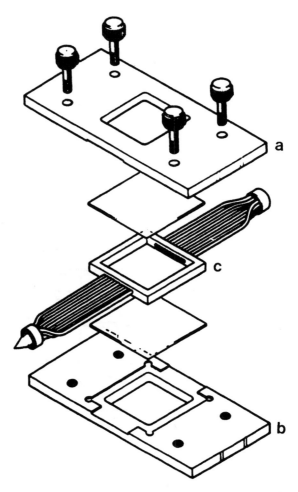

Fig. 3 An exploded view of a "sandwich"-type perfusion chamber designed by Pearce *et al.* (1981) to study neurotransmitter release from CNS cells maintained in monolayer culture. The two coverslips are inserted, cell side facing inward, into the recesses of the outer elements *a* and *b*. Component *c* is positioned between *a* and *b* and held in place by four securing screws. In this design, which uses a nonflexible metal spacer (*c*), the glass-to-metal joints must be sealed with silicone grease (from Pearce, Currie, Dutton, Hussey, Beale and Pigott; Reproduced from the *Journal of Neurosciences Methods,* 1981, **3:**255–259, by copyright permission of Elsevier Sciences Inc.).

al., 1988). The components of the chamber are usually held together by double-stick tape and/or VALAP (an equal part mixture of Vaseline, lanolin, and paraffin). Perfusion is accomplished by gently applying the perfusate to a hole (for inverted microscopes; Fig. 2) or an open end (for upright microscopes) with a syringe or pipette and then drawing it through the chamber with a filter-paper wick placed at an opposing hole or open end. These "microscope slide"-based

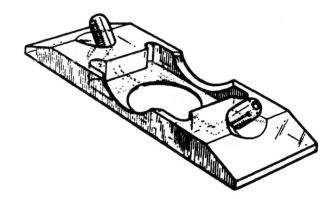

Fig. 4 Perspective drawing of an early "planchet"-type perfusion chamber, milled from polished plexiglass, that was designed and used by Christiansen *et al.* (1953). This sophisticated design meets many of the requirements needed of a perfusion chamber, including the ability to view specimens under continuous perfusion at the highest possible resolution. Its primary weakness is its lack of laminar perfusate flow (from Christainsen, Danes, Allen and Leinfelder; Reproduced from *Experimental Cell Research*, 1953, **5**:10–15, by copyright permission of Academic Press).

perfusion chambers can be quickly constructed and allow rapid exchange of fluids. However, their use requires considerable diligence (and sometimes more than two hands) and they are limited to short-term observations. In addition, since the chamber is always open to the atmosphere, care must be taken to prohibit specimen drying. Finally, because the perfusion process induces subtle fluctuations in the distance between the coverslip and slide, the specimen "bounces," which makes it difficult or impossible to maintain focus.

In the "sandwich" type of perfusion chamber the two glass planes are separated by a spacer, usually composed of silicone rubber or Teflon but sometimes steel (Fig. 3) that has a hole in it through which the specimen is viewed. The glass planes themselves are held securely against the spacer by sandwiching the ensemble between two metal or plastic plates or rings, both of which contain holes that superimpose over the hole in the spacer. The metal plates, in turn, are then clamped or screwed together (e.g., Richter and Woodward, 1955; Poyton and Branton, 1970; Dvorak and Stotler, 1971; Pouchelet and Moncel, 1974; Sakariassen *et al.*, 1983; Finch and Stier, 1988; Ince *et al.*, 1990; Pentz and Horler, 1992; Salih *et al.*, 1992; Kaplan *et al.*, 1996). Most of these chambers are useful for long-term studies because they seldom leak. However, as a rule the viewing area is restricted and with few exceptions (e.g., Pearce *et al.*, 1981) they are not designed for laminar flow.

The third type of chamber is based on a flat, rectangular plexiglass (e.g., Christiansen *et al.*, 1953; Fig. 4), brass (e.g., Mole-Bajer and Bajer, 1968), or stainless steel (e.g., Eijgentstein and Prop, 1970; McGee-Russell and Allen, 1971; Berg and Block, 1984) rectangular plate or "planchet" in which a rectangular hole and inflow/outflow perfusion ports have been milled and soldered. Two coverslips, one of which supports the specimen, are mounted over this hole and sealed in place using VALAP or other sealants (e.g., silicone cement, double-stick tape, or tackiwax). When made of stainless steel these planchet-based chambers are rugged and many are designed for laminar flow (see below). However, they all have a restricted viewing area and are susceptible to leakage when not properly assembled.

IV. General Considerations When Selecting or Building a Perfusion Chamber

Probably the foremost consideration in designing or selecting a perfusion chamber is the resolution required by the study, because this will define the maximum chamber thickness. For high-resolution performance the light micro-scope must be aligned for Köhler illumination, which involves focusing an image of the field diaphragm on the specimen plane with the condenser lens. The maximum distance that the two glass planes can be separated for perfusion chambers, while still allowing for Köhler alignment, will be dictated by the position of the specimen and the numerical apertures (NA) of the condenser and objective. In general, the lower the NAs, the further the two parallel glass planes can be separated, but the poorer the resolution. High-resolution video-enhanced LM studies require a condenser and objective NAs of 1.3–1.4, which restrict the total separation distance between the two glass plane surfaces (i.e., the chamber thickness) to near that of a single microscope slide and coverslip (i.e., ≤1.2 mm).

Because perfusion chambers for high-resolution LM are so thin, they hold very small volumes of fluid, usually ≤1 ml and often only 50–250 μl (see Finch and Stier, 1988 for a perfusion chamber with a 5-μl volume). For some studies, even relatively long-term ones, this may not be a problem. We have found, for example, that cultures of amphibian (newt) epithelial cells remain healthy in our perfusion chambers, which hold only 250 μl of fluid, for up to 2 days at 23°C without a media exchange. On the other hand, we cannot maintain monolayers of mammalian cells in these same chambers at the required 35–37°C unless we perfuse them with fresh media every 20–30 min (and then we can keep them healthy ≥2 days—see Jensen *et al.*, 1994). The need in some studies to frequently replenish (or sample) the media around the specimen has led to construction of perfusion chambers that are designed to maintain a continuous long-term flow of fluid across the cell surface (e.g., Pouchelet and Moncel, 1974; Sprague *et al.*, 1987; Salih *et al.*, 1992).

The surface area of the chamber over which the specimen(s) can be viewed will again depend on the type of investigation. In general studies on random populations of cells that are distributed relatively densely and evenly across the specimen plane may require a chamber containing only a nominal glass surface area (e.g., Berg and Block., 1984). By contrast, a much larger surface area may be needed, for example, if the study involves very large cells of reticulopodial networks (e.g., Allogromia; see Rupp *et al.,* 1986), searching through many cells to find one in a particular stage of mitosis (e.g., Mole-Bajer and Bajer, 1968; Rieder and Borisy, 1981), or if specimen density is low. Regardless of the surface area, it is important that the two glass planes forming the specimen cavity be maintained in a strain-free, optically flat, and parallel array which, when combined with the thin nature of the chamber, makes chambers with large viewing areas more susceptible to breakage and/or leakage.

When considering perfusion chambers it is also important to remember that the barrel of the objective lens has a considerable diameter and that when the specimen is in focus the lens is extremely close to the glass plane supporting the specimen (and connected to it by an oil interface). These features severely restrict the usable viewing area of a chamber if the specimen-bearing (glass) substrate is rimmed by a border of metal with an appreciable thickness, as is commonly the case in sandwich- and planchet-based chambers (e.g., Fig. 3; also see discussion in Finch and Stier, 1988). To minimize this problem some chambers are designed so that the metal rim supporting the viewing substrate is thinner near the viewing area (e.g., Bajer and Mole-Bajer, 1968; Eijgenstein and Prop, 1970; Poyton and Branton, 1970; McGee-Russell and Allen, 1971; Finch and Stier, 1988).

Obviously, the components of the chamber that contact the specimen and the fluid surrounding the specimen cannot be toxic. For the most part modern chambers are constructed from glass, stainless steel, silicone rubber, and/or an adhesive (silicone glue, grease, or VALAP). When new, the first two components can contain toxic traces of heavy metals and organic solvents that should be removed by a thorough cleaning. The other components, including silicone-based rubbers and glues and VALAP, are generally nontoxic to cells.

A frequently overlooked (e.g., Finch and Stier, 1988; Ince *et al.,* 1990; Pentz and Horler, 1992; Sevcik *et al.,* 1993) but often important consideration is that the chamber be designed for laminar flow so that solutions within it can be exchanged rapidly and with relatively little mixing (e.g., see Section II). For some studies laminar flow may not be of critical importance (e.g., Kaplan *et al.,* 1996), but for others it clearly is (e.g., see Wetzels *et al.,* 1991; Inoué and Spring, 1997). For the characteristics of laminar flow, and how it is achieved in perfusion chamber designs, the reader is referred to Inoué and Spring (1997).

The fluid in a perfusion chamber is usually exchanged by passing new fluid into the chamber through an inlet port, which in turn forces the old fluid out through an exit port. In all chambers the specimen bounces out of focus to some degree as the perfusion process is initiated and/or terminated. This is especially true of microscope-slide-type chambers in which perfusion cannot be carefully controlled. Depending on the goal of the study, how often the specimen is photographed (i.e., the time-lapse framing rate), and the frequency of perfusions, specimen bouncing may not be important. However, it can seriously affect those studies that require very high temporal resolution (≤ 2 sec) as, for example, when assaying a rapid response (e.g., Strange and Spring, 1986) or fixing cells for a subsequent EM analysis (e.g., Rieder and Alexander, 1990; McEwen *et al.*, 1996; Fig. 1).

In order to minimize specimen bounce, pressure changes within the chamber must be minimized when the perfusion process is started and stopped. To achieve this the diameter of the inflow port should be significantly smaller than that of the outflow port (so that hydrostatic pressure does not build up during fluid flow). In addition, instead of using a peristaltic pump to force fluid through the chamber (e.g., Soll and Herman, 1983; Berg and Block, 1984; Varnum *et al.*, 1985), it should be gravity-fed from a reservoir above the microscope stage and the effluent exiting the chamber directed into another "catch basin" reservoir below the chamber (e.g., Mole-Bajer and Bajer, 1968; McGee-Russell and Allen, 1971; Spring and Hope, 1978). If necessary the effluent can then be filtered, treated, and pumped into the upper reservoir to be recycled (e.g., Pouchelet and Moncel, 1974; Salih *et al.*, 1992). The rate of perfusion can be controlled by varying the vertical distance between the two reservoirs or by a stopcock that controls the flow from the upper reservoir (see Fig. 5).

In order to avoid air bubbles during the perfusion process, which can rapidly ruin an experiment, the perfusate and the perfusion chamber should form a closed system (see Mole-Bajer and Bajer, 1968; McGee-Russell and Allen, 1971; Fig. 5). Air hidden in the system, or potential leaks around the chamber or at tube/chamber junctions, should be eliminated after the specimen is mounted but prior to initiating the viewing process. For gravity-fed systems the easiest way to check for and prevent air bubbles is to fill the system and then shut it down with a stopcock when it is fully loaded with fluid, once all the air and leaks have been eliminated. Indeed, the system should be completely closed and filled with fluid prior to the first perfusion in order to minimize specimen bounce as perfusions are started and stopped (see Fig. 5). This includes the effluent tube that conveys fluid from the exit port into the lower catch basin, which should extend into the reservoir and terminate in fluid. The goal here is to prevent drops of fluid from breaking away and falling from the end of the effluent tube during the perfusion.

When choosing a perfusion chamber the position of the inflow and outflow ports should also be considered since their placement can complicate rapid changes of objectives and also limit the type of microscope that the chamber can be used on. Most sandwich-type chambers are designed so that the inflow and outflow ports are positioned on the same or opposing chamber *edges* (see Fig. 3). These types of designs do not hinder the rotation of objectives around the nose turret, and they can be used interchangeably on upright and inverted microscopes if the top and bottom aspects are the same (e.g., Dvorak and Stotler, 1971; Finch and Stier, 1988; Pentz and Horler, 1992; Ince *et al.*, 1990). Unfortunately, few of the chambers designed with side-mounted perfusion ports deliver a laminar flow during fluid exchange (see above). Instead, most laminar-flow chambers are designed so that the inflow/outflow ports are positioned on the top of the chamber which, depending on their design may (e.g., Christiansen *et*

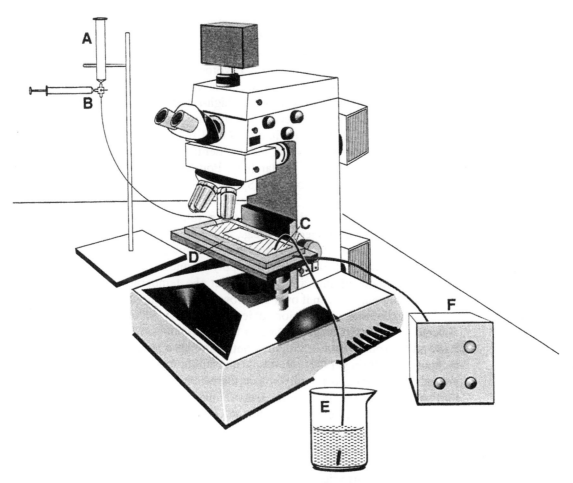

Fig. 5 A schematic diagram showing how perfusate can be gravity-fed into and through our perfusion chamber with minimum specimen bounce. The upper fluid reservoir (A) is connected with the inflow port of the chamber. Perfusate is then gravity-fed to the chamber, and the rate of flow is controlled by a three-way stopcock (B) which can also be used to introduce a different solution into the chamber. Fluid passing through the chamber then exits the chamber via the outflow port (C), where it is fed by gravity into a lower reservoir (E). The chamber temperature is controlled (F) by mounting it in a heating block (D) on the microscope stage. See text for details.

al., 1953; Salih *et al.*, 1992) or may not (e.g., McGee-Russell and Allen, 1971; Inoué and Tilney, 1982) interfere with changes in objective lenses. Also, because the perfusion ports protrude away from and not into the microscope stage (e.g., Fig. 4), these chambers cannot be used interchangeably between upright and inverted microscopes unless they are modified so that the specimen can be mounted interchangeably on either glass plane (e.g., see Section V,A).

For many studies it is important to maintain the temperature of the specimen above the ambient temperature and a number of methods have been developed for this task (e.g., Pouchelet and Moncel, 1974; Chabala *et al.,* 1985; Lowndes and Hallett, 1986; Payne *et al.,* 1987; Toyotomi and Momose, 1989; Ince *et al.,* 1990; Sevcik *et al.,* 1993). Few of these, however, considered the influence of the objective (or condenser) lens, which in high-resolution studies resides ≤1 mm from the specimen and is connected to it by glass and oil. Under these conditions both lenses act as effective radiators to cool the specimen. Early attempts to overcome this problem involved independently regulating the temperature of the objective with a heating collar (e.g., Pouchelet and Moncel, 1974; Rieder and Bajer, 1977; see also Inoué and Spring, 1997). A simpler and more popular solution is to use a heating element and fan (e.g., an air-curtain incubator or hair drier) to blow air of the desired temperature across a thermistor placed near the objective and specimen (e.g., Soll and Herman, 1983). Temperature near the specimen can then be automatically regulated, within narrow limits, by a feedback control that turns the air supply on and off at the appropriate time. The disadvantage of this approach is that the temperature of the specimen constantly fluctuates, even if over a narrow range of temperatures. Also it cannot be used with water-immersion objectives and it induces severe specimen "bounce" and focus changes whenever the air stream is switched on or off (which leads to the production of less than esthetic time-lapse sequences). As emphasized by Inoué and Spring (1997), the surest and simplest way to control the temperature of the specimen is to simply place the whole microscopic system, including the loaded chamber, in a room (e.g., Tilney and Porter, 1967; Hiraoka and Haraguchi, 1996) or box (e.g., Salih *et al.,* 1992; Pentz and Horler, 1992) that is maintained at or near the desired temperature. If needed, the chamber temperature can then be fine tuned by the use, for example, of Peltier devices (e.g., Chabala *et al.,* 1985; Delbridge *et al.,* 1990; Inoue and Spring, 1997)

Special problems arise when studying organisms that need to be maintained at very low temperatures or when examining the effects of sudden cooling on a specimen. For long-term studies the method-of-choice is again to conduct the experiment in a cold room set at the desired temperature (e.g., Stephens, 1973; Rieder and Borisy, 1981). In addition to its simplicity this approach eliminates the plaguing problem of water condensation that forms on cold surfaces that reside in warm rooms. On the other hand this method is not amenable to delivering sudden and reversible cold shocks to specimens usually studied at or above ambient temperatures. For these studies "cooling" perfusion chambers are used most of which are based on flowing chilled alcohol solutions through a separate glass chamber which is constructed so that one of its walls is the opposite side of the glass supporting the specimen (e.g., Inoué *et al.,* 1970; Stephens, 1973; Lambert and Bajer, 1977). Again, it is important to consider the influence of the objective and condenser, which in these studies can effectively raise the temperature of the cell (Lambert and Bajer, 1977). Frosting can be prevented by coating all external glass surfaces with Photo-Flo (Eastman Kodak Co., Rochester, NY) prior to initiating the experiment (e.g., Stephens, 1973).

In many investigations, including those conducted in our laboratory, maintaining an exact temperature is not as important as consistently achieving and maintaining the same temperature. Thus, we keep the microscope room at a constant 20–23°C and control the specimen temperature by maintaining the perfusion chamber at 39–40°C as measured by a thermistor near the specimen (e.g., see Section VB). Under these conditions the temperature of the cells approximates 37°C, and there is little temperature variability between experiments.

V. A Perfusion Chamber for Video–LM of Cultured Cell Monolayers

In this section we describe a perfusion chamber used in our laboratory for correlative microscopic studies on vertebrate cell monolayers. It holds ~250μl of fluid, allows for video-enhanced LM studies of the highest resolution (e.g., Hayden *et al.,* 1990; Fig. 6, and solutions can be changed with laminar flow very rapidly ($\leq$1 sec with flow velocities approaching 1 ml/sec). It is rugged, inexpensive to build, and easy to assemble and use. Although originally designed for use on an upright microscope (Fig. 7), it can be adapted for use on an inverted system (Fig. 8). The primary disadvantage of our chamber is that it must be heated to mount the nonspecimen-containing coverslip. In practice this means that it cannot be assembled rapidly from scratch in an emergency situation (which sometimes occurs when one is trying to catch a rare event). However, this problem can be eliminated by simply keeping a stock of partly assembled chambers available at all times. Another disadvantage is that, like all planchet-based chambers, it can spring leaks at the most inopportune times if not properly assembled.

Our perfusion chamber is modified from one built by McGee-Russell and Allen (1971) to study reticulopodial networks in marine foraminifera, which in turn was adapted from one designed by Mole-Bajer and Bajer (1968) to study mitosis in *Haemanthus* endosperm. For our work we increased the mass of the chamber in order to make it more durable and to reduce warping during the machining process. We also enhanced its laminar-flow characteristics and strengthened the otherwise fragile inflow/outflow ports by modifying their placement. Finally, we designed an adaptation for use on an inverted LM.

A. Design and Fabrication

Both chambers, whether for use on an upright (Fig. 7) or inverted (Fig. 8) LM, are milled from 0.25 × 1.25-inch-type 304 stainless steel bar stock. During fabrication the stock is cut into 80-mm-long planchets that approximate the chamber length, and the top and bottom surface of each planchet are then milled

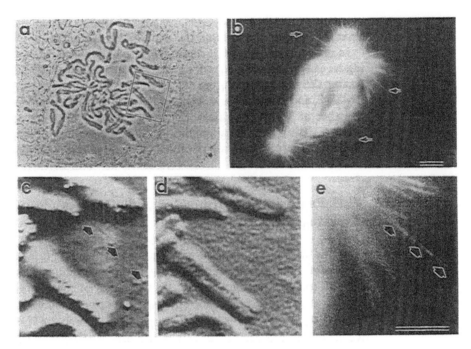

Fig. 6 Phase-contrast (a) and fluorescence (b) micrographs of an early prometaphase newt pneumocyte after antitubulin staining. This cell contains a number of monooriented chromosomes and the spindle is surrounded by a clear area of cytoplasm (asterisks in a). Note that some of the longer astral microtubules (arrowheads in b) terminat between the distal ends of the monooriented chromosomes and the periphery of the clear area. High-magnification video-enhanced DIC (c), conventional DIC (d), and immunofluorescence (e) micrographs of the region outlined by the box in a, just before (c) and after (d and e) perfusion fixation and antitubulin staining. Note that the thin linear element seen in the living cell corresponds to a single fluorescent line; *that is,* microtubule (compare c and d, arrows). Bars: (a and b) 10 μm; (c–e) 5 μm. (From Hayden, Bowser and Rieder; Reproduced from the *Journal of Cell Biology,* 1990, **111:**1039–1045, by copyright permission of the Rockefeller University Press.)

until the planchet is the appropriate thickness (4.75 mm). During this milling step it is critical that these surfaces are made perfectly flat and parallel. Each planchet is then milled to the correct length (76 mm). Next a large "lens clearance" depression is milled out of each planchet (see Fig. 7). Electrical discharge machining (EDM) is then used to create a continuous 0.25-mm-deep groove or trough on that side of the planchet opposite the objective depression (Fig. 7). This groove, which traces a long, thin diamond shape will prevent VALAP, used to secure a coverslip to this side during the assembly process, from invading the chamber reservoir and perfusion ports. EDM is then used to create a very shallow (0.25-mm-deep) elongated diamond-shaped depression on this same side (Fig. 7), which will become the fluid reservoir between the two coverslips. EDM is the method of choice for making these cuts because end-milling introduces excessive heat, which can warp the thinner areas of the chamber. To increase

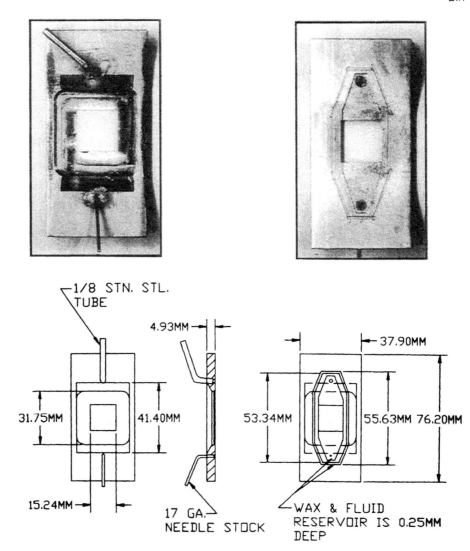

Fig. 7 The objective-facing top (left) and condenser-facing bottom (right) of a laminar-flow perfusion chamber designed for use with monolayer cell cultures and an upright microscope. (Bottom) Schematic diagram of the chamber viewed from various aspects including the top (left), side (middle), and bottom (right). All measurements are in millimeters. See text for details.

their life spans the EDM electrodes should be milled from copper graphite rather than pure carbon.

The final fabrication stages vary, depending on whether the chamber is to be used on an inverted or upright microscope. For chambers used on upright microscopes (Fig. 7), inflow and outflow holes are drilled completely through the planchet and counter-bored on the side containing the fluid reservoir, which

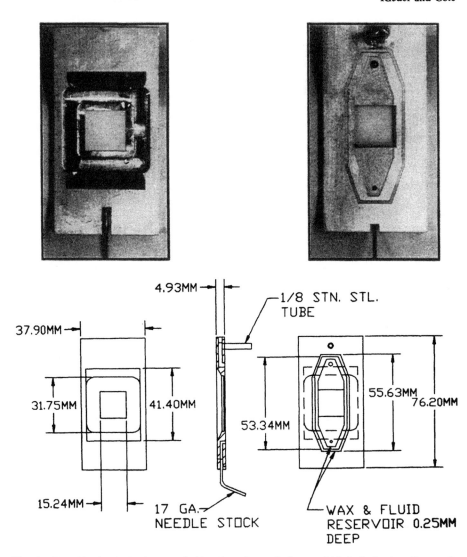

Fig. 8 The objective-facing bottom (left) and condenser-facing top (right) of a laminar-flow perfusion chamber designed for use on an inverted microscope. (Bottom) Schematic diagram of the chamber viewed from various aspects including the bottom (left), side (middle), and top (right). All measurements are in millimeters. See text for details.

will ultimately lie against the microscope stage. For the inverted microscope design three holes are drilled for each port (Fig. 8). Initially two sets of adjacent holes, each of which will form a port, are drilled from the surface opposing the objective depression. These holes are on opposite ends of the chamber and must be the same depth. Next, the adjacent holes are connected by drilling another hole of the same size through each end of the chamber. The ends of these holes are then plugged with soft silver (i.e., lead-free) solder.

Stainless steel inlet (17-gauge needle stock; O.D. = 1.28 mm) and outlet (stock, O.D. = 3.175 mm; Small Parts Inc., Miami Lake, FL 33014) tubes are then soft-silver soldered to the chamber so that they extend away and slant from that chamber surface facing the condenser (Figs. 7 and 8). Finally, a square window is cut through the chamber for viewing the specimen (Figs. 7 and 8).

B. Temperature Regulation

We control the temperature of the specimen in our system by adjusting the temperature of the perfusion chamber, which has a considerable thermal mass. We do this by mounting the fully assembled chamber inside a hollowed metal block, milled from type 606 aluminum, that sits flush on the microscope stage (Fig. 9). This manner of thermal regulation eliminates focus shifts visible at moderate-to-high magnification that are generated by air curtain incubators/ heaters. We control the temperature of the heating block by using a proportional temperature controller (Nexus Corp., Brandon, VT) that regulates the electrical

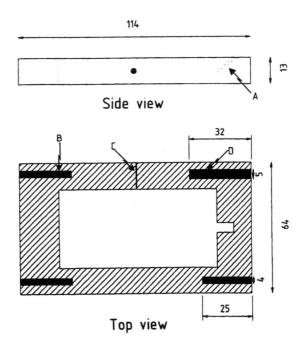

Fig. 9 Schematic diagram of the aluminum block heater used to maintain the temperature of the perfusion chambers pictured in Figs. 7 and 8. A, cutout for perfusion chamber inflow tube; B, one of three cartridge heaters; C, thermocouple; D, thermistor. See text for details. All measurements are in millimeters.

current delivered to three 15-watt heater cartridges (Prime Inc., Paramus, NJ) mounted in the block (see Fig. 9). This controller uses a thermistor as part of a feedback loop for temperature regulation. A digital thermistor (Omega Engineering, Stamford, CT) can also be attached to the block for verifying temperature visually.

For reasons noted in Section IV we make no attempt to control the temperature of the objective lens. When a high-NA immersion objective at room temperature is focused on the cells, the temperature difference between the cells and the 39–40°C heating block remains relatively constant when room temperature is maintained between 19–23°C. By placing a very small (0.0005-inch diameter) type K thermocouple (Omega Engineering, Stamford, CT, model chal-0005) into the oil between the coverslip and specimen we have determined that, when the heating block is 39–40°C, the center of the specimen bearing coverslip is ~37°C.

Although we have yet to explore the utility of our system for maintaining specimens at temperatures below ambient, it is possible that the heaters could be replaced by a Peltier device (i.e., a semiconductor device that functions similarly to a heat pump), mounted on top of the chamber, that allows for controlled cooling (or heating; e.g., see Delbridge *et al.*, 1990.)

C. Assembly

In this section we describes how to assemble the perfusion chamber shown in Fig. 7, which is designed for use on an upright microscope. Assembly of the chamber for use on inverted microscopes (Fig. 8) will not be detailed since only a few obvious modifications are required from the method used to assemble chambers used on upright LMs. Prior to assembly the chamber should be carefully cleaned of adhering oils, chemicals, VALAP, and glass fragments by soaking it in xylene for 20 min in a fume hood. The chambers should then be boiled or sonicated for 20–40 min in a 10% w/v sparkleen (Fischer Scientific Co., Pittsburgh, PA) solution and then sonicated in three changes of distilled H_2O for 5 min each.

For assembly the clean chamber is placed flat-surface down on a hot plate maintained at ~85°C. Once hot it should be inverted and placed on a platform that will hold it stationary with its flat surface up (as shown in Fig. 7, upper right). The surface of the chamber peripheral to the outer trough or groove should then be immediately but carefully covered with a thin layer of VALAP, which will remain molten as long as the chamber is hot (see Fig. 7). [We use VALAP (MP = 50°C) for assembling chambers to be used at room temperature, but substitute a 50/50 mixture of VALAP/tackiwax (Boekel Ind., Philadelphia, PA) when assembling chambers for use at 37°C.] Next a clean number 1.5 coverslip (24 × 60 mm) should be placed flush on the VALAP-covered chamber surface. (Although coverslips thicker than 1.5 are more durable and thus less susceptible to breakage, they should not be used in constructing the chamber because they will compromise final resolution.) The molten VALAP will then quickly form, by capillary motion, a thin and complete interface between the chamber surface and coverslip. Care must be taken to ensure that enough VA-

LAP is used so that a complete seal forms between the coverslip surface contacting the metal surface of the chamber. However, too much VALAP will induce the coverslip to ride unevenly on the chamber surface and, most likely, some of it will flow past the outer groove (which was originally made during chamber fabrication to prevent the migration of VALAP into the chamber cavity and perfusion ports) and into the perfusate reservoir when the coverslip is applied. Once this coverslip is properly mounted the chamber should be allowed to cool undisturbed. It can then be sterilized by UV irradiation or washing with 70% ethanol.

The chamber is now ready to accept cells, which should be grown on 22- to 25-mm square number 1.5 coverslips. For this procedure place the chamber bottom-side down on a clean flat surface (as seen in Fig. 7, upper left). Next, using a syringe with an 18-gauge needle connected to a short length of 1.2-mm I.D. tubing, fill the chamber cavity through the inlet port almost to the top with media (careful; this will require ≤ 250 μl of fluid). Partly filling the chamber with media prevents the cells from drying out or overheating during the subsequent steps. Next, thoroughly dry that surface of the specimen-containing coverslip that does not contain the cells. This is important since any moisture on this surface will inhibit the formation of a good VALAP seal. Next, carefully place the specimen-containing coverslip over the chamber space specimen-side down. Using slivers of filter paper blot away all excess media from around the edges of the coverslip and dry the regions of the chamber surrounding the coverslip. Next, using a paintbrush, carefully seal the coverslip to the chamber by applying several thin coats of molten VALAP around its perimeter. After the VALAP hardens, remove all residual air trapped in the chamber by completely filling it with media via the syringe while holding it in a vertical position with the outlet port up. At this time check for and repair any leaks around the specimen coverslip/chamber seal.

The chamber is now ready to be mounted on the microscope stage and coupled to the perfusion system. As outlined in Section III we deliver various perfusates by gravity feeding them via intramedic tubing (I.D. = 1.20 mm; O.D. = 1.70 mm; number 427436, Becton Dickinson, Sparks, MD 21152) from a reservoir held above the chamber. For an upper reservoir we generally use a large-capacity (50–100 ml) syringe without its plunger. Fluid from this reservoir is then fed via an 18-gauge needle and intramedic tubing to the inflow port of the chamber. Flow is controlled and regulated by placing a stopcock in the tubing between the reservoir and chamber. Finally, tygon tubing is then used to convey fluid leaving the chamber into a bottom reservoir (see Section IV).

After use, both coverslips and most of the VALAP should be removed from the chamber with a razor blade. They should then be cleaned for reuse as described above.

Acknowledgments

This paper is dedicated to a pioneer of perfusion chamber designs, Dr. Sam McGee-Russell, who retired from academia on July 25, 1996. The authors thank Dr. Sam Bowser and Dr. Ken Spring for providing material, thoughtful discussions, and valuable comments on the manuscript. We also thank Mr. George Matuszek of the Wadsworth Center machine shop for assistance with the design and milling of the chambers described in this paper, and Drs. Bob Palazzo and Roger Sloboda for generously making writing space available in their MBL/Woods Hole laboratory. The original work described in this chapter was supported by NIH Grant GMS 40198 to C.L.R. and by NIH Grant RR 01219, awarded by the Biomedical Research Technology Program (NCRR/DHHR-PHS) to support the Wadsworth Center's Biological Microscopy and Image Reconstruction facility as a National Biotechnological Resource.

References

Adler, K. B., Schwartz, J. E., Whitcutt, M. J., and Wu, R. (1987). A new chamber system for maintaining differential guinea pig respiratory epithelial cells between air and liquid phases. *BioTechniques* **5**, 462–465.

Bajer, A. S. (1969). Time-lapse microcinematography in cell biology. *Theoret. Exper. Phys.* **3**, 217–267.

Bajer, A. S., and Jensen, C. (1969). Delectability of mitotic spindle microtubules with the light and electron microscopes. *J. Microsc.* **8**, 343–354.

Bajer, A. S., Cypher, C., Mole-Bajer, J., and Howard, H. M. (1982). Taxol-induced anaphase reversal: Evidence that elongating microtubules can exert a pushing force in living cells. *Proc. Natl. Acad. Sci. USA* **79**, 6569–6573.

Berg, H. C., and Block, S. M. (1984). A miniature flow cell designed for rapid exchange of media under high-power microscope objectives. *J. Gen. Microbiol.* **130**, 1915–1920.

Burrows, M. T. (1912). A method of furnishing a continuous supply of new medium to a tissue culture *in vitro*. *Anat. Rec.* **6**, 141–144.

Cande, W. Z. (1982). Permeabilized cell models for studying chromosome movement in dividing PtK$_1$ cells. *Methods Cell Biol.* **25**, 57–68.

Chabala, L. D., Sheridan, R. E., Hodge, D. C., Power, J. N., and Walsh, M. P. (1985). A microscope stage temperature controller for the study of whole-cell or single-channel currents. *Pflugers Arch.* **404**, 374–377.

Cole, R. W., Ault, J. G. Hayden, J. H., and Rieder, C. L. (1991). Crocidolite asbestos fibers undergo size-dependent microtubule-mediated transport after endocytosis in vertebrate lung epithelial cells. *Cancer Res.* **51**, 4942–4947.

Christiansen, G. S., Danes, B., Allen, L., and Leinfelder, P. J. (1953). A culture chamber for the continuous biochemical and morphological study of living cells in tissue culture. *Exp. Cell Res.* **5**, 10–15.

Datyner, N. B., Gintant, G. A., and Cohen, I. S. (1985). Versatile temperature controlled tissue bath for studies of isolated cells using an inverted microscope. *Pflugers Arch.* **403**, 318–323.

Delbridge, L. M., Harris, P. J., Pringle, J. T., Dally, L. J., and Morgan, T. O. (1990). A perfusion bath for single-cell recording with high-precision optical depth control, temperature regulation, and rapid solution switching. *Pflugers Arch.* **416**, 94–97.

Dvorak, J. A., and Stotler, W. F. (1971). A controlled environment culture system for high resolution light microscopy. *Exp. Cell Res.* **68**, 144–148.

Eijgenstein, L. H., and Prop, F. J. A. (1970). A simple versatile large-surface tissue culture chamber for phase contrast photography and microcinematography. *Exp. Cell Res.* **60**, 464–466.

Finch, S. A. E., and Stier, A. (1988). A perfusion chamber for high-resolution light microscopy of cultured cells. *J. Microsc.* **151**, 71–75.

Forscher, P., Kaczmarek, L. K., Buchanan, J., and Smith, S. J. (1987). Cyclic AMP induces changes in distribution of organelles within growth cones of *Aplysia* bag cell neurons. *J. Neurosci.* **7**, 3600–3611.

Forscher, P., and Smith, S. J. (1988). Actions of cytochalasins on the organization of actin filaments and microtubules in a neuronal growth cone. *J. Cell Biol.* **107**, 1505–1516.

Friend, J. V. (1971). Improvements to the functional design of the Roberts and Trevan type culture chamber. *J. Microsc.* **94**, 79–82.

Haas, H. L., Schaerer, B., and Vosmansky, M. (1979). A simple perfusion chamber for the study of nervous tissue slices *in vitro*. *J. Neurosci. Methods* **1**, 323–325.

Harrison, R. G. (1907). Observations on the living developing nerve fiber. *Proc. Soc. Exp. Biol. Med.* **4**, 140–143.

Hayden, J. H., Bowser, S. S., and Rieder, C. L. (1990). Kinetochores capture astral microtubules during chromosome attachment to the mitotic spindle: Direct visualization in live newt lung cells. *J. Cell Biol.* **111**, 1039–1045.

Hiraoka, Y., and Haraguchi, T. (1996). Fluorescence imaging of mammalian living cells. *Chromosome Res.* **4**, 173–176.

Ince, C., van Dissel, J. T., and Diesselhoff, M. M. C. (1985). A Teflon culture dish for high-magnification microscopy and measurements in single cells. *Pflugers Arch.* **403**, 240–244.

Ince, C., Beekman, R. E., and Verschragen, G. (1990). A micro-perfusion chamber for single-cell fluorescence measurements. *J. Immunol. Methods* **128**, 227–234.

Inoué, S., Ellis, G. W., Salmon, E. D., and Fuseler, J. W. (1970). Rapid measurement of spindle birefringence during controlled temperature shifts. *J. Cell Biol.* **47** (2), 95a–96a.

Inoué, S., and Tilney, L. G. (1982). Acrosomal reaction of *Thyone* sperm. 1. Changes in the sperm head visualized by high resolution video microscopy. *J. Cell Biol.* **93**, 812–819.

Inoué, S., and Spring, K. R. (1997). "Video Microscopy: The Fundamentals," 2nd edition, p. 741. Plenum, NY.

Jensen, C. G., Jensen, L. C. W., Ault, J. G., Osorio, G., Cole, R., and Rieder, C. L. (1994). Time-lapse video light microscopic and electron microscopic observations of vertebrate epithelial cells exposed to crocidolite asbestos. *In* "Cellular and Molecular Effects of Mineral and Synthetic Dusts and Fibers" (J. M. Davis and M.-C. Jurand, Eds.), Vol. H85, pp. 63–78. Springer-Verlag, Berlin.

Kaplan, D., Bungay, P., Sullivan, J., and Zimmerberg, J. (1996). A rapid-flow perfusion chamber for high-resolution microscopy. *J. Microsc.* **181**, 286–297.

Lambert, A.-M., and Bajer, A. S. (1977). Microtubule distribution and reversible arrest of chromosome movements induced by low temperature. *Cytobiologie* **15**, 1–23.

Lowndes, R. H., and Hallett, M. B. (1985). A versatile light microscope heating stage for biological temperatures. *J. Microsc.* **142**, 371–374.

McEwen, B. F., Heagle, A. B., Cassels, G. O., Buttle, K. F., and Rieder, C. L. (1997). Kinetochore fiber maturation in PtK_1 cells and its implications for the mechanisms of chromosome congression and anaphase onset. *J. Cell Biol.* **137**, 1567–1580.

McGee-Russell, S. M., and Allen R. D. (1971). Reversible stabilization of labile, microtubules in the reticulopodial network of *Allogromia*. *Adv. Cell Mol. Biol.* **1**, 153–184.

Mole-Bajer, J., and Bajer, A. S. (1968). Studies of selected endosperm cells with the light and electron microscope: The technique. *Cellule* **67**, 257–265.

Nicklas, R. B., Brinkley, B. R., Pepper, D. A., Kubai, D. F., and Rickards, G. K. (1979). Electron microscopy of spermatocytes previously studied in life: Methods and some observations on micro-manipulated chromosomes. *J. Cell Sci.* **35**, 87–104.

Nomarski, G. (1955). Microinterferometre differential à ondes polarizés. *J. Phys. Radium* **16**, 59–513.

Payne, J. N., Cooper, J. D., MacKeown, S. T., and Horobin, R. W. (1987). A temperature controlled chamber to allow observation and measurement of uptake of fluorochromes into live cells. *J. Microsc.* **147**, 329–335.

Pearce, B. R., Currie, D. N., Dutton, G. R., Hussey, R. E. G., Beale, R., and Pigott, R. (1981). A simple perfusion chamber for studying neurotransmitter release from cells maintained in monolayer culture. *J. Neurosci. Methods* **3**, 255–259.

Pentz, S., and Horler, H., (1992). A variable cell culture chamber for "open" and "closed" cultivation, perfusion and high microscopic resolution of living cells. *J. Microsc.* **167**, 97–103.

Pouchelet, M., and Moncel, C. (1974). Mise au point, pour observation microscopique prolongée de cellules vivantes, d'un système de thermostatisation et de perfusion continue pour microchambres de culture. *Microscopica Acta* **75**, 352–360.

Poyton, R. O., and Branton, D. (1970). A multipurpose microperfusion chamber. *Exp. Cell Res.* **60**, 109–114.

Richter, K. M., and Woodward, N. W. (1955). A versatile type of perfusion chamber for long-term maintenance and direct microscopic observation of tissues in culture. *Exp. Cell Res.* **9**, 585–587.

Rieder, C. L., and Bajer, A. S. (1977). Effect of elevated temperatures on spindle microtubules and chromosome movements in cultured newt lung cells. *Cytobios* **18**, 201–234.

Rieder, C. L., and Borisy, G. G. (1981). The attachment of kinetochores to the prometaphase spindle in PtK₁ cells: Recovery from low temperature treatment. *Chromosoma* **82**, 693–717.

Rieder, C. L., and Alexander, S. P. (1990). Kinetochores are transported poleward along a single astral microtubule during chromosome attachment to the spindle in newt lung cells. *J. Cell Biol.* **110**, 81–95.

Roberts, D. C., and Trevan, D. J. (1961). A versatile microscope chamber for the study of the effects of environmental changes on living cells . *J. Royal Microsc. Soc.* **79**, 361–366.

Romeis, B. (1912). Ein verbesserter Kulturapparat fär Explantate. *Z. Wiss. Mikroskop.* **29**, 530–534.

Rose, G. (1954). A separate and multipurpose tissue culture chamber. *Texas Rept Biol. Med.* **12**, 1075–1081.

Rose, G., PomeraC. M., Shindler, T. O., and Trunnell, J. B. (1958). A cellophane-strip technique for culturing tissue in multipurpose culture chambers. *Biophys. Biochem. Cytol.* **4**, 761–769.

Roth, K. E., Rieder, C. L., and Bowser, S. S. (1988). Flexible-substratum technique for viewing cells from the side: Some *in vivo* properties of primary (9 + 0) cilia in cultured kidney epithelia. *J. Cell Sci.* **89**, 457–466.

Rupp, G., Bowser, S. S., Mannella, C. A., and Rieder, C. L. (1986). Naturally occurring tubulin-containing paracrystals in *Allogromia:* Immunocytochemical identification and functional significance. *Cell Motil. Cytoskel.* **6**, 363–375.

Sakariassen, K. S., Aarts, P. A., de Groot, P. G., Houdijk, W. Ṗ M., and Sixma, J. J. (1983). A perfusion chamber developed to investigate platelet interaction in flowing blood with human vessel wall cells, their extracellular matrix, and purified components. *J. Lab. Clin. Med.* **102**, 522–535.

Salih, V., Greenwald, S. E., Chong, C. F., Coumbe, A., and Berry, C. L. (1992). The development of a perfusion system for studies on cultured cells. *Int. J. Exp. Pathol.* **73**, 625–632.

Schwartzkroin, P. A. (1975). Characteristics of CA1 neurons recorded intracellularly in the hippocampal *in vitro* slice preparations. *Brain Res.* **85**, 423–436.

Sevcik, G., Guttenberger, H., and Grill, D. (1993). A perfusion chamber with temperature regulation. *Biotechnic Histochem.* **68**, 229–236.

Soll, D. R., and Herman, M. A. (1983). Growth and the inducibility of mycelium formation in *Candida albicans:* A single-cell analysis using a perfusion chamber. *J. Gen. Microbiol.* **29**, 2809–2824.

Sprague, E. A., Steinbach, B. L., Nerem, R. M., and Schwartz, C. J. (1987). Influence of a laminar steady-state fluid-imposed wall shear stress on the binding, internalization, and degradation of low-density lipoproteins by cultured arterial endothelium. *Circulation* **76**, 648–656.

Spring, K. R., and Hope, A. (1978). Size and shape of the lateral intercellular spaces in a living epithelium. *Science* **200**, 54–58.

Stephens, R. E. (1973). A thermodynamic analysis of mitotic spindle equilibrium at active metaphase. *J. Cell Biol.* **57**, 133–147.

Strange, K., and Spring, K. R. (1986). Methods for imaging renal tubule cells. *Kidney Int.* **30**, 192–200.

Sykes, J. A., and Moore, E. B. (1959). A new chamber for tissue culture. *Proc. Soc. Exp. Biol. Med.* **100**, 125–127.

Sykes, J. A., and Moore, E. B. (1960). A simple tissue culture chamber. *Texas Rep. Biol. Med.* **18**, 288–297.

Tilney, L. G., and Porter, K. R. (1967). Studies on the microtubules in heliozoa. The effect of low temperature on these structures in the formation and maintenance of the axopodià *J. Cell Biol.* **34**, 327–343.

Toy, B. L., and Bardawil, W. A. (1958). A simple plastic perfusion chamber for continuous maintenance and cinematography of tissue cultures. *Exp. Cell Res.* **14**, 97–103.

Toyotomi, S., and Momose, Y. (1989). Temperature-controlled perfusion apparatus for microscope using transparent conducting film heater. *Am. J. Physiol.* **256**, C214–C217.

Varnum, B., Edwards, K. B., and Soll, D. R. (1985). *Dictyostelium* amebae alter motility differently in response to increasing versus decreasing temporal gradients of cAMP. *J. Cell Biol.* **101**, 1–5.

Wetzels, J. F. M., Kribben, A., Burke, T. J., and Schrier, R. W. (1991). Evaluation of a closed perfusion chamber for single cell fluorescence measurements. *J. Immunol. Methods* **141**, 289–291.

White, P. R. (1966). Versatile perfusion chamber for living cells and organs. *Science* **152**, 1758–1760.

Willmer, E. N. (1954). "Tissue Culture: Methuen's Monographs on Biological Subjects," 2nd edition. Methuen and Co., London.

Zernike, F. (1942). Phase contrast, a new method for the microscopic observation of transparent objects. *Physica* **9**, 686–698.

CHAPTER 14

Fluorescence Lifetime Imaging Techniques for Microscopy

Todd French, * **Peter T. C. So,** † **Chen Y. Dong,** ‡
Keith M. Berland, ‖ **and Enrico Gratton** ‡

* LJL Bio-Systems
Sunnyvale, California 94089

† Department of Mechanical Engineering
Massachusetts Institute of Technology
Cambridge, Massachusetts 02139

‡ Laboratory for Fluorescence Dynamics
Department of Physics
University of Illinois at Urbana-Champaign
Urbana, Illinois 61801

‖ IBM
Yorktown Heights, New York 10598

METHODS IN CELL BIOLOGY, VOL. 56
Copyright © 1998 by Academic Press. All rights of reproduction in any form reserved.
0091-679X/98 $25.00

I. Introduction

Fluorescence is an especially sensitive technique that is a vital contrast enhancement mechanism for microscopy. Fluorescence microscopy has excellent background rejection and highly specific staining (particularly with the use of antibody-conjugated probes). It provides the high contrast needed to spatially resolve microscopic structures such as cellular organelles. Although the spatial relationship of the organelles is important, the functional properties of the organelles are also vital to the understanding of cellular life. Local properties such as pH or molecular concentration can reveal much about the organelle. Fluorescent probes are normally very sensitive to the local environment. For example, some probes are fluorescent only in a particular pH or polarity condition. Fluorescence intensity measurements, however, are unsuitable for quantitative work as the measured intensity is not just dependent on the environment but also dependent on the local probe concentration which cannot easily be determined. Quantitative measurements require a parameter that is independent of the local probe concentration. Spectral and temporal excited state properties of a fluorescent probe are not dependent upon the probe concentration. Therefore, techniques which exploit either spectral or lifetime-sensitive probes will be able to measure quantitatively environmental factors that affect the excited state. Consequently ratiometric (spectral) and fluorescence lifetime (temporal) techniques coupled with a microscope can provide localized quantitative information in a cellular environment.

Fluorescence reports the molecular state of a chromophore. Spectral and polarization changes in the emitted light reveal the excited-state properties and the orientation of the transition dipole moment. To investigate more completely the nature of the excited state, time-resolved measurements of these quantities are required. For instance, fluorescence quenching may occur due to a ground-state reaction or an excited-state reaction. Only by measuring the fluorescence lifetime may one determine which is the case. The excited state is also affected by the molecular environment through effects such as solvent relaxation, energy transfer, and conformational changes. This sensitivity makes fluorescent probes excellent monitors of such factors as pH, viscosity, molecular concentration, and local order.

The sensitivity of fluorescence lifetime to the microenvironment has been used to measure pH (Sanders *et al.*, 1995a), metal ion concentration (Piston *et al.*, 1994; Lakowicz *et al.*, 1994a), fluorescence resonance energy transfer (Oida *et al.*, 1993; Gadella and Jovin, 1995), cellular photostress (König *et al.*, 1996), and antigen processing (French *et al.*, 1997). Other quantities that can be measured are molecular oxygen concentration, environmental polarity, and local order. The lifetime itself also can be used as a contrast-enhancing mechanism. Several distinct fluorophores with different lifetimes can be imaged in one picture and the fraction of each fluorophore can be resolved within a single image element (So *et al.*, 1995; Sanders *et al.*, 1995b).

Time-resolved fluorescence microscopy has advanced greatly from the first single-pixel measurements (Dix and Verkman, 1990; Keating and Wensel, 1990). These first experiments deduced important cellular information such as calcium concentration or cytoplasm matrix viscosity at selected points inside the cell. The next step in fluorescence lifetime resolved microscopy was to extend the single point measurements to obtain lifetime information across the entire cell. Two different approaches have been applied. The first method uses CCD cameras equipped with gain-modulated image intensifiers to collect data simultaneously over the whole image. The other approach modifies traditional confocal laser scanning microscopes and obtains time resolved information on a point-by-point basis.

The laser scanning microscope uses a single point detector to acquire images one point at a time, serially. Serial acquisition systems consist of a single point detector and a mechanism to move the sample or the excitation light. Fluorescence lifetime imaging can be achieved by adapting normal lifetime resolved detectors and electronics to the new optical setup. Laser scanning microscopy has been used to measure three-dimensionally resolved lifetime images with confocal detection (Buurman et al., 1992; Morgan et al., 1992), two-photon excitation (Piston et al., 1992; So et al., 1995), and time-dependent optical mixing (Dong et al., 1995; Müller et al., 1995). In addition, nearfield scanning optical microscopy has been used to image fluorescence lifetimes in a plane (Ambrose et al., 1994; Xie and Dunn, 1994).

Time-resolved imaging of the full field simultaneously is harder to implement. Full-field imaging generally employs a two-dimensional array detector (such as a CCD) to capture all points in an image in parallel (rather than one at a time). The problem to solve with two-dimensional detectors is how to achieve nanosecond time resolution (required for fluorescence) across a large spatial area. The basic approaches are to use a gated image intensifier coupled to a camera or a position-sensitive photomultiplier. Both frequency domain (Wang et al., 1989; Marriott et al., 1991; Lakowicz and Berndt, 1991Œ So et al., 1994; Morgan et al., 1995) and time domain (Wang et al., 1991; Oida et al., 1993; McLoskey et al., 1996) techniques have been developed.

This chapter describes the development and application of fluorescence lifetime imaging microscopy. Three different fluorescent lifetime imaging microscopes that use the frequency domain heterodyning technique are described: one based on a CCD, one using laser scanning with two-photon excitation, and one using laser scanning with a pump-probe (stimulated emission) technique.

II. Time-Resolved Fluorescence Methods

A. Time Domain and Frequency Domain Measurements

Time-resolved fluorescence measurements are performed in two functionally equivalent ways: using the time domain or the frequency domain. In the time domain, the impulse response of a system is probed. A fluorescent sample is illuminated with a narrow pulse of light and the resulting fluorescence emission decay captured with a fast recorder. A common method to reconstruct the decay profile is time-correlated single-photon counting. With this method, the time delay between the emitted photon and the excitation pulse is recorded. For a low-enough emission rate (less than 10^5 photons/sec) every photon's delay can be recorded. An entire decay curve can be built by plotting the number of photons versus delay time. To properly reconstruct the decay curve, the excitation profile must also be measured so that it may be used to deconvolve the finite width of the excitation pulse from the emission profile.

The frequency domain method measures the harmonic response of the system. The sample is excited by a sinusoidally modulated source. An arbitrarily complex modulation waveform can be decomposed into sinusoidal components so that only sinusoidal modulation need be discussed. The emission signal appears as a sine wave that is demodulated and phase shifted from the source (Fig. 1). The phase shift and modulation are used to obtain the lifetime of the fluorophore. For a single exponential decay, the excited state lifetime (τ) is related to the phase (ϕ) and modulation (M) as given in equations (1) and (2), where ω is the angular modulation frequency or 2π times the modulation frequency. The angular modulation frequency should be roughly the inverse of the lifetime to obtain the maximum sensitivity. Since typical lifetimes are 1–10 ns, typical modulation frequencies are 20–200 MHz.

$$\omega\tau = \tan(\phi) \tag{1}$$

$$\omega\tau = \sqrt{\frac{1}{M^2} - 1} \tag{2}$$

Because the emission signal is too fast to sample continuously, the high-frequency signal is converted to a lower frequency. This conversion, called heterodyning, is accomplished with a gain-modulated detector. The detector is modulated at a frequency close to or of the same order as the source frequency. The result is a beating of the emission and detector modulation signals that yields a signal at the difference of the frequencies, called the cross-correlation frequency.

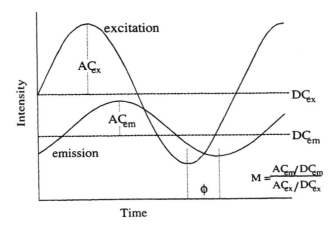

Fig. 1 Definitions of phase (ϕ) and modulation (M) in frequency domain measurements.

If the source and detector frequencies are the same, the method is called homodyning. Homodyning, by definition, results in a DC signal. The intensity is proportional to the sine of the difference of the phase between the detector and the emission. To acquire the entire phase and modulation information of the emission signal, it is necessary to systematically step the phase difference between the source and detector modulation signals. Homodyning is commonly used with a fixed phase difference to collect phase-resolved data. By properly choosing the phase of the detector, one can suppress or enhance certain lifetimes. A disadvantage of homodyning is that it is more sensitive than heterodyning to spurious DC components such as stray light. Heterodyning recovers all of the high frequency phase and modulation information in each cross-correlation period (typically 10–100 ms) and can be used for phase-resolved data also.

For the three different microscopes we apply the same form of heterodyned data acquisition. First, we apply a synchronous averaging filter. Several cross-correlation waveforms are collected and combined to form an average waveform. This filter removes random noise by rejecting signals that are not at the cross-correlation frequency or its harmonics. The reduction of the bandwidth along with the averaging causes the signal-to-noise ratio to increase linearly with the number of waveforms averaged (Feddersen *et al.*, 1989). The next step in data reduction is a fast Fourier transform (FFT). The cross-correlation signal is uniformly sampled in time so an FFT can be used to calculate the phase and modulation. In the manner that we apply the FFT, it is equivalent to (but simpler than) a least-squares fit of multiple sine waves to the data. Additional averaging of sequential phase and modulation images is sometimes used in the scanning microscopes to further reduce the noise. This is a less efficient filter and the noise is reduced as expected from Poisson statistics.

The sampling frequency is generally chosen such that four samples are collected in one cross-correlation period (Fig. 2). To measure a signal at a given frequency, the signal must be sampled at least at twice that frequency (Nyquist theorem). Especially because noise appears as a high-frequency signal, it is better to sample at a higher rate to produce a more accurate measurement. More points sampled results in better discrimination of the various harmonics. Thus, by increasing the number of samples per cross-correlation period, higher-frequency signals (generally noise) can be more easily separated from the fundamental.

B. Simultaneous Multiple-Lifetime Component Measurement

To discriminate more harmonics and achieve better noise rejection, more points can be sampled per period. For the first-generation microscope instruments presented in this chapter, we chose to use four points per cross-correlation period because it was faster and the higher harmonics were small compared to the fundamental frequency. However, there are situations that can benefit from measuring higher harmonics.

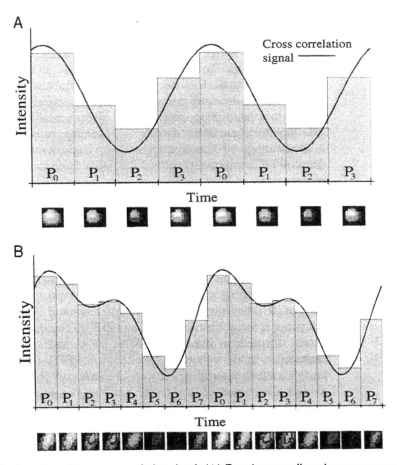

Fig. 2 Sampling of the cross-correlation signal. (A) Four images collected per cross-correlation period. (B) Eight images can reduce the error in the measurement (harmonic distortion).

One such case is the simultaneous measurement of two probes with dissimilar lifetimes. To measure the two lifetimes, the phase and modulation would have to be collected in separate measurements at each of the frequencies appropriate for each of the fluorophores. However, it is possible to map the two modulation frequencies to two different cross-correlation frequencies and collect the phase and modulation in one measurement. If the two cross-correlation frequencies are related by an integer multiple, they can be separated by an FFT.

Consider the case of a sample that consists of two probes whose lifetimes are 1 and 1000 ns. Two appropriate modulation frequencies are 100 MHz and 100 kHz. If the laser is modulated at a mixture of 100 MHz and 100 kHz and the detector is modulated at a mixture of 100 MHz + 10 Hz and 100 kHz + 20 Hz, the two signals can be separated. The result of the heterodyning is two signals at 10 and 20 Hz that correspond to the fluorescence decay at 100 MHz and 100 kHz, respectively.

To achieve the mixing of the 100-MHz and 100-kHz signals and still produce a strong fluorescence signal, the two signals should both be sine waves. When combined with a high-frequency mixer, the result is the high-frequency signal enclosed in an envelope of the intermediate frequency signal (Fig. 3A). Upon heterodyning, the low-frequency resultant signal will be composed mainly of the two cross-correlation frequencies (Fig. 3B).

After measuring the phase and modulation at each frequency, it is still necessary to separate the lifetime components. The most common approach is to fit a decay scheme to the acquired data. For the example outlined in this section it is possible to use algebraic methods to separate the two components. The algebraic method is possible whenever the two lifetimes are known to be different by several orders of magnitude.

To describe the example system, the following conventions are used. The shorter lifetime component (1 ns) will be referred to with the subscript 1. The

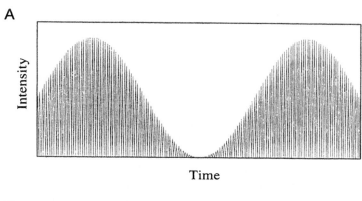

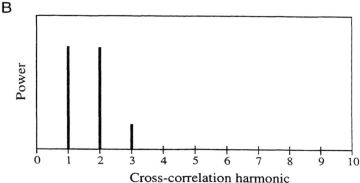

Fig. 3 Dual modulation signal. (A) the source or detector signal; (B) the low-frequency power spectrum of the heterodyned signal.

longer component (1000 ns) will have the subscript 2. The fractional intensity of the fluorescence contributed by the short lifetime probe is indicated by f $(1 - f$ will be contributed by the other component).

In the frequency domain, lifetime components add like vectors. The vector sum of the phase and modulation times the fractional intensity of each component results in the measured phase and modulation of a multiple component system. The cartesian components of the resultant lifetime vector for a two-component system are given in equations (3) and (4).

$$M_r(\omega)\cos[\phi_r(\omega)] = fM_1(\omega)\cos[\phi_1(\omega)] + (1 - f)M_2(\omega)\cos[\phi_2(\omega)] \quad (3)$$

$$M_r(\omega)\sin[\phi_r(\omega)] = fM_1(\omega)\sin[\phi_1(\omega)] + (1 - f)M_2(\omega)\sin[\phi_2(\omega)] \quad (4)$$

By measuring the phase and modulation at the two frequencies appropriate for the two components ($\omega_1 \approx 1/\tau_1$, $\omega_2 \approx 1/\tau_2$), equations (3) and (4) can be rearranged and combined with equations (1) and (2) to solve for each of the lifetime values and the fractional intensity.

$$f = \frac{M_r(\omega_1)}{\cos[\phi_r(\omega_1)]} \quad (5)$$

$$\tau_1 = \frac{\tan[\phi_r(\omega_1)]}{\omega_1} \quad (6)$$

$$\tau_2 = \frac{M_r(\omega_2) \cdot \sin[\phi_r(\omega_2)]}{\omega_2 \cdot [M_r(\omega_2) \cdot \cos[\phi_r(\omega_2)] - f]} \quad (7)$$

Table I lists the values that would be measured in three different two-component systems. The first system consists of 20% of the short component, the second 50%, and the third 80%. These three configurations demonstrate the ability of the algebraic method. As can be seen in Table II there is some error involved in using this technique. The error, however, is less than the typical error of the fluorescence lifetime imaging systems presented in this chapter. For an integration time of several seconds, an error of 0.4° phase and 0.005 modulation is normal. This magnitude of error results in about a 1% error in the lifetime at

Table I
Phase and Modulation Values of Three Two-Component Systems at Two Sample Frequencies[a]

Short/long (%)	ϕ_r (ω_1)	M_r (ω_1)	ϕ_r (ω_2)	M_r (ω_2)
20/80	32.5°	0.170	25.0°	0.853
50/50	32.2°	0.424	14.7°	0.888
80/20	32.1°	0.678	5.5°	0.948

[a] $\omega_1 = 2\pi \cdot 100$ MHz, $\omega_2 = 2\pi \cdot 100$ kHz.

Table II
Results of the Numerical Separation of Different Lifetime Components in the Three Sample Systems

Value	20% τ_1		50% τ_1		80% τ_1	
	Result	Error (%)	Result	Error (%)	Result	Error (%)
f	0.2016	0.8	0.5011	0.2	0.8004	0.05
τ_1	1.014 ns	1	1.002 ns	0.2	0.998 ns	0.2
τ_2	1.004 μs	0.4	1.002 μs	0.2	1.010 μs	1

the appropriate frequency. If smaller errors are required, the algebraic method could be performed iteratively or the measured phase and modulation could be fit to a two-component model using standard fitting routines.

C. Photobleaching Effects

Fluorescence photobleaching (photon-induced destruction of the fluorophore) is a major problem in fluorescence microscopy. This can make quantitatively reproducible intensity measurements next to impossible. Lifetime measurements, on the other hand, should be insensitive to such a loss of fluorophore. However a problem can still arise if the concentration changes significantly during a lifetime measurement. In this section, lifetime measurement errors due to photobleaching are compared for three frequency domain data acquisition methods. In the comparison each method was given the same amount of time to observe the sample and all methods collected four phase samples per integration cycle.

The first method investigated is heterodyning. The cross-correlation frequency was 32 Hz and integration was performed by synchronous averaging. The second method is normal homodyning. The detector was modulated at the four phase angles 0°, 90°, 180°, and 270°. Each phase was integrated for one-quarter of the total integration time. The third method is modified homodyning. For this method the detector was modulated at the same four phase angles but the phase was first stepped up and then back to counteract trends due to photobleaching. The phase steps and order were 0°, 90°, 180°, 270° and then 270°, 180°, 90°, and 0°. Each phase step was integrated one-eighth of the total time and intensities at corresponding phase angles were averaged.

To compare the three acquisition methods, two samples were produced: one that exhibited strong photobleaching and one that had insignificant photobleaching. For both samples, DAPI (4,6-diamidino-2-phenylindole; Sigma Chemical, St. Louis, MO), a common fluorescent probe, was chosen. This probe is quite easily photobleached. The nonphotobleaching sample was a solution of DAPI loaded into a hanging drop slide. The bleaching in this sample was negligible due to the diffusion of fresh DAPI molecules from outside the excitation volume.

For the photobleaching sample, the effects of diffusion were limited by using a polyacrylimide gel commonly used for gel chromatography (Bio-Gel P6, exclusion limit 6 KDa, Bio-Rad, Richmond, CA). When a sample volume of the gel was photobleached, no return of the fluorescence intensity to its original value was observed even after several minutes. The lifetime and photobleaching rate were constant in these samples.

Figure 4 shows a representative photobleaching decay in the photobleaching-sensitive system. The intensity decrease was fit to a double-exponential model that resulted in a short photobleaching time of 9 sec and a long photobleaching time of 140 sec. The double-exponential fit suggests a dual-rate process. The origin of the processes is not known. To avoid the effects of the fast component, all of the photobleaching samples were illuminated for at least 30 sec before lifetime data was acquired.

Phase and modulation errors of the three data-collection methods are shown in Fig. 5. For the bleaching measurements, the errors for the various integration times were calculated by comparing the longer integration measurements to the average of reference measurements. As a reference for each method, the phase and modulation were measured for 10 sec repeatedly. During this short integration time, photobleaching caused little change in average intensity (as shown in the inset of Fig. 4). Consequently photobleaching should have little effect on the phase and modulation during the reference measurements using any method. The errors presented are the average errors of 4–8 different samples at each integration time. The graph clearly shows that the heterodyning acquisition method is the least affected by photobleaching. This is an important factor in our choice of the heterodyning acquisition method for fluorescence lifetime imaging microscopy.

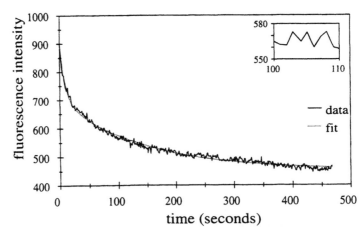

Fig. 4 Fluorescence intensity of DAPI as a function of time showing photobleaching. The bleaching effects were fit to a double-exponential decay with decay times of 9 and 140 sec.

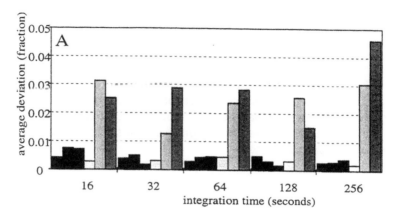

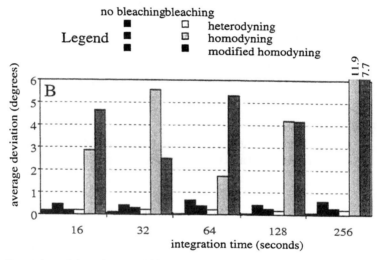

Fig. 5 Comparison of three data acquisition methods in the presence and absence of photobleaching. The average deviation is the average of the magnitude of the error of several measurements. (A) Modulation; (B) phase.

III. Fluorescence Lifetime–Resolved Camera

The fluorescence lifetime-resolved camera uses a CCD detector to collect nanosecond time-resolved data in parallel. The camera system can simultaneously measure fluorescence lifetime at every point in a microscopic image. It uses a high-speed CCD camera modified to collect lifetime data. The heterodyning technique coupled with a fast camera yields several unique advantages over previously reported camera systems (McLoskey *et al.,* 1996; Morgan *et al.,* 1995; Gadella *et al.,* 1993; Oida *et al.,* 1993; Lakowicz and Berndt, 1991; and Wang *et al.,* 1991). First, this instrument is capable of generating lifetime images in a few seconds. Second, it minimizes the effect of photobleaching, a common problem in microscopy.

This system can operate in the presence of high background fluorescence and can quantitatively measure lifetimes of microscopic samples. Time-resolved imaging cannot only effectively and selectively enhance the contrast of microscope fluorescence images, but can also quantitatively measure lifetimes within cellular compartments to monitor their microenvironment. With an acquisition rate as short as a few seconds, kinetic studies of lifetime changes are also possible with this camera.

A. Instrumentation

Our first camera system was constructed of a CCD camera coupled to an image intensifier (So *et al.*, 1994). Subsequently, a second-generation lifetime-resolved camera was constructed of a scientific-grade high-speed CCD and a custom image intensifier. The second-generation instrument addressed problems with the CCD, frame rate, modulation frequency, and other troubles. Throughout the development, the optics and data acquisition remained essentially the same. A general schematic of the instrument is illustrated in Fig. 6.

1. Camera and Image Intensifier

The two most important components of the lifetime-resolved camera system are the CCD and the microchannel plate image intensifier. The second generation

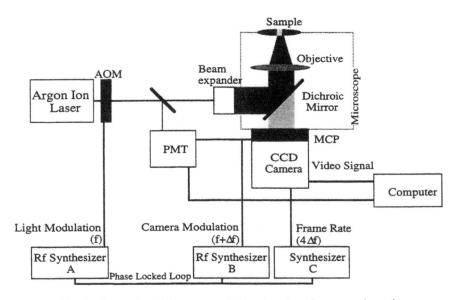

Fig. 6 Camera-based fluorescence lifetime imaging microscope schematic.

camera includes a scientific grade CCD (CA-D1-256, Dalsa, Waterloo, Ontario). This CCD operates in the progressive scan mode as opposed to the interlaced mode. Its frame rate is variable (derived from an external signal) and can be as high as 200 Hz. The new CCD also has a signal to noise ratio of 70 dB. The resolution is 256 × 256 pixels, each 16 × 16 μm. Although we could have chosen a camera with higher resolution, we chose one with less resolution but a fast frame rate. The microchannel plate is coupled via fiber optic to the CCD. It was custom built by Hamamatsu (model V6390U, Bridgewater, NJ) for high-speed modulation. It has a double stage design with a gain of approximately 10^5. The camera is gain modulated by modulating the cathode of the microchannel plate (So *et al.*, 1994). We have tested the modulation to 500 MHz.

2. Light Source and Optics

To illuminate the sample, a laser is directed into an epifluorescence microscope (Axiovert 35, Zeiss Inc., Thronwood, NY). The laser beam is coupled via fiber optic to a beam expander and then delivered to the microscope for wide field illumination. Shown in Fig. 6 is an argon ion laser (model 2025, Spectra Physics, Mountain View, CA) modulated by an acoustooptical modulator (AOM; Intra-Action, Belwood, IL). The argon ion laser can provide illumination at several selected blue-green wavelengths, most notably 514 and 488 nm. The acoustooptical modulator can optically modulate light at discrete frequencies from 30–120 MHz. One particular benefit of an AOM is its high throughput (up to 50%). As an alternative to the AOM, a Pockels cell (ISS Inc., Champaign, IL) has also been used. The Pockels cell has the advantage of modulating at any frequency from 0–300 MHz. The advantage of continuous wideband modulation is offset by the relatively poor throughput (~5%). Another alternative modulated light source used in our experiments was a cavity-dumped DCM dye laser (model 700, Coherent Inc., Palo Alto, CA) synchronously pumped by a Nd:YAG laser (Antares, Coherent Inc.). The dye laser produces 10 ps pulses at a range of frequencies. The frequency can be chosen by the cavity dumper to be any integer divisor (4–256) of the Nd:YAG fundamental frequency (76.2 MHz). The narrow pulse width insures that there will be harmonics measurable to many gigahertz. The dye laser provides a tunable source in the near UV region (300–340 nm).

3. Data Acquisition

The data acquisition scheme plays an important part in the performance of the camera system. The hardware determines the sensitivity and time response of the system while the software provides a user interface and noise/artifact reduction. The camera has an 8-bit digital output. The output signal is sent to an Image-1280 digitizer (MATROX Electronic Systems Ltd., Dorval, Quebec). Four frames are collected per cross-correlation period. The cross-correlation frequency of the system is from 1–50 Hz. Each frame is synchronously averaged with the corresponding frame of the previous cross-correlation period. The Image-1280 can average the frames and transfer the averaged frames to the host computer without interrupting image acquisition. Hence there is no limit on the number of cross-correlation periods that can be averaged. After collecting the average cross-correlation images, the phase and modulation are independently calculated for each point in the image. Real measurements of typical fluorescence probes generally have an uncertainty of about 200 ps which can be achieved in about 300 cross-correlation periods (30 sec at 10 Hz).

Although the camera can record phase and modulation images that are internally consistent, there is no guarantee that the values in different images are comparable. Generally, the high-frequency modulation signals change slightly between measurements. Slow phase and modulation drifts are common in acoustooptical and electrooptical modulators. To account for these small variations, a second detector is used to monitor the laser light source. All values calculated for an image are referenced to the monitor values. The detector that monitors the laser is a photomultiplier tube (PMT; R928, Hamamatsu,). Data from the PMT is digitized with a second digitizer (A2D-160, DRA Laboratories, Sterling, VA).

B. Camera-Based Microscope Examples

Lifetime images of some simple microscopic systems were obtained to test the performance of this fluorescence lifetime imaging microscope. The following examples demonstrate the ability of our microscope to selectively enhance image contrast and to quantitatively measure fluorescence lifetime.

One of the most important features of the lifetime microscope is its ability to measure minute differences in cellular microenvironments where fluorescence intensity measurements do not have sufficient sensitivity. A string of live spirogyra cells is shown in Fig. 7. Spirogyra cells organize as long strands, cells connected end-to-end. The most prominent features in each of these cells are the large, double-helical chloroplasts. The chlorophyll in these organelles is brightly fluorescent when illuminated at 514 nm. The string of cells was imaged with the time resolved microscope at a modulation frequency of 81 MHz.

This image was taken at the junction between two distinct spirogyra cells. The break in the helical structures corresponds to the separating cell walls between the two cells. The cell walls (which have been drawn for clarity) are not visible since they are not fluorescent. Judging from the fluorescence intensity picture alone, these two cells appear to be identical. However, a significant difference in chlorophyll lifetimes in these two cells can be observed in the phase- and modulation-resolved pictures (Figs. 7c and 7d). Note that the chloroplast within each individual cell has a uniform lifetime that is distinct from the other one. The observation shows that the nonuniformity in the phase and modulation pictures reflects the environment of the chloroplast in the cells and is not caused

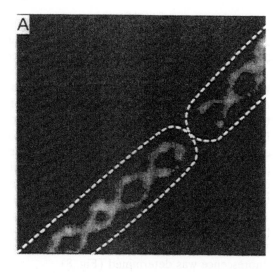

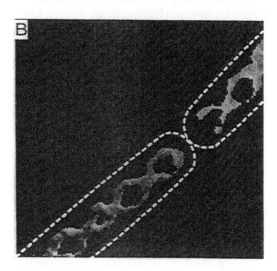

Fig. 7 Frequency domain pictures of chloroplast fluorescence of two spirogyra cells. (A) DC intensity (arbitrary units); (B) phase (0–90°) at 81 MHz. The gray scale is linear from the minimum (black) to the maximum (white). The dotted line illustrates the cell walls.

by experimental artifacts of the lifetime camera. Comparing the phase and modulation-resolved pictures, the cell at the upper right-hand corner has a higher phase value and is more demodulated while the other cell has a lower phase value and is better modulated. That is consistent with the hypothesis that the chlorophyll of the cell at the upper right corner has a longer lifetime. Since the growth of these cells was not controlled, the cause of the lifetime difference is unknown. This example demonstrates the promise of the time-resolved microscopy technique in quantitatively monitoring cellular functions.

To demonstrate quantitative measurement of lifetime, a two-layer sample slide was constructed. The well of a hanging drop slide was filled with a POPOP–ethanol solution and sealed with a coverslip. The second layer, placed on top of the first, consisted of 15 μm blue fluorescent spheres (Molecular Probes, Inc., Eugene, OR) in water sealed with another coverslip. POPOP exhibits a single exponential decay in ethanol of 1.3 ns. It served as an internal fluorescence lifetime reference. The two solutions were not allowed to mix together because different solvents were required for each probe. Frequency domain pictures were taken at multiple frequencies from 8 to 140 MHz using the DCM dye laser. The phase differences and modulation ratios between the sphere and the dye background were measured at each frequency. Using the known lifetime of the POPOP background, the absolute phase shift and modulation of the sphere fluorescence was determined (Fig. 8).

The absolute phase shift and modulation values were analyzed using GLO-BALS Unlimited (University of Illinois, Urbana, IL). The data obtained fits best to a double-exponential decay as expected by the two-fluorophore system. One component was fixed to have the lifetime of POPOP while the other was free to vary. From the intensity images, the POPOP was expected to contribute about 0.7 of the total intensity. The calculated fractions were 76% for POPOP and 24% ($\pm$3%) for the sphere in close agreement with the intensity estimate. The lifetime of the spheres was calculated to be 2.5 ns ($\pm$0.3 ns).

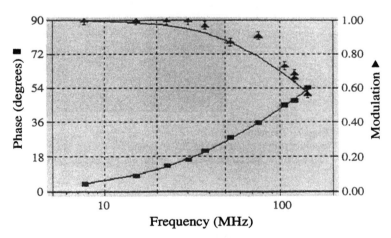

Fig. 8 Phase shift and demodulation of 15-μm fluospheres as a function of frequency. The data are the average value of all of the spheres in one image. The fit (solid line) is a double-exponential fit. The phase error was 0.4° (error bars are smaller than the symbols) and the modulation error was 0.05.

⟳ IV. Two-Photon Fluorescence Lifetime Microscopy

The second fluorescence lifetime microscope design uses two-photon excitation and laser scanning to produce three-dimensionally resolved fluorescence lifetime images. Two-photon excitation was first introduced to microscopy by Denk *et al.* (1990). Chromophores can be excited by the simultaneous absorption of two photons each having half the energy needed for the excitation transition (Friedrich, 1982; Birge, 1983, 1985). The two-photon excitation probability is proportional to the square of the excitation power. The key to efficient two-photon excitation is a high temporal and spatial concentration of photons. Only in the region of a high photon density will there be appreciable two-photon excitation. The high spatial concentration of photons can be achieved by focusing laser light with a high numerical aperture objective to a diffraction limited spot. It is more difficult to obtain a high temporal concentration of photons. This was only recently made possible by the development of high-peak power mode-locked lasers.

Depth discrimination is the most important advantage of two-photon excitation. For one photon excitation in a spatially uniform fluorescent sample, equal fluorescence intensities are contributed from each z-section above and below the focal plane, assuming negligible excitation attenuation. This is a consequence of the conservation of energy (Wilson and Sheppard, 1984). On the other hand, in the two-photon case and for objectives with a numerical aperture of 1.25, over 80% of the total fluorescence intensity comes from a 1-μm-thick region about the focal point (So *et al.*, 1995).

The spatial resolution and depth discrimination of the two-photon microscope are comparable to conventional confocal systems. For excitation of the same chromophore, the resolution is roughly half the one-photon confocal resolution (Sheppard and Gu, 1990). This reduction in spatial resolution is due to the larger diffraction-limited spot of the longer-wavelength two-photon excitation source (double the wavelength of the one-photon source). For our microscope at 960 nm excitation, the FWHM of the point spread function is 0.3 μm radially and 0.9 μm axially (So *et al.*, 1995).

Two-photon excitation also has a unique advantage in that photobleaching and photodamage are localized to a submicron region at the focal point. In contrast, a conventional scanning confocal microscope causes photobleaching and photodamage in the entire sample. In addition, no detection pinhole is required with a two-photon microscope to achieve 3D resolution, improving the overall light throughput of the microscope. Since more fluorescence photons are required to generate high-quality time-resolved images than intensity imaging, the light-gathering efficiency is an important factor in the successful implementation of a time-resolved microscope. This requirement is critical in three-dimensional imaging where the sample is exposed to a long period of laser excitation. Therefore, maximizing microscope detector efficiency and minimizing sample photobleaching are important design considerations.

Our instrument operates at a high cross-correlation frequency that provides lifetime information on a per-pixel basis. The time resolution of this system is about 400 ps which is slightly inferior to standard fluorometers which have a resolution of about 25 ps. This result is expected given the data acquisition time at each pixel is typically a million times shorter than the time used by a conventional fluorometer. Due to the heterogeneous nature of the cellular environment, it is seldom necessary to determine specimen lifetime to an accuracy comparable to the standard fluorometer.

A. Instrumentation

Figure 9 presents the schematics of the two-photon time-resolved microscope. It essentially consists of a laser scanning microscope and a high-peak power laser. This system operates efficiently at the normal light levels of microscopy. It also can operate at very low light levels. The system can detect events with as little as 2–3 photons and has single-molecule detection capabilities. The background noise is equivalent to about one photon per pixel per frame at the highest sensitivity.

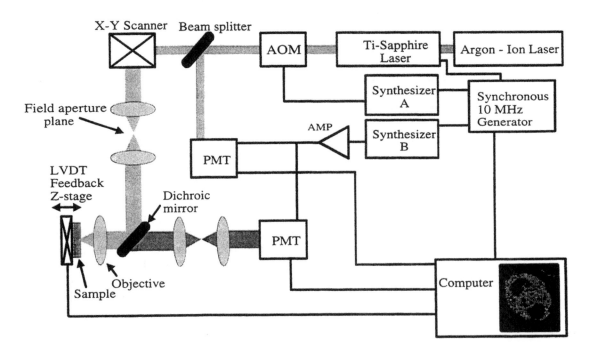

Fig. 9   Two-photon fluorescence lifetime imaging microscope schematic.

1. Light Source

The laser used in this microscope was a mode-locked titanium sapphire (Ti-Sapphire) laser (Mira 900, Coherent Inc.) pumped by an Argon-Ion laser (Innova 310, Coherent Inc.). The features of this laser system are a high average power of up to 1 watt, a high pulse repetition rate of 80 MHz, and a short pulse width of 150 fs. Additionally, the intensity output of this laser is very stable with typical fluctuations of less than 0.5%. Such stability is important for uniform excitation in the scanning system. The Ti-Sapphire laser has a tuning range from 720 to 1000 nm corresponding to one-photon excitation wavelengths of 360 to 500 nm. This wide tuning range allows most common near UV, blue and green fluorescent probes to be excited.

As a frequency domain light source, the pulse train of the Ti-Sapphire laser has modulation frequency content of 80 MHz and its harmonics. The harmonic content of the femtosecond pulse extends to terahertz. The lifetime determination of many common fluorescent probes requires the availability of modulation frequencies below 80 MHz. An acoustooptical modulator (Intra-Action, Belwood, IL) driven by the amplified signal of a phase-locked synthesizer is used to generate these additional frequencies not available directly from the laser. This AOM has the desirable properties of low-pulse-width dispersion an high-intensity transmittance (typically 50%). Additional modulation frequencies can be generated by beating the AOM modulation with the 80 MHz laser pulse train and its harmonics. The mixing between the laser repetition frequency and the AOM modulation frequency generates new components at the sum and difference frequencies. This scheme allows one to generate intensity modulation from kilohertz to tens of terahertz (Piston *et al.*, 1989).

2. Optics and Scanner

An *x-y* scanner (Cambridge Technology, Watertown, MA) directs the laser light into the microscope. A low-cost single-board computer (New Micros, Inc., Dallas, TX) is the synchronization and interface circuit between the scanner and the master data acquisition computer. The master computer synchronizes the scan steps by delivering a TTL trigger pulse to the interface circuit which outputs the required 16-bit number to move the scanner to the next pixel position. The normal region is a square of 256 × 256 scanned in a raster fashion. The scanner can scan over its full range with a maximum scan rate of 500 Hz. This rate limits the scan time needed for a typical 256 × 256 pixel region to be about 0.5 sec which corresponds to a pixel residence time of 8 μs. For normal operation, the pixel spacin in the images was 0.14 μm but could be adjusted as needed.

The excitation light enters the microscope (Axiovert 35, Zeiss Inc.) via a modified epiluminescence light path. The light is reflected by the dichroic mirror to the objective. The dichroic mirror is a custom-made short-pass filter (Chroma Technology Inc., Brattleboro, VT) which maximizes reflection in the infrared and transmission in the blue-green region of the spectrum. Since tight focusing increases the photon density, a high numerical aperture objective is critical to obtain efficient two-photon excitation. The objective normally used is a well-corrected Zeiss 63× Plan-Neofluar with a numerical aperture of 1.25. The objective delivers the excitation light and collects the fluorescence signal. The fluorescence signal is transmitted through the dichroic mirror and refocused on the detector. Since two-photon excitation has the advantage that the excitation and emission wavelengths are well separated (by 300–400 nm), suitable short-pass filters eliminate most of the residual scatter with a minimum attenuation of the fluorescence signal. In the excitation region of 900–1000 nm, we use 3 mm of BG39 Schott filter glass. It has an optical density of over 1— for the excitation wavelength while retaining over 80% of the transmittance in the fluorescence wavelength range.

The z-axis position of the microscope stage is controlled by a second single-board computer (Iota System Inc., Incline Village, NV). The distance between the objective and the sample is monitored by a linear variable differential transformer (LVDT; Schaevitz Engineering, Camden, NJ). The LVDT output is digitized and the single-board computer compares the measured position with the preset position. The position of the objective is dynamically maintained by the computer which drives a geared stepper motor coupled to the standard manual height adjustment mechanism of the microscope. This control system is designed to have a position resolution of 0.1 μm over a total range of 200 μm. This dynamic feedback control system has a bandwidth of about 10 Hz.

3. Detector and Data Acquisition

A gain-modulated photomultiplier tube (model R3896, Hamamatsu) detects the fluorescence signal from each position. A second gain-modulated photomultiplier tube (model R928, Hamamatsu) monitors the laser beam to correct for frequency and modulation drift of the laser pulse train. The analog outputs of the two PMTs are digitized by the data acquisition computer.

One unique feature of this microscope is its high cross-correlation frequency. A high frequency ensures that complete cross-correlation waveforms can be collected within a minimum pixel residence time. A short pixel residence time is essential to achieve an acceptable frame refresh rate. The absolute minimum pixel residence time is limited by the mechanical response of the scanner to about 8 μs. However, the minimum residence time is actually limited by the digitizer and the number of photons available. The digitizer used in this instrument can operate at 60 kHz in simultaneous dual channel mode (12-bit resolution; A2D-160, DRA Laboratories, Sterling, VA). Data is acquired at four points per cross-correlation period. Thus, the digitizer limit is 70 μs/pixel (15 kHz). In order to make the timing easier, the actual maximum frequency is chosen to be 12.5 kHz (80 μs/pixel). At this rate an entire 256 × 256 frame can be acquired in just over 6 sec. For fluorescence photon fluxes less than 10^7 s^{-1}, the pixel residence time should be increased to reduce statistical noise. As has been discussed, the most efficient way to reduce noise is to average over a few waveforms at each pixel. In this instrument, at least two waveforms are averaged, which brings the pixel residence time to 160 μs with a reasonable frame rate of about 13 sec.

B. Two–Photon Microscopy Examples

The time resolution of this microscope was studied by imaging latex spheres labeled with chromophores of different lifetimes (Molecular Probes, Eugene, OR). Orange fluorescent latex spheres (2.3 μm diameter) were mixed with Nile red fluorescent latex spheres (1.0 μm diameter) in the sample. The lifetime of these spheres were independently measured in a conventional cuvette lifetime fluorometer (ISS). The orange and red spheres each show a single exponential decay with a lifetime of 4.3 and 2.9 ns, respectively.

An image of two orange spheres (the larger ones) and three red spheres (the smaller ones) is presented in Fig. 10. Note that the small red sphere at the right edge of the picture has the same intensity as that of the larger orange spheres and is significantly less intense than the other red spheres. This observation stresses the fact that intensity is a poor parameter for discerning the property of the specimen in a fluorescent image. Intensity differences can be due to differences in concentration and many experimental artifacts such as the uniformity of the illumination and detector response. In contrast, the lifetime picture shows lifetime values of 4.0 + 0.2 ns and 2.6 ± 0.2 ns, in good agreement with the data obtained in the conventional fluorometer. The sphere on the right is correctly identified only in the lifetime picture.

Fluorescence lifetime imaging is most valuable in cellular systems. As an example, FITC-dextran (FITC, fluorescein-5-isothiocyanate isomer I) was used to observe the pH in the vacuoles of mouse macrophages. The three-dimensional resolution of the two-photon microscope proved to be crucial in distinguishing closely packed vacuoles. Out-of-focus vacuoles contributed minimal fluorescence to the measured lifetime images due to the three-dimensional resolution.

Macrophages were incubated with 1 mg/ml FITC-dextran (Sigma Chemical; 50.7 kDa, 1.2 mol FITC per mol dextran) for 24 hr before observation. The cells were not washed before examination. Fluorescence was observed only in vacuoles and in the extracellular media (Fig. 11). The average lifetime measured in the extracellular fluid was 3.9 ns corresponding to a neutral pH while the lifetime measured in the vacuoles was 3.1 ns on average indicating an average intravacuolar pH of 4.0, consistent with previous reports (Aubry *et al.*, 1993; Straubinger *et al.*, 1990; Murphy *et al.*, 1984; Tycko and Maxfield, 1982; Ohkuma and Poole, 1978). All vacuoles have roughly the same lifetime, indicating the vacuoles are at the same pH even though the intensity varied dramatically even within individual cells.

V. Pump-Probe Microscopy

The third example of a fluorescence lifetime imaging microscope uses an asynchronous pump-probe technique. We have adapted a frequency domain pump-probe technique (Elzinga *et al.*, 1987) for use in microscopy. Pump-probe spectroscopy uses a pump pulse of light to promote the sample to the excited state and a probe pulse to monitor the relaxation back to the ground state. The most common implementation is transient absorption (Fleming, 1986). Recently, applications using pump-probe spectroscopy with stimulated emission have also been reported (Lakowicz *et al.*, 1994b; Kusba *et al.*, 1994). With stimulated emission, one can either measure the stimulated emission directly or the resulting change in fluorescence emission intensity. In our microscope, we measure the modulation in fluorescence emission from chromophores excited by a pump laser source and stimulated to emit by a probe laser.

In microscopy, the asynchronous pump-probe technique offers spatial resolution equal to conventional microscopy and it provides high frequency response of a fluorescence system without a fast photodetector. The principle of this technique is illustrated in Fig. 12. Two pulsed laser beams are overlapped at the sample. Their wavelengths are chosen such that one beam (pump) is used to excite the molecules under study and another beam (probe) is used to stimulate emission from the excited-state chromophores. The two laser pulse trains are offset in frequency by a small amount. This arrangement generates a fluorescence signal at the cross-correlation frequency and corresponding harmonics. Since the

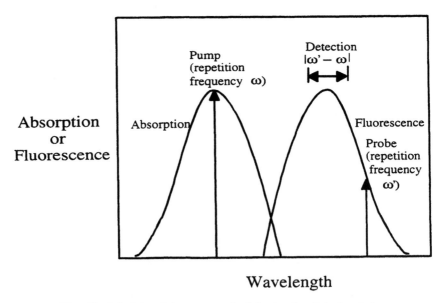

Fig. 12 Principles of the pump-probe (stimulated emission) technique.

cross-correlation signal depends strongly on the efficient overlap of the two laser sources (Dong *et al.*, 1995), superior spatial resolution can be obtained by detecting the fluorescence at the beat frequency or its harmonics. Furthermore, pulsed laser systems can have high harmonic content. The low-frequency cross-correlation signal harmonics correspond to the high-frequency laser harmonics. Therefore, a high-speed detector is not required to record the high-frequency components of the fluorescence signal.

A. Instrumentation

The experimental arrangement for our pump-probe fluorescence microscope is shown in Figure 13. The design is quite similar to the two-photon microscope. The differences are mainly due to the use of two laser beams instead of one. For example, the light sources are different along with the excitation and emission filters. Also the detector does not need to be gain modulated because the pump-probe signal is not at high frequency. Otherwise, the majority of the components are the same (such as the microscope, the *x-y* scanner, *z* (focus)-positioner and the data acquisition software and hardware).

1. Light Sources

A master synthesizer which generates a 10-MHz reference signal is used to synchronize two mode-locked neodymium-YAG and -YLF (Nd-YAG and Nd-YLF, Antares, Coherent Inc.) lasers and to generate a clock for the digitizer and scanner. The 532-nm output of the Nd-YAG laser is used for excitation.

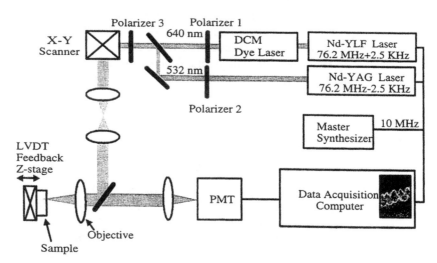

Fig. 13 Pump-probe fluorescence lifetime imaging microscope schematic.

The probe Nd-YLF laser pumps a DCM dye laser (Model 700, Coherent Inc.) tuned to 640 nm and is used to induce stimulated emission. The pulse width (FWHM) of the pump laser is 150 ps and the DCM probe laser has a pulse width of 10 ps. Combinations of polarizers are used to control the laser power reaching the sample. The average power of the pump and probe beams, at the sample, are about 10 μW and 7 mW, respectively. For time-resolved fluorescence microscopy, the pump source is operated at 76.2 MHz − 2.5 kHz and the probe laser's repetition frequency is 5 kHz away at 76.2 MHz + 2.5 kHz.

2. Optics

The two lasers are combined at a dichroic mirror (Chroma Technology Inc.) before reaching the *x-y* scanner (Cambridge Technology, Watertown, MA). As with the two-photon system, the images are typically 256 × 256 pixels with a 0.14-μm pixel spacing. The beams are reflected into the microscope objective by a second dichroic mirror. To align the pump and probe lasers, we found it convenient to overlap their projections on the laboratory's ceiling.

Because tight focusing increases the photon density and localizes the pump-probe effect, a high-numerical-aperture objective is used. The objective used in these studies is a well corrected, Zeiss 63× Plan-Neofluar with numerical aperture of 1.25. With this objective, the optical point spread function has a FWHM of 0.2 μm radially and 0.5 μm axially for excitation at 532 nm (Dong *et al.*, 1995). The fluorescence signal is collected by the same objective, transmitted through the second dichroic mirror and two 600 ± 20-nm bandpass filters.

3. Detector and Data Acquisition

A photomultiplier tube (model R928 or R1104, Hamamatsu) detects the fluorescence signal from each position. The analog PMT signal is electronically filtered by a preamplifier (Stanford Research, Sunnyvale, CA) to isolate the 5-kHz, cross-correlation signal before being delivered to the data acquisition hardware and software. Typically we use a pixel dwell time of 800 μs with a resulting frame acquisition time of 65 sec.

B. Pump–Probe Microscopy Examples

We obtained lifetime-resolved images of a mixture of 2.3-μm orange and 1.1-μm Nile-red (absorption maximum: 520 nm, emission maximum: 580 nm) fluorescent latex spheres (Molecular Probes). The two types of spheres were known to have different lifetimes. The measured lifetimes using a standard frequency-domain phase fluorometer are 2.70 ns for 1.1-μm spheres and 4.28 ns for 2.3-μm spheres. In our images, the first harmonic amplitude and phase are measured (Fig. 14). The phase image was referenced to that of a 4.16-mM rhodamine-B slide for the purpose of lifetime calculations. From the histograms of lifetime values, the lifetimes of the spheres were determined to be 3.2 ± 1.0 ns (1.1 μm) and 4.2 ± 1.4 ns (2.3 μm). These values agree within error to the results from frequency domain phase fluorometry.

We also examined mouse fibroblast cells doubly labeled with the nucleic acid stain ethidium bromide and the membrane stain rhodamine DHPE (Molecular Probes). The pump-probe image is shown in Fig. 15. These cells were grown on a coverslip and then fixed with acetone. The cells were stained first with ethidium bromide (1 mM in PBS, 0.1% Triton X-100) for 30 min and then stained by rhodamine DHPE (10 μg/ml in PBS, 0.1% Triton X-100) for another 30 min before it was rinsed twice in PBS and mounted for viewing. The lifetimes of the cytoplasmic and nuclear region were determined from the phase image. The reference phase was obtained from a slide of 4.16 mM rhodamine B in water. It was found that the average of lifetime histograms in the cytoplasm and nucleus are 2.0 ± 0.5 ns and 6.6 ± 4.8 ns, respectively. For comparison, the lifetime of rhodamine B in water was determined in a standard frequency domain phase fluorometer to be 1.44 ns. Furthermore, the lifetimes of the unbound ethidium bromide and bound ethidium bromide to nucleic acid are known to be 1.7 and 24 ns, respectively (So *et al.*, 1995). Our measurements of lifetime in cytoplasm show that there was significant staining of cytoplasmic structures by rhodamine DHPE. The average lifetime in the nucleus is between that of bound and unbound ethidium bromide indicative of the fact that both populations of the chromophores exist in the nucleus. Nonetheless, the lifetime contribution from bound ethidium bromide is sufficient to distinguish the different lifetimes in the nucleus and cytoplasm as demonstrated by the phase image.

This example demonstrates one advantage of lifetime-resolved imaging. From intensity imaging, it is difficult to distinguish the cytoplasmic and nuclear regions since these chromophores have similar emission spectra. With lifetime imaging, sharp contrast between the two species of chromophores can be generated.

VI. Conclusion

Fluorescence lifetime imaging provides the unique opportunity to study functional structures of living cellular systems. The three implementations of the lifetime-resolved microscope described in this chapter offer complimentary ways to solve similar problems. The lifetime camera captures all points in an image simultaneously which is better for fast events and phosphorescence imaging. The laser scanning systems can image arbitrarily shaped regions of interest. Three-dimensional resolution is inherently part of both the two-photon and pump-probe implementations. The lifetime camera images, on the other hand, must be mathematically deconvolved to achieve 3D resolution.

Each microscope presented has unique beneficial features. All three microscopes benefit from the frequency domain heterodyning technique. The lifetime camera achieves high-speed imaging, high temporal resolution, and much more photobleaching resistance than existing camera alternatives. Compared to confocal detection, the two-photon microscope has better separation of excitation and emission light, better spatial confinement of photodamage, and similar resolution. The pump-probe microscope provides equal spatial resolution and effective off-focal background rejection while eliminating the need for high-speed detectors.

Fluorescence lifetime imaging is a powerful technique that has far-ranging applications. Some projects that may be explored further are microscopic thermal imaging, multiple structure labeling of cells, and imaging and fluorescence in highly scattering media. Any process that needs nanosecond temporal resolution across an extended object is a good candidate. Fluorescence lifetime imaging microscopy has the potential to dramatically enhance the field of fluorescence microscopy.

Acknowledgments

We thank Dr. Matt Wheeler, Dr. Laurie Rund, Ms. Linda Grum and Ms. Melissa Izard in the Department of Animal Sciences, University of Illinois for providing us with mouse fibroblast cells. We also thank Dr. Edward Voss, Dr. Jenny Carrero, Ms. Anu Cherukuri, and Mr. Donald Weaver in the Department of Microbiology, University of Illinois for providing us with mouse macrophage cells. This work was supported by the National Institutes of Health Grant RR03155.

References

Ambrose, W. P., Goodwin, P. M., Martin, J. C., and Keller, R. A. (1994). Alterations of single molecule fluorescence lifetimes in near-field optical microscopy. *Science* **265**, 364–367.

Aubry, L., Klein, G., Martiel, J.-L., and Satre, M. (1993). Kinetics of endosomal pH evolution in *Dictyostelium discoideum* amoebae. *J. Cell Sci.* **105**, 861–866.

Birge, R. R. (1983). One-photon and two-photon excitation spectroscopy. In *"Ultrasensitive Laser Spectroscopy,"* pp. 109–174. Academic Press, New York.

Birge, R. R. (1985). Two-photon spectroscopy of protein-bound chromophores. *Acc. Chem. Res.* **19**, 138–146.

Buurman, E. P., Sanders, R., Draaijer, A., Gerritsen, H. C., van Veen, J. J. F., Houpt, P. M., and Levine, Y. K. (1992). Fluorescence lifetime imaging using a confocal laser scanning microscope. *Scanning* **14**, 155–159.

Denk, W., Strickler, J. H., and Webb, W. W. (1990). Two-photon laser scanning fluorescence microscopy. *Science* **248**, 73–76.

Dix, J. A., and Verkman, A. S. (1990). Pyrene eximer mapping in cultured fibroblasts by ratio imaging and time-resolved microscopy. *Biochemistry* **29**, 1949–1953.

Dong, C. Y., So, P. T. C., French, T., and Gratton, E. (1995). Fluorescence lifetime imaging by asynchronous pump-probe microscopy. *Biophys. J.* **69**, 2234–2242.

Elzinga, P.A., Kneisler, R. J., Lytle, F. E., King, G. B., and Laurendeau, N. M. (1987). Pump/probe method for fast analysis of visible spectral signatures utilizing asynchronous optical sampling. *Appl. Opt.* **26**, 4303–4309.

Feddersen, B. A., Piston, D. W., and Gratton, E. (1989). Digital parallel acquisition in frequency domain fluorimetry. *Rev. Sci. Instrum.* **60**, 2929–2936.

Fleming, G. R. (1986). *"Chemical applications of ultrafast spectroscopy."* Oxford Univ. Press, New York.

French, T., So, P. T. C., Weaver, D. J., Jr., Coelho-Sampaio, T., Gratton, E., Voss, E. W., Jr., and Carrero, J. (1997). Two-photon fluorescence lifetime imaging microscopy of macrophage mediated antigen processing. *J. Microsc.* (in press).

Friedrich, D. M. (1982). Two-photon molecular spectroscopy. *J. Chem. Education* **59**, 472–481.

Gadella, T. W. J., Jr., and Jovin, T. M. (1995). Oligomerization of epidermal growth factor receptors on A431 cells studied be time-resolved fluorescence imaging microscopy: A stereochemical model for tyrosine kinase receptor activation. *J. Cell Biol.* **129**, 1543–1548.

Gadella, T. W. J., Jr., Jovin, T. M., and Clegg, R. M. (1993). Fluorescence lifetime imaging microscopy (FLIM): Spatial resolution of microstructures on the nanosecond time scale. *Biophys. Chem.* **48**, 221–239.

Keating, S. M., and Wensel, T. G. (1990). Nanosecond fluorescence microscopy: Emission kinetics of Fura-2 in single cells. *Biophys. J.* **59**, 186–202.

König, K., So, P. T. C., Mantulin, W. W., Tromberg, B. J., and Gratton, E. (1996). Two-photon excited lifetime imaging of autofluorescence in cells during UVA and NIR photostress. *J. Microsc.* **183**, 197–204.

Kusba, J., Bogdanov, V., Gryczynski, I., and Lakowicz, J. R. (1994). Theory of light quenching: Effects on fluorescence polarization, intensity, and anisotropy decays. *Biophys. J.* **67**, 2024–2040.

Lakowicz, J. R., Szmacinski, H., Lederer, W. J., Kirby, M. S., Johnson, M. L., and Nowaczyk, K. (1994a). Fluorescence lifetime imaging of intracellular calcium in COS cells using QUIN-2. *Cell Calcium* **15**, 7–27.

Lakowicz, J. R., Gryczynski, I., Bogdanov, V., and Kusba, J (1994b). Light quenching and fluorescence depolarization of rhodamine B. *J. Phys. Chem.* **98**, 334–342.

Lakowicz, J. R., and Berndt, K. W. (1991). Lifetime-selective fluorescence imaging using an RF phase-sensitive camera. *Rev. Sci. Instrum.* **62**, 1727–1734.

Marriott, G., Clegg, R. M., Arndt-Jovin, D. J., and Jovin, T. M. (1991). Time resolved imaging microscopy. *Biophys. J.* **60**, 1374–1387.

McLoskey, D., Birch, D. J. S., Sanderson, A., Suhling, K., Welch, E., and Hicks, P. J. (1996). Multiplexed single-photon counting: 1. A time correlated fluorescence lifetime camera. *Rev. Sci. Instrum.* **67**, 2228–2237.

Morgan, C. G., Murray, J. G., and Mitchell, A. C. (1995). Photon correlation system for fluorescence lifetime measurements. *Rev. Sci. Instrum.* **66**, 3744–3749.

Morgan, C. G., Mitchell, A. C., and Murray, J. G. (1992). Prospects for confocal imaging based on nanosecond fluorescence decay time. *J. Microsc.* **165**, 49–60.

Müller, M., Ghauharali, R., Visscher, K., and Brakenhoff, G. (1995). Double pulse fluorescence lifetime imaging in confocal microscopy. *J. Microsc.* **177**, 171–179.

Murphy, R. F., Powers, S., and Cantor, C. R. (1984). Endosome pH measured in single cells by dual fluorescence flow cytometry: Rapid acidification of insulin to pH 6. *J. Cell Biol.* **98**, 1757–1762.

Oida, T., Sako, Y., and Kusumi, A. (1993). Fluorescence lifetime imaging microscopy (flimscopy). *Biophys. J.* **64**, 676–685.

Ohkuma, S., and Poole, B. (1978). Fluorescence probe measurement of the intralysosomal pH in living cells and the perturbation of pH in various agents. *Proc. Natl. Acad. Sci. USA* **75**, 3327–3331.

Piston, D. W., Kirby, M. S., Cheng, H., Lederer, W. J., and Webb, W. W. (1994). Two photon excitation fluorescence imaging of three-dimensional calcium ion activity. *Appl. Optics* **33**, 662–669.

Piston, D. W., Sandison, D. R., and Webb, W. W. (1992). Time-resolved fluorescence imaging and background rejection by two-photon excitation in laser scanning microscopy. *Proc. SPIE* **1640**, 379–389.

Piston, D. W., Marriott, G., Radivoyevich, T., Clegg, R. M., Jovin, T. M., and Gratton, E. (1989). Wide-band acousto-optic light modulator for frequency domain fluorometry and phosphorimetry. *Rev. Sci. Instrum.* **60**, 2596–2560.

Sanders, R., Draaijer, A., Gerritsen, H. C., Houpt, P. M., and Levine, Y. K. (1995a). Quantitative pH imaging in cells using confocal fluorescence lifetime imaging microscopy. *Analytical Biochem.* **227**, 302–308.

Sanders, R., Draaijer, A., Gerritsen, H. C., and Levine, Y. K. (1995b). Selective imaging of multiple probes using fluorescence lifetime contrast. *Zoological Studies* **34**, 173–174.

Sheppard, C. J. R., and Gu, M. (1990). Image formation in two-photon fluorescence microscopy. *Optik* **86**, 104–106.

So, P. T. C., French, T., Yu, W. M., Berland, K. M., Dong, C. Y., and Gratton, E. (1995). Time-resolved fluorescence microscopy using two-photon excitation. *Bioimaging* **3,** 49–63.

So, P. T. C., French, T., and Gratton, E. (1994). A frequency domain time-resolved microscope using a fast-scan CCD camera. *Proc. SPIE* **2137,** 83–92.

Straubinger, R. M., Papahadjopoulos, D., and Hong, K. (1990). Endocytosis and intracellular fate of liposomes using pyranine as a probe. *Biochem.* **29,** 4929–4939.

Tycko, B., and Maxfield, F. R. (1982). Rapid acidification of endocytic vesicles containing α_2-macro-globulin. *Cell* **28,** 643–551.

Wang, X. F., Uchida, T., Coleman, D. M., and Minami, S. (1991). A two-dimensional fluorescence lifetime imaging system using a gated image intensifier. *Appl. Spect.* **45,** 360–366.

Wang, X. F., Uchida, T., and Minami, S. (1989). A fluorescence lifetime distribution measurement system based on phase-resolved detection using an image dissector tube. *Appl. Spect.* **43,** 840–845.

Wilson, T., and Sheppard, C. (1984). *Theory and Practice of Scanning Optical Microscopy.* (Academic: New York).

Xie, X. S., and Dunn, R. C. (1994). Probing single molecule dynamics. *Science* **265,** 361–364.

CHAPTER 15

Digital Deconvolution of Fluorescence Images for Biologists

Yu-li Wang

Cell Biology Group
Worcester Foundation for Biomedical Research
Shrewsbury, Massachusetts 01545

I. Introduction

The performance of optical microscopes is limited both by the the aperture of the lens, which causes light from a point source to spread (or diffract) over a finite volume, and by the cross-contamination of light that originates from out-of-focus planes. To overcome these limitations, approaches have been developed in recent years both to improve the microscope design, as exemplified by confocal scanning microscopy, and to reverse mathematically the degrading effects of the conventional microscope. This latter approach is commonly referred to as "deconvolution," since the degradation effects of the microscope can be described mathematically as the convolution of input signals by the point spread function of the optical system (see below; Young, 1989; Russ, 1994).

A number of deconvolution algorithms have been tested for restoring fluorescence images. The most straightforward approach, 3D inverse filtering (Agard et al., 1989; Holmes and Liu, 1992), attempts to reverse the effects of image degradation through direct calculations. Unfortunately, it usually suffers from excessive computational artifacts. The two methods currently in wide use are constrained iterative deconvolution (Agard, 1984; Agard *et al.*, 1989; Shaw, 1993; Holmes and Liu, 1992) and nearest-neighbor deconvolution (Castleman, 1979; Agard, 1984; Agard *et al.*, 1989). The purpose of this chapter is to introduce the basic rationale of these two methods in languages easily understood by biologists. It will also provide details for the implementation of nearest-neighbor deconvolution using readily available hardware and software. Readers who wish to have a concise introduction of convolution and Fourier transformation for imaging are referred to the book by Russ (1994).

II. Rationale of Constrained Iterative Deconvolution

The two-dimensional convolution operation is defined mathematically as:

$$i(x,y) \otimes s(x,y) = \sum_{u,v} i(u,v)s(x - u, y - v) \tag{1}$$

This equation can be easily understood when one looks at the effects of convolving a matrix i with a 3×3 matrix:

$$
\begin{array}{cc}
\begin{array}{l}
\cdots\cdots\cdots\cdots\cdots\cdots \\
\cdots\cdots i_1 i_2 i_3 \cdots\cdots \\
\cdots\cdots i_4 i_5 i_6 \cdots\cdots \\
\cdots\cdots i_7 i_8 i_9 \cdots\cdots \\
\cdots\cdots\cdots\cdots\cdots\cdots
\end{array}
&
\otimes \quad
\begin{array}{l}
s_1 s_2 s_3 \\
s_4 s_5 s_6 \\
s_7 s_8 s_9
\end{array}
\end{array}
$$

Following the calculation, the element i_5 is replaced by $(i_1 \times s_9) + (i_2 \times s_8) + (i_3 \times s_7) + (i_4 \times s_6) + (i_5 \times s_5) + (i_6 \times s_4) + (i_7 \times i_3) + (i_8 \times i_2) + (i_9 \times i_1)$. That is, each element is now "contaminated" by contributions from the surrounding elements to an extent specified by the values in the s matrix. In an optical system, the degree of "contamination" is measured as the point spread function (the output image of a point source).

The process of image formation in a microscope can be described as the original distribution of intensities convolved by the 3D point spread function of the optical system (Agard *et al.*, 1989):

$$o(x,y,z) = i(x,y,z) \otimes s(x,y,z) \tag{2}$$

where $i(x,y,z)$ is a 3D matrix describing the signal originating from the sample and $s(x,y,z)$ is a matrix describing the 3D point spread function.

Alternatively, it is equally valid to write the equation using a series of 2D point spread functions (Agard *et al.*, 1989):

$$o(x,y) = i_0(x,y) \otimes s_0(x,y) + i_{-1}(x,y) \otimes s_{-1}(x,y) + i_{+1}(x,y) \otimes s_{+1}(x,y) \tag{3}$$
$$+ i_{-2}(x,y) \otimes s_{-2}(x,y) + i_{+2}(x,y) \otimes s_{+2}(x,y) + \cdots$$

where $i(x,y)$s are 2D matrices describing the signal originating from the plane of focus (i_0) and from planes above ($i_{+1}, i_{+2}, \ldots$) and below ($i_{-1}, i_{-2}, \ldots$), $s(x,$

y)s are matrices of 2D point spread functions that describe how point sources on the plane of focus (s_0) or planes above (s_1, s_2, . . .) and below (s_{-1}, s_{-2}. . . .) spread out when they reach the image plane.

Constrained iterative deconvolution uses a trial and error process to look for signal distribution $i(x,y,z)$ that satisfies equation (2). It usually starts with the assumption that $i(x,y,z)$ equals the measured stack of optical sections $o(x,y,z)$. As expected, when $o(x,y,z)$ is plugged into the right hand side of equation (2) in place of $i(x,y,z)$, it generates a matrix $o'(x,y,z)$ that deviates from $o(x,y,z)$ on the left-hand side. To decrease this deviation, adjustment is made to the initial matrix $o(x,y,z)$, voxel by voxel, based on the deviation of $o'(x,y,z)$ from $o(x,y,z)$ and on constraints such as nonnegativity of voxel values. Various approaches have been developed to determine how adjustments should be made to the trial image and how voxel values should be "constrained" (Agard *et al.*, 1989; Holmes and Liu, 1992). The modified $o(x,y,z)$ is then plugged back into the right-hand side of equation (2) to generate a new matrix, $o''(x,y,z)$, which resembles more closely $o(x,y,z)$. This process is repeated at least 20–30 times until there is no further improvement or until the calculated image matches closely the actual image.

III. Rationale of Nearest–Neighbor Deconvolution

The nearest-neighbor algorithm uses equation (3) as the starting point. The equation is simplified by introducing three assumptions:

1. Out-of-focus light from planes other than those adjacent to the plane of focus is negligible (i.e., terms containing s_{-2}, s_{+2}, and beyond are insignificant).
2. Light originating from planes immediately above or below the plane of focus can be approximated by images taken while focusing on these planes (i.e., $i_{-1} \approx o_{-1}$ and $i_{+1} \approx o_{+1}$).
3. Point spread functions for planes immediately above and below the focal plane, s_{-1} and s_{+1}, are equivalent (hereafter denoted as s_1).

Together, these approximations simplify equation (3) into:

$$o = i_0 \otimes s_0 + (o_{-1} + o_{+1}) \otimes s_1 \qquad (4)$$

Rearranging the terms and taking advantage of the mathematical fact that if $a \otimes b = c$, then $F(a) \times F(b) = F(c)$, where F represents Fourier transformation and "$\times$" represents multiplication of corresponding elements in the matrices, it can be shown that

$$i_0 = [o - (o_{-1} + o_{+1}) \otimes s_1] \otimes F^{-1}(1/F(s_0)) \qquad (5)$$

where F^{-1} represents reverse Fourier transformation. This equation can be understood in a simple, intuitive way: it states that the unknown signal distribution,

i_0, can be obtained by taking the in-focus image, o, subtracting out estimated contributions from planes above and below the plane of focus, $(o_{-1} + o_{+1}) \otimes s_1$, followed by convolution with the matrix $F^{-1}(1/F(s_0))$, which reverses diffraction-induced spreading of signals on the plane of focus.

Among the three approximations, the second is the most serious. On the one hand, since images taken from planes immediately above or below the plane of focus can include significant contributions of signals from the plane of focus, the use of these images leads to oversubtraction and erosion of structures. On the other hand, due to the diffraction of light, these out-of-focus images also somewhat underrepresent the true contribution from the corresponding planes.

In practice, nearest neighbor deconvolution is performed with a modified form of equation (5):

$$i_0 = [o - (o_{-1} + o_{+1}) \otimes (c_1 \cdot s_1)] \otimes F^{-1}(F(s_0)/ (F(s_0)^2 + c_2)), \qquad (6)$$

where constants c_1 and c_2 are empirical factors. c_1 is used to offset errors caused by oversubtraction as described above. c_2 is required to deal with problems associated with the calculation of reciprocals at the end of equation (5): the error could become astronomical when the value of the matrix element is small. The use of constant c_2 keeps the reciprocal value from getting too large when the matrix element is small compared to c_2. However it does not significantly affect the outcome when the matrix element is large compared to c_2.

IV. Implementation of Nearest-Neighbor Deconvolution

Nearest-neighbor deconvolution can be performed with readily available equipment: a conventional fluorescence microscope, a stable light source, a stepping motor coupled to the microscope focusing mechanism, a cooled slow-scan CCD camera, and a personal computer. According to equation (6), the calculation of i_0 requires the collection of only in-focus image o and images immediately above and below the focal plane (Fig. 1). These images are convolved with two matrices, $c_1 \cdot s_1$ and $F^{-1}(F(s_0)/(F(s_0) + c_2))$, which are determined by the point spread functions of the microscope system (alternatively, similar calculations can be performed in the frequency space, by replacing matrices in equation (6) with corresponding Fourier transformations and convolution operations with element-by-element multiplication of the matrices).

The easiest way to obtain the two matrices, s_0 and s_1, is to take serial optical sections, at a spacing equal to that used for collecting images of the sample, of fluorescent beads of ~ 0.1 μm diameter. The image with the highest intensity at the center of the bead is identified as the in-focus image. To obtain matrix s_0, this image is trimmed to an appropriate size, made radially symmetric by averaging, and normalized such that the sum of all elements equals 1 (Fig. 1). We found that the optimal size of the matrix lies between 11×11 and 17×17. This matrix is used as the input for Fourier transformation and matrix multiplication/

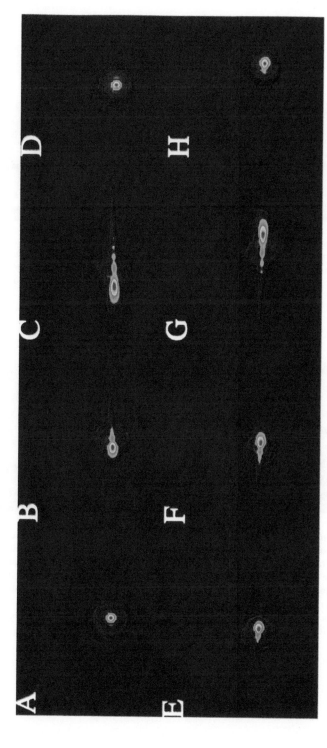

Ch.6, Fig. 5 The three-dimensional point spread function. Pseudocolor images of the point spread function of a microscope. A Z-series was taken of a 0.2-mm bead in the microscope and then reconstructed with pseudocolor. (A) The view looking down the optical axis of the microscope; (B) 40 degrees to the axis; (C) 100 degrees; (D) 190 degrees; (E) 210 degrees; (F) 260 degrees; (G) 300 degrees; and (H) 360 degrees (images were prepared by Ted Inoue and Scott Blakely of Universal Imaging Corporation. Westchester, PA).

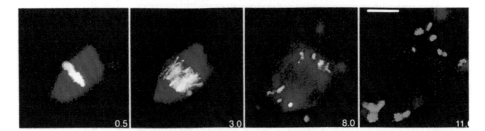

Ch.10, Fig. 11 Anaphase recorded for mitotic spindles assembled in the test-tube from *Xenopus* egg extracts as described (Murray *et al.*, 1996). At 30-sec intervals, an X-rhodamine-tubulin image and a DA PI-stained chromosome image were recorded. Red and blue image stacks were overlaid with MetaMorph and printed in Adobe Photoshop. Time in minutes is indicated on each frame taken from the time-lapse sequence.

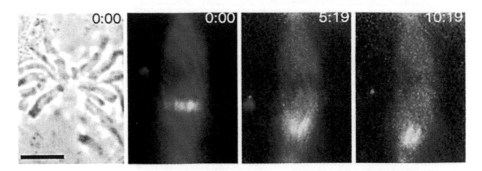

Ch.10, Fig. 12 Multimode imaging and local fluorescence photoactivation to see the dynamics of kinetochore fiber microtubules in metaphase newt lung epithelial cells. Cell were injected in early mitosis with X-rhodamine-labeled tubulin and C2CF caged-fluorescein-labeled tubulin. At each time point, a phase image showed chromosome position, an X-rhodamine image showed spindle microtubule assembly, and the C2CF fluorescein image recorded the poleward flux and turnover of photoactivated fluorescent tubulin within the kinetochore fiber microtubules. In the color frames, the green photoactivated C2CF fluorescein fluorescence is overlaid on the red image of the X-rhodamine tubulin in the spindle fiber microtubules. Time is given in minutes after photoactivation. See Waters *et al.* (1996) for details. Scale = 10 μm.

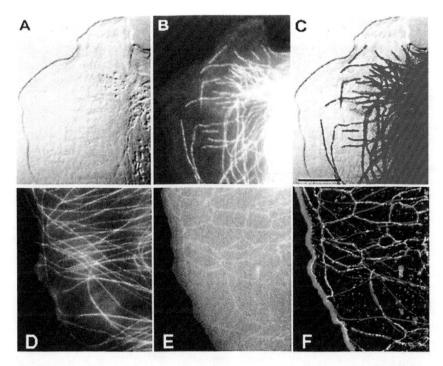

Ch.10, Fig. 13 Microtubules in the lamella of a live migrating newt lung epithelial cells. Microtubules are fluorescently labeled by microinjection of X-rhodamine-labeled tubulin into the cytoplasm and ER-labeled with $DiOC_6$. (A) DIC image of leading edge of lamella. (B) X-rhodamine microtubule image. (C) RGB color overlay of A and B. (D) X-rhodamine tubulin image of another cell. (E) $DiOC_6$ image of the ER. (F) RGB color overlay of E and F. See text for details.

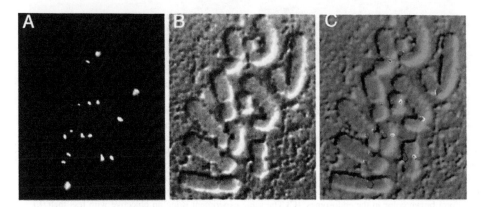

Ch.10, Fig. 15 The mitotic arrest protein MAD2 localizes to kinetochores on chromosomes in mitotic PtK1 cells when they are unattached from spindle microtubules. The cell was treated with nocodazole to induce spindle microtubule disassembly and then lysed and fixed for immunofluorescence staining with antibodies to the MAD2 protein (Chen *et al.*, 1996). (A) A single optical fluorescence section. (B) The corresponding DIC image. (C) The fluorescent MAD2 image and the DIC image combined in an RGB image to show the localization of MAD2 on either side of the centomeric constriction of the chromosomes where the sister kinetochore are located. Some kinetochores are above or below the plane of focus and not visible.

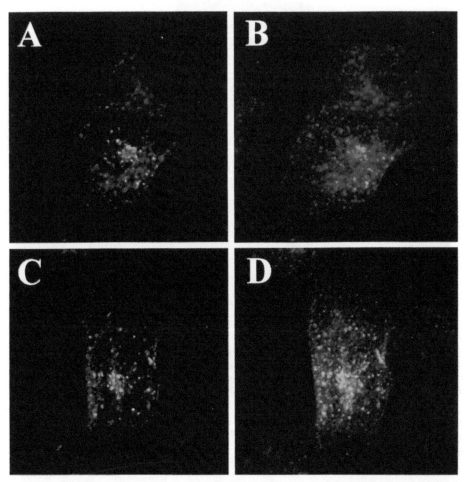

Ch.11, Fig. 1 Effects of axial chromatic aberration on ratio measurements. Living cells were incubated with a mixture of transferrin conjugated to either fluorescein or to Cy-5 and then fixed and imaged by confocal microscopy. As the two conjugates are expected to proportionally label the same endocytic compartments, one would expect the ratio between the two to be constant in all labeled endosomes. Fluorescein fluorescence was excited by the 488-nm line and Cy-5 fluorescence excited by the 647-nm line of a Krypton-Argon laser. (A) An image of a single focal plane using an objective in which a disparity of 0.4 microns in the focal plane of the 488- and 647-nm excitation lines results in the production of variously colored endosomes ranging from green to red. That each endosome has the same ratio of fluorescein to Cy-5 fluorescence is shown when one combines the vertical series of images of this field (B) and one obtains a uniform brown color in the endosomes. (C) The same experiment using an objective with less than 0.1 microns of axial chromatic aberration between the 488- and 647-nm lines. In this image, the constant ratio between the two different transferrin conjugates in each endosome is apparent in the uniform yellow color of the endosomes. This same color is obtained when one combines all the images of a vertical series into a single projected image (D), indicating that both colors of fluorescence are accurately collected in each single focal plane.

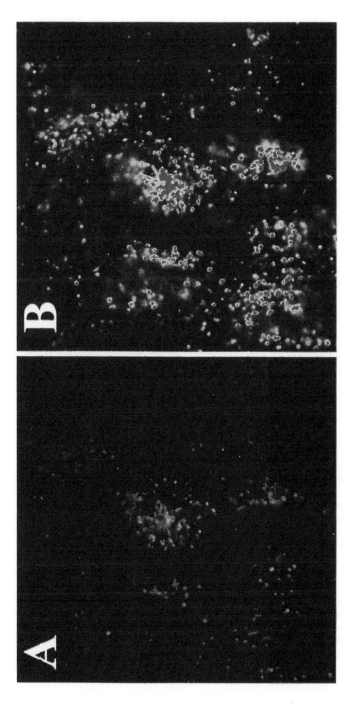

Ch.11, Fig. 2 Effects of limited dynamic range on ratio measurements. Living cells were incubated with dextran conjugated to both fluorescein and rhodamine which is delivered to lysosomes. Since the fluorescence of fluorescein is quenched at low pH, but the fluorescence of rhodamine is pH insensitive, the ratio of red-to-green fluorescence can be used to sensitively measure the pH of lysosomes. (A) A field of living cells whose lysosomes appear yellow due to the quenched fluorescein fluorescence at the low pH lysosomes. (B) The same field of cells after the lysosomes were alkalinized by the addition of an ionophore. In addition to the endosomes become greener, one also sees that many, if not most, of the detectable lysosomes in A now show saturating signal levels of green fluorescence (depicted in red).

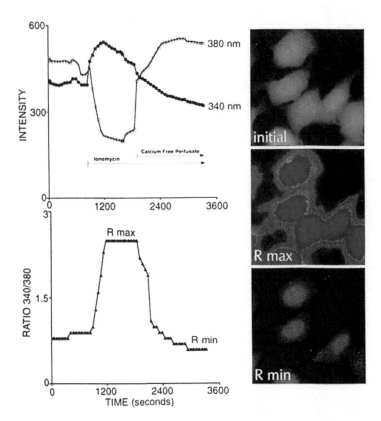

Ch.12, Fig. 7 *In situ* calibration of fura-2 in HeLa cells using ratiometric excitation digital imaging techniques. These data were generated using a Universal Imaging (Metafluor) work station and a cooled CCD camera with a frame transfer chip (EEV-37) and 12-bit readout (Princeton Instruments) interfaced to an inverted epifluorescence microscope (Nikon Diaphot) and viewed with a Nikon CF Fluor 100X/1.3 nA oil immersion objective. The graphs of data obtained during a protocol for one individual cell in the field at 340 nm (closed circles), 380 nm (open circles), and the 340/380 ratio (closed triangles) are shown as a function of time. At the first arrow (~1000 seconds), ionomycin (10 μM) is added to the perfusate bathing the cells which is a HEPES-buffered Ringer's solution containing 2 mM CaCl$_2$, to generate an R_{max} value. At 1900 sec (second arrow) the solution is switched to an ionomycin containing calcium free solution with 2 mM EGTA, in order to measure an R_{min} value. In addition three representative pseudocolor ratio images of this field of HeLa cells corresponding to three time points (initial, R_{max} and R_{min}) are shown. The pseudocolor scale is the same for all three images with blue being the lowest ratio corresponding to the lowest Ca$_i$ values and red being the highest ratio associated with maximal Ca$_i$ values.

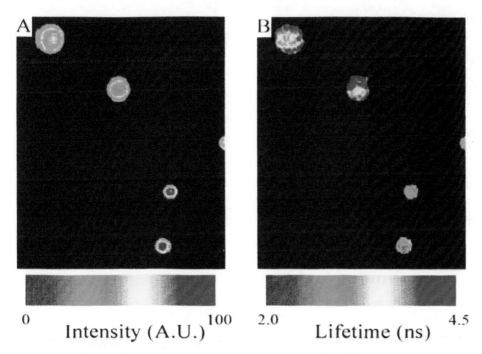

Ch.14, Fig. 10 Time-resolved image of orange (2.3 μm diameter) and red (1.0 μm diameter) fluorescent latex spheres. (A) Intensity: (B) lifetime.

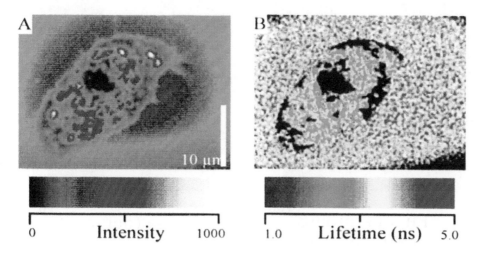

Ch.14, Fig. 11 Unwashed macrophages incubated 24 hr with FITC-dextran. (A) Intensity: (B) lifetime. (Cells were provided by the laboratory of Dr. E. W. Voss, Dept. of Microbiology, University of Illinois.)

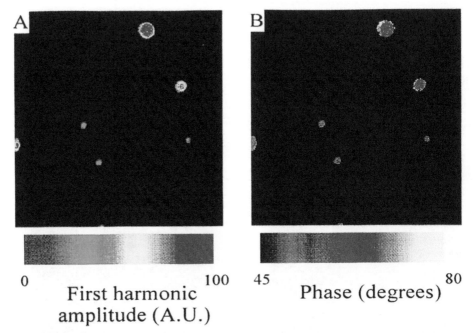

0   100   45   80

First harmonic
amplitude (A.U.)   **Phase (degrees)**

Ch.14, Fig. 14   Fluorescent lifetime images of 1.1 µm nile-red (τ_p = 3.20±1.0 ns) and 2.3 µm orange (τ_p = 4.19±1.4 ns) fluorescent latex spheres. (A) First harmonic amplitude; (B) phase.

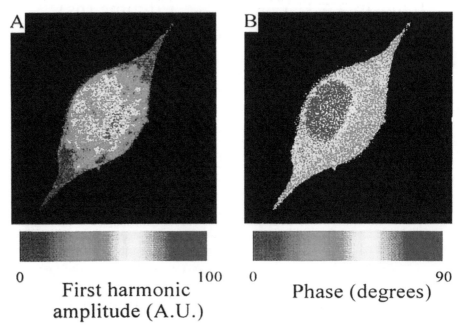

0   100   0   90

First harmonic
amplitude (A.U.)   **Phase (degrees)**

Ch.14, Fig. 15   Fluorescent lifetime images of a mouse fibroblast cell labeled with rhodamine DHPE and ethidium bromide (membrane and cytoplasm: τ_p = 2.00±0.54 ns, nucleus: τ_p = 6.62±4.8 ns). (A) First-harmonic amplitude; (B) phase. (Cells were provided by the laboratory of Dr. M. Wheeler, Department of Animal Sciences, University of Illinois.)

division to generate $F^{-1}(F(s_0)/(F(s_0)^2 + c_2))$. The calculations can be streamlined using readily available PC programs such as Mathcad (Fig. 1). The generation of s_1 is more straightforward. Images of beads immediately above and below the plane of sharp focus are averaged, followed by trimming, symmetrization, and normalization as for the generation of s_0. The constant c_1 is then incorporated into the s_1 matrix as shown in the last equation of Fig. 1.

The performance of nearest-neighbor deconvolution is highly sensitive to the values c_1 and c_2 in equation (6). In general, c_1 falls in the range of 0.45 to 0.50: the optimal value varies with the optical condition and the separation between adjacent optical slices. Too large a value causes erosion and discontinuity of in-focus structures (Fig. 3c), while too small a value would lead to high residual background due to the incomplete removal of out-of-focus noises (Fig. 3a). The value of c_2 is generally in the range of 0.0001 to 0.001. Too large a c_2 value causes the loss of details, yielding blurry images (Fig. 3e), while too small a value would cause the amplification of random noises into bright spots, rings, or patches (Fig. 3d). The optimal values for c_1 and c_2 can be found only through systematic trials.

The convolution matrices and sample images are then fed into equation (6) to generate the desired image i_o. This requires a computer program that can perform image/matrix subtraction and convolution with floating-point matrices (or fast Fourier transformation for calculations in the frequency space). These functions again can be programmed into common mathematical packages such as Mathcad 6.0 Plus. To improve the speed of calculation, these operations are performed in this laboratory with a DSP board installed in a personal computer (AL860-40MHz, single processor; Alacron, Inc., Nashua, NH), which with dedicated software can perform convolution of a 512×384 image with a 17×17 floating-point matrix within 2 sec. The entire computations of equation (6) can be completed within 10 sec. After computation, the image needs to be scaled properly to yield gray values suitable for display on a computer monitor. A stack of processed optical sections can then be used for 3D reconstruction or visualization (Fig. 4).

V. Evaluation of Digital Deconvolution Methods

For both constrained iterative deconvolution and nearest-neighbor deconvolution, images are collected with a conventional microscope, which has advantages such as limited photobleaching and versatility in the choice of excitation wavelengths. While it is essential to use a high-quality cooled CCD camera, the rest of the system can be installed with limited cost using readily available equipment and software. In addition, it is possible to implement both nearest-neighbor and constraint iterative programs in the same system, using nearest-neighbor for preliminary evaluation of images and constrained iterative deconvolution for more precise restorations.

File "onfile" contains the in-focus intensity distribution of fluorescent beads in a 17x17 matrix, obtained by trimming and averaging a number of in-focus images.

ON := READPRN(onfile)

$$
ON = \begin{bmatrix}
25 & 25 & 25 & 25 & 27 & 26 & 26 & 27 & 25 & 24 & 22 & 21 & 21 & 23 & 20 & 22 & 20 \\
26 & 24 & 27 & 28 & 27 & 29 & 28 & 29 & 29 & 28 & 26 & 24 & 25 & 21 & 22 & 21 & 20 \\
27 & 28 & 28 & 31 & 32 & 33 & 35 & 34 & 32 & 29 & 27 & 25 & 24 & 25 & 22 & 21 & 22 \\
26 & 27 & 29 & 34 & 36 & 39 & 41 & 42 & 41 & 36 & 31 & 30 & 26 & 25 & 24 & 23 & 23 \\
28 & 30 & 33 & 35 & 43 & 48 & 53 & 56 & 53 & 47 & 41 & 32 & 28 & 25 & 25 & 22 & 22 \\
29 & 31 & 35 & 42 & 54 & 59 & 69 & 82 & 86 & 79 & 60 & 39 & 33 & 26 & 24 & 24 & 22 \\
32 & 33 & 37 & 49 & 60 & 71 & 104 & 157 & 190 & 163 & 107 & 61 & 36 & 29 & 27 & 26 & 23 \\
33 & 36 & 41 & 55 & 65 & 90 & 172 & 296 & 345 & 294 & 176 & 90 & 45 & 34 & 28 & 22 & 23 \\
32 & 37 & 44 & 60 & 68 & 107 & 225 & 383 & 451 & 370 & 220 & 102 & 56 & 33 & 28 & 26 & 24 \\
34 & 36 & 44 & 59 & 70 & 98 & 202 & 332 & 395 & 322 & 193 & 93 & 50 & 35 & 29 & 26 & 23 \\
33 & 36 & 43 & 55 & 68 & 82 & 129 & 204 & 236 & 188 & 122 & 66 & 44 & 35 & 30 & 27 & 23 \\
34 & 36 & 40 & 50 & 63 & 69 & 81 & 99 & 104 & 92 & 70 & 51 & 44 & 33 & 29 & 25 & 23 \\
35 & 35 & 40 & 45 & 52 & 58 & 62 & 59 & 57 & 52 & 47 & 43 & 37 & 33 & 28 & 25 & 25 \\
32 & 35 & 36 & 38 & 46 & 47 & 48 & 49 & 45 & 43 & 39 & 37 & 33 & 30 & 28 & 25 & 23 \\
29 & 31 & 34 & 37 & 37 & 38 & 40 & 40 & 35 & 35 & 35 & 32 & 27 & 27 & 25 & 24 & 22 \\
27 & 29 & 32 & 32 & 33 & 32 & 34 & 31 & 31 & 29 & 30 & 30 & 26 & 26 & 23 & 24 & 22 \\
27 & 26 & 28 & 29 & 27 & 30 & 31 & 30 & 30 & 28 & 26 & 26 & 24 & 25 & 23 & 23 & 22
\end{bmatrix}
$$

File "offfile" contains the out-of-focus intensity distribution of fluorescent beads in a 17x17 matrix, obtained by trimming and averaging a number of images immediately above and below the plane of focus.

OFF := READPRN(offfile)

$$
OFF = \begin{bmatrix}
24 & 27 & 27 & 28 & 28 & 30 & 31 & 30 & 27 & 25 & 25 & 26 & 22 & 22 & 20 & 21 & 20 \\
26 & 27 & 28 & 30 & 30 & 35 & 34 & 32 & 31 & 31 & 26 & 25 & 22 & 22 & 22 & 21 & 20 \\
26 & 28 & 31 & 34 & 37 & 36 & 38 & 36 & 35 & 30 & 32 & 27 & 25 & 24 & 24 & 23 & 21 \\
28 & 28 & 34 & 39 & 41 & 41 & 43 & 45 & 42 & 40 & 35 & 33 & 28 & 25 & 21 & 25 & 22 \\
29 & 30 & 34 & 36 & 44 & 50 & 56 & 59 & 59 & 56 & 45 & 35 & 30 & 27 & 26 & 23 & 22 \\
30 & 32 & 38 & 41 & 51 & 59 & 74 & 93 & 100 & 90 & 70 & 49 & 37 & 28 & 27 & 25 & 24 \\
34 & 34 & 40 & 46 & 58 & 72 & 113 & 168 & 195 & 172 & 120 & 70 & 45 & 32 & 28 & 26 & 25 \\
33 & 36 & 44 & 51 & 59 & 88 & 166 & 276 & 312 & 269 & 177 & 98 & 54 & 36 & 31 & 25 & 24 \\
32 & 38 & 44 & 54 & 62 & 97 & 195 & 306 & 368 & 314 & 200 & 110 & 59 & 38 & 32 & 28 & 27 \\
35 & 37 & 43 & 55 & 61 & 88 & 162 & 261 & 304 & 256 & 172 & 100 & 57 & 38 & 32 & 29 & 26 \\
35 & 34 & 43 & 51 & 60 & 69 & 103 & 152 & 178 & 160 & 115 & 77 & 50 & 37 & 32 & 29 & 28 \\
36 & 37 & 39 & 49 & 55 & 61 & 67 & 80 & 87 & 83 & 69 & 54 & 43 & 37 & 30 & 30 & 27 \\
34 & 37 & 38 & 41 & 47 & 53 & 51 & 52 & 53 & 53 & 46 & 45 & 39 & 35 & 29 & 28 & 25 \\
32 & 34 & 38 & 38 & 41 & 41 & 43 & 43 & 43 & 40 & 39 & 37 & 34 & 31 & 28 & 27 & 25 \\
28 & 31 & 32 & 33 & 35 & 37 & 38 & 35 & 35 & 36 & 36 & 34 & 31 & 29 & 27 & 24 & 24 \\
28 & 28 & 30 & 32 & 31 & 32 & 32 & 33 & 34 & 31 & 30 & 30 & 28 & 28 & 25 & 24 & 25 \\
27 & 26 & 29 & 29 & 29 & 30 & 30 & 31 & 30 & 29 & 28 & 28 & 26 & 25 & 24 & 23 & 24
\end{bmatrix}
$$

Fig. 1 Mathcad program used for the calculation of s_0 and s_1. The averaged in-focus image of fluorescent beads was stored in an ASCII file "onfile." Out-of-focus images collected above and below the plane of sharp focus were averaged and stored in an ASCII file "offile." These images are shown as two 17×17 matrices near the beginning of the figure. The calculations then symmetrize and normalize the matrices as indicated in the comments. The resulting matrices, denoted as "ON" and "OFF," are then used for the calculation of s_0 and s_1, which are output as ASCII files at the end of the program. Note that parameter c_1 (0.50) is incorporated into the s_1 matrix near the end of the figure.

310

Symmetrize the matrices

$ii := 0..8$

$jj := 0..8$

$$ON_{ii,jj} := \frac{ON_{ii,jj} + ON_{16-ii,jj} + ON_{ii,16-jj} + ON_{16-ii,16-jj} + ON_{jj,ii} + ON_{16-jj,ii} + ON_{jj,16-ii} + ON_{16-jj,16-ii}}{8}$$

$$\begin{bmatrix} ON_{jj,ii} \\ ON_{16-jj,ii} \\ ON_{jj,16-ii} \\ ON_{16-jj,16-ii} \\ ON_{16-ii,jj} \\ ON_{ii,16-jj} \\ ON_{16-ii,16-jj} \end{bmatrix} := \begin{bmatrix} ON_{ii,jj} \\ ON_{ii,jj} \\ ON_{ii,jj} \\ ON_{ii,jj} \\ ON_{ii,jj} \\ ON_{ii,jj} \\ ON_{ii,jj} \end{bmatrix}$$

$$OFF_{ii,jj} := \frac{OFF_{ii,jj} + OFF_{16-ii,jj} + OFF_{ii,16-jj} + OFF_{16-ii,16-jj} + OFF_{jj,ii} + OFF_{16-jj,ii} + OFF_{jj,16-ii} + OFF_{16-jj,16-ii}}{8}$$

$$\begin{bmatrix} OFF_{jj,ii} \\ OFF_{16-jj,ii} \\ OFF_{jj,16-ii} \\ OFF_{16-jj,16-ii} \\ OFF_{16-ii,jj} \\ OFF_{ii,16-jj} \\ OFF_{16-ii,16-jj} \end{bmatrix} := \begin{bmatrix} OFF_{ii,jj} \\ OFF_{ii,jj} \\ OFF_{ii,jj} \\ OFF_{ii,jj} \\ OFF_{ii,jj} \\ OFF_{ii,jj} \\ OFF_{ii,jj} \end{bmatrix}$$

Use the corner value as background intensity, subtract from all elements

$i := 0..16$

$j := 0..16$

$Back := if(ON_{0,0} > OFF_{0,0}, OFF_{0,0}, ON_{0,0})$

$ON_{i,j} := ON_{i,j} - Back$

$OFF_{i,j} := OFF_{i,j} - Back$

Normalize the matrices

$$SUMON := \sum_i \sum_j ON_{i,j} \qquad SUMOFF := \sum_i \sum_j OFF_{i,j}$$

$$ON_{i,j} := \frac{ON_{i,j}}{SUMON} \qquad OFF_{i,j} := \frac{OFF_{i,j}}{SUMOFF}$$

Calculate the Fast Fourier Transform, ASSUMING c2 = 0.0005

$FTON := cfft(ON)$

$$FTS0_{i,j} := \frac{FTON_{i,j}}{\left(FTON_{i,j} \cdot \overline{FTON_{i,j}} + 0.0005\right)}$$

Perform Inverse FFT

$S0 := Re(icfft(FTS0))$

Normailze s0 such that the sum of elements equals 10. This makes the output decimal numbers more readable but does not affect the final results. Parameter c1 is multiplied into the s1 matrix.

$$SUMG := \sum_i \sum_j \frac{S0_{i,j}}{10}$$

$PRNCOLWIDTH := 12$

$$WRITEPRN(s0) := \frac{S0}{SUMG} \qquad\qquad WRITEPRN(s1) := OFF \cdot 0.50$$

Fig. 1—*Continued*

Fig. 2 Original images used for nearest-neighbor deconvolution. NRK epithelial cells were stained with antibodies against β-tubulin and rhodamine-conjugated secondary antibodies. Images were recorded with a Zeiss Axiovert microscope, a 100X/N.A. 1.30 neofluar lens, and a cooled slow-scan CCD camera from Princeton Instruments (Trenton, NJ). The image is 576×384 pixels, with each pixel corresponding to a sample area of 0.085×0.085 μm. The three images are 0.25 μm apart in focus from one another. During the calculation, b is used as the in-focus image and a and c are used as its nearest neighbor.

Compared to confocal laser scanning microscopy, one disadvantage of both methods is their limited ability to penetrate thick samples. In addition, when the sample is very weak, input images become limited in S/N ratio, which could seriously jeopardize the calculation for image restoration. Both deconvolution methods are also sensitive to instabilities in image intensity, caused by fluctuating lamp intensities or sample photobleaching. Thus samples should be mounted in antibleaching solutions whenever possible and microscope lamps should be driven with stabilized DC power supplies. Another common characteristic of the deconvolution techniques is their sensitivity to the vertical distance among optical sections. The distance should be close enough such that the stack includes focused or close-to-focused images of all structures. However, too close a distance would cause increases in the calculation time for the constrained iterative method, and for the nearest-neighbor method, serious errors due to assumption 2 discussed above. Typically, the optimal distance falls in the range of 0.25–0.50 μm.

Under ideal conditions, constrained iterative deconvolution can achieve a resolution that goes beyond what is predicted by the optical theory and provide 3D intensity distribution of photometric accuracy (Carrington *et al.,* 1995). In addition, it can generate accurate image slices between collected slices. With confocal microscopes or nearest-neighbor deconvolution, such information can be generated only through mathematical interpolation. The main drawback of constrained iterative deconvolution is its computational demand. While it is possible to perform the calculation with a PC, the restoration of one stack of images could take many hours. Thus for practical purposes, constrained iterative deconvolution has to be performed at least with a powerful workstation. A second disadvantage with constrained iterative deconvolution is its requirement for a complete stack of images for deconvolution, even if the experiment requires only one optical slice. Iterative constrained deconvolution is also very sensitive to the condition of the microscope (dust, alignment, etc.) and the quality of the

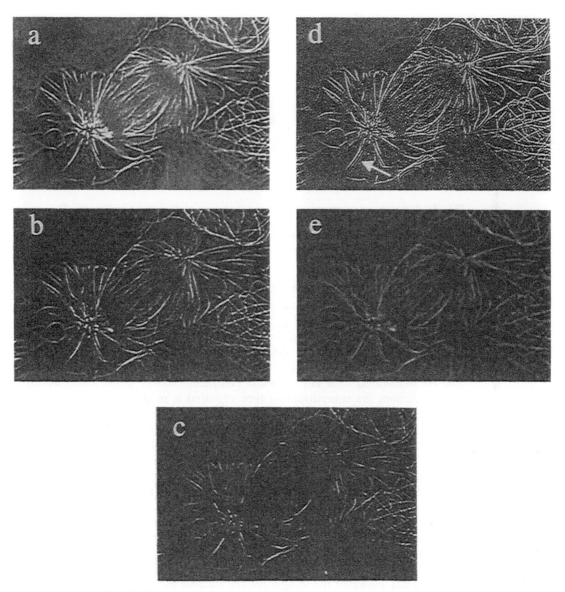

Fig. 3 Nearest-neighbor deconvolution of images shown in Fig. 2 with different values of c_1 and c_2. (b) The optimal setting, with $c_1 = 0.50$ and $c_2 = 0.0005$. When c_1 is decreased to 0.46 (a), the removal of out-of-focus noises becomes incomplete, resulting in high diffuse background. When c_1 is increased to 0.54 (c), out-of-focus noises are overcorrected, resulting in fragmented structures. The parameter c_2 controls the sharpness of the final image. When it is too small (d; $c_2 = 0.00004$), the image suffers from excessive spotty noise. In addition, some structures become double (arrow). When c_2 is too large (e; $c_2 = 0.01$), the structures become fuzzy and faint.

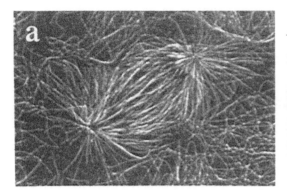 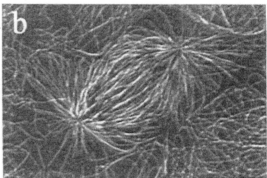

Fig. 4 A stereo pair constructed from a complete stack of 25 deconvolved images. The images were deconvolved as in Fig. 3b, with values of 0.50 and 0.0005 for c_1 and c_2, respectively. Although the stereo pair is generated with custom-written software, similar results can be achieved with commercial packages such as Metamorph (Universal Imaging, West Chester, PA).

objective lens. In addition, with a poorly corrected lens the point spread function can vary significantly with the position in the field and with the focal plane, introducing serious position-dependent errors in the output images.

The most significant advantage of nearest-neighbor deconvolution is its simplicity. With various parameters appropriately tuned, it yields noticeably better images than those provided by high-pass filtering or unsharp masking, yet it can be performed with any personal computer. Unlike constrained iterative deconvolution, the computational time required is in the order of seconds to minutes rather than hours. In addition, when the experiment is focused on a single plane of focus, images need to be acquired only from the plane of interest plus two adjacent planes, and the processed image can be obtained with a single cycle of calculation using equation (6). Unlike constrained iterative deconvolution, the approach is relatively insensitive to the quality of the lens or the point spread function, and visually satisfactory results could be generated with either actual or computer calculated point spread functions.

The most notable limitation with nearest-neighbor deconvolution is its precision, associated with the approximations described above. It performs well when only qualitative images are required and when the sample consists of discrete structures such as microtubules, vesicles, bundles of actin filaments, and chromosomal bands. However the images are not suitable for quantitative analysis such as the determination of 3D distribution of fluorophores. The limitation also becomes apparent when the sample involves continuous grades of intensities or consists of dark objects imbedded in an otherwise uniform block of fluorescence.

VI. Prospectus

With the improvements in computer performance, digital deconvolution is becoming increasingly feasible for average cell biology laboratories. Even with the advancement in confocal scanning microscopy such as two-photon excitation, computation methods will continue to serve their unique purposes. For example, nearest-neighbor deconvolution will remain as an efficient, economical method for samples of discrete structures and limited thickness (5–20 μm). The iterative constraint deconvolution, on the other hand, will remain as the method of choice for obtaining precise 3D fluorescence distribution. Most importantly, since confocal scanning microscopy and computational deconvolution work under independent principles, these methods can be easily combined to obtain resolution and photometric precision far beyond what was feasible with individual approaches.

References

Agard, D. A. (1984). *Annu. Rev. Biophys. Bioeng.* **13**, 191–219.

Agard, D. A., Hiraoka, Y., Shaw, P., and Sedat, J.W. (1989). *Methods Cell Biol.* **30**, 353–377.

Carrington, W. A., Lynch, R. M., Moore, E. D., Isenberg, G., Fogarty, K. E., and Fay, F. S. (1995). *Science* **268**, 1483–1487.

Castleman, K. R. (1979). "Digital Image Processing." Prentice-Hall, Englewood Cliffs, NJ.

Holmes, T. J., and Liu, Y.-H. (1992). *In* "Visualization in Biomedical Microscopies" (A. Kriete, ed.), pp. 283–327. VCH Press, New York.

Russ, J. C. (1994). "The Image Processing Handbook." CRC Press, Boca Raton, FL.

Shaw, P. J. (1993). *In* "Electronic Light Microscopy" (D. Shotton, ed.), pp. 211–230. Wiley-Liss, New York.

Young, I. T. (1989). *Methods Cell Biol.* **30**, 1–45.

INDEX

VOLUMES IN SERIES

Founding Series Editor
DAVID M. PRESCOTT

Volume 1 (1964)
Methods in Cell Physiology
Edited by David M. Prescott

Volume 2 (1966)
Methods in Cell Physiology
Edited by David M. Prescott

Volume 3 (1968)
Methods in Cell Physiology
Edited by David M. Prescott

Volume 4 (1970)
Methods in Cell Physiology
Edited by David M. Prescott

Volume 5 (1972)
Methods in Cell Physiology
Edited by David M. Prescott

Volume 6 (1973)
Methods in Cell Physiology
Edited by David M. Prescott

Volume 7 (1973)
Methods in Cell Biology
Edited by David M. Prescott

Volume 8 (1974)
Methods in Cell Biology
Edited by David M. Prescott

Volume 9 (1975)
Methods in Cell Biology
Edited by David M. Prescott

Volume 10 (1975)
Methods in Cell Biology
Edited by David M. Prescott

Volume 11 (1975)
Yeast Cells
Edited by David M. Prescott

Volume 12 (1975)
Yeast Cells
Edited by David M. Prescott

Volume 13 (1976)
Methods in Cell Biology
Edited by David M. Prescott

Volume 14 (1976)
Methods in Cell Biology
Edited by David M. Prescott

Volume 15 (1977)
Methods in Cell Biology
Edited by David M. Prescott

Volume 16 (1977)
Chromatin and Chromosomal Protein Research I
Edited by Gary Stein, Janet Stein, and Lewis J. Kleinsmith

Volume 17 (1978)
Chromatin and Chromosomal Protein Research II
Edited by Gary Stein, Janet Stein, and Lewis J. Kleinsmith

Volume 18 (1978)
Chromatin and Chromosomal Protein Research III
Edited by Gary Stein, Janet Stein, and Lewis J. Kleinsmith

Volume 19 (1978)
Chromatin and Chromosomal Protein Research IV
Edited by Gary Stein, Janet Stein, and Lewis J. Kleinsmith

Volume 20 (1978)
Methods in Cell Biology
Edited by David M. Prescott

Advisory Board Chairman
KEITH R. PORTER

Volume 21A (1980)
Normal Human Tissue and Cell Culture, Part A: Respiratory, Cardiovascular, and Integumentary Systems
Edited by Curtis C. Harris, Benjamin F. Trump, and Gary D. Stoner

Volume 21B (1980)
Normal Human Tissue and Cell Culture, Part B: Endocrine, Urogenital, and Gastrointestinal Systems
Edited by Curtis C. Harris, Benjamin F. Trump, and Gary D. Stoner

Volume 22 (1981)
Three-Dimensional Ultrastructure in Biology
Edited by James N. Turner

Volume 23 (1981)
Basic Mechanisms of Cellular Secretion
Edited by Arthur R. Hand and Constance Oliver

Volume 24 (1982)
The Cytoskeleton, Part A: Cytoskeletal Proteins, Isolation and Characterization
Edited by Leslie Wilson

Volume 25 (1982)
The Cytoskeleton, Part B: Biological Systems and *in Vitro* Models
Edited by Leslie Wilson

Volume 26 (1982)
Prenatal Diagnosis: Cell Biological Approaches
Edited by Samuel A. Latt and Gretchen J. Darlington

Series Editor

LESLIE WILSON

Volume 27 (1986)
Echinoderm Gametes and Embryos
Edited by Thomas E. Schroeder

Volume 28 (1987)
***Dictyostelium discoideum*: Molecular Approaches to Cell Biology**
Edited by James A. Spudich

Volume 29 (1989)
Fluorescence Microscopy of Living Cells in Culture, Part A: Fluorescent Analogs, Labeling Cells, and Basic Microscopy
Edited by Yu-Li Wang and D. Lansing Taylor

Volume 30 (1989)
Fluorescence Microscopy of Living Cells in Culture, Part B: Quantitative Fluorescence Microscopy—Imaging and Spectroscopy
Edited by D. Lansing Taylor and Yu-Li Wang

Volume 41 (1994)
Flow Cytometry, Second Edition, Part A
Edited by Zbigniew Darzynkiewicz, J. Paul Robinson,
  and Harry A. Crissman

Volume 42 (1994)
Flow Cytometry, Second Edition, Part B
Edited by Zbigniew Darzynkiewicz, J. Paul Robinson,
  and Harry A. Crissman

Volume 43 (1994)
Protein Expression in Animal Cells
Edited by Michael G. Roth

Volume 44 (1994)
***Drosophila melanogaster:* Practical Uses in Cell and Molecular Biology**
Edited by Lawrence S. B. Goldstein, and Eric A. Fyrberg

Volume 45 (1994)
Microbes as Tools for Cell Biology
Edited by David G. Russell

Volume 46 (1995)
Cell Death
Edited by Lawrence M. Schwartz, and Barbara A. Osborne

Volume 47 (1995)
Cilia and Flagella
Edited by William Dentler, and George Witman

Volume 48 (1995)
***Caenorhabditis elegans:* Modern Biological Analysis of an Organism**
Edited by Henry F. Epstein, and Diane C. Shakes

Volume 49 (1995)
Methods in Plant Cell Biology, Part A
Edited by David W. Galbraith, Hans J. Bohnert, and Don P. Bourque

Volume 50 (1995)
Methods in Plant Cell Biology, Part B
Edited by David W. Galbraith, Don P. Bourque, and Hans J. Bohnert

Volume 51 (1996)
Methods in Avian Embryology
Edited by Marianne Bronner-Fraser

Volume 52 (1997)
Methods in Muscle Biology
Edited by Charles P. Emerson, Jr. and H. Lee Sweeney

Volume 53 (1997)
Nuclear Structure and Function
Edited by Miguel Berrios

Volume 54 (1997)
Cumulative Index

Volume 55 (1997)
Laser Tweezers in Cell Biology
Edited by Michael P. Sheez

Volume 56 (1998)
Video Microscopy
Edited by Greenfield Sluder and David E. Wolf

ISBN 0-12-564158-3

90018

Printed and bound by CPI Group (UK) Ltd, Croydon, CR0 4YY

08/05/2025

01864963-0001